illustrate just how relevant math is to your everyday life and beyond.

Based on this plan, calculate the amount of your down payment: _____
Calculate the amount of each monthly payment: _____

3. The fine print!

   *Until January next year, interest will accrue at 23.4%.
   If the balance is paid in full by end of deferral period,
   all finance charges will be credited, resulting in no interest to you.*

   If you do not pay off this loan by the end of December of this year, then you will pay interest on the entire balance due at the time of purchase. Using the simple interest formula and the 24 months of financing available, how much interest would you pay if you weren't able to pay off the total balance due by the end of the year? _____

4. Use the Annual Percentage Rate (APR) Table (Table 5-3) to answer these questions.

   (a) You plan to pay 10% down on your purchase. How much do you pay at the time of purchase? _____ How much is left to finance? _____
   (b) If you finance the balance due at 7.5% for 36 months, how much interest will you owe? _____
   (c) What will your monthly payment be? _____

5. Using the installment loan table, calculate the monthly payments for your purchase if you decide to finance the purchase for 4 years at 8.5% APR.

   Total interest due _____
   Monthly payment _____

Fundamentals of Algebraic Modeling

An Introduction to Mathematical Modeling with Algebra and Statistics

Fundamentals of Algebraic Modeling

6e

An Introduction to Mathematical Modeling with Algebra and Statistics

Daniel L. Timmons
Alamance Community College

Catherine W. Johnson
Alamance Community College

Sonya M. McCook
Alamance Community College

Australia • Brazil • Japan • Korea • Mexico • Singapore • Spain • United Kingdom • United States

Fundamentals of Algebraic Modeling: An Introduction to Mathematical Modeling with Algebra and Statistics, 6th Edition
Timmons, Johnson, McCook

Executive Editor: Charlie Van Wagner

Acquisitions Editor: Marc Bove

Developmental Editor: Stefanie Beeck

Assistant Editor: Lauren Crosby

Editorial Assistant: Ryan Furtkamp

Media Editor: Bryon Spencer

Senior Brand Manager: Gordon Lee

Senior Market Development Manager: Danae April

Content Project Manager: Tanya Nigh

Art Director: Vernon Boes

Manufacturing Planner: Rebecca Cross

Rights Acquisitions Specialist: Tom McDonough

Production and Composition: MPS Limited

Photo Researcher: Scott Rosen, Q2A/Bill Smith

Text Researcher: Pablo D'Stair

Copy Editor: Martha Williams

Text Designer: Rokusek Design

Interior Designer: Kim Rokusek

Cover Designer: Larry Didona

Cover photo credits: House keys on blueprint: Anthony Bradshaw/Getty Images; Ballpoint pen: Tsoi Hoi Fung/Getty Images; Radio telescopes: Robert Llewellyn/Corbis; Golfers: Robert Michael/Corbis; Credit card on keyboard: Whisson/Jordan/Corbis; Financial and global business: Henry Choi/Corbis

© 2014, 2010 Brooks/Cole, Cengage Learning

Unless otherwise noted, all art is © Cengage Learning.

ALL RIGHTS RESERVED. No part of this work covered by the copyright herein may be reproduced, transmitted, stored, or used in any form or by any means graphic, electronic, or mechanical, including but not limited to photocopying, recording, scanning, digitizing, taping, Web distribution, information networks, or information storage and retrieval systems, except as permitted under Section 107 or 108 of the 1976 United States Copyright Act, without the prior written permission of the publisher.

> For product information and technology assistance, contact us at
> **Cengage Learning Customer & Sales Support, 1-800-354-9706.**
>
> For permission to use material from this text or product, submit all requests online at **www.cengage.com/permissions**.
>
> Further permissions questions can be e-mailed to
> **permissionrequest@cengage.com**

Library of Congress Control Number: 2012938474

ISBN-13: 978-1-133-62777-7
ISBN-10: 1-133-62777-3

Brooks/Cole
20 Davis Drive
Belmont, CA 94002-3098
USA

Cengage Learning is a leading provider of customized learning solutions with office locations around the globe, including Singapore, the United Kingdom, Australia, Mexico, Brazil, and Japan. Locate your local office at **www.cengage.com/global**.

Cengage Learning products are represented in Canada by Nelson Education, Ltd.

To learn more about Brooks/Cole, visit
www.cengage.com/brooks/cole
Purchase any of our products at your local college store or at our preferred online store **www.CengageBrain.com**.

Printed in China
2 3 4 5 6 7 16 15 14 13

Contents

Preface ix

CHAPTER R — A REVIEW OF ALGEBRA FUNDAMENTALS   1

R-1 **Real Numbers and Mathematical Operations** 2
R-2 **Solving Linear Equations** 9
R-3 **Percents** 13
R-4 **Scientific Notation** 19
    Chapter Summary 23
    Chapter Review Problems 23
    Chapter Test 25
    Suggested Laboratory Exercises 25

CHAPTER 1 — FUNDAMENTALS OF MATHEMATICAL MODELING   27

1-1 **Mathematical Models** 28
1-2 **Formulas** 30
1-3 **Ratio and Proportion** 34
1-4 **Word Problem Strategies** 40
    Chapter Summary 47
    Chapter Review Problems 47
    Chapter Test 48
    Suggested Laboratory Exercises 49

CHAPTER 2 — APPLICATIONS OF ALGEBRAIC MODELING   53

2-1 **Models and Patterns in Plane Geometry** 54
2-2 **Models and Patterns in Triangles** 60
2-3 **Models and Patterns in Right Triangles** 67
2-4 **Models and Patterns in Art, Architecture, and Nature** 73
2-5 **Models and Patterns in Music** 82
    Chapter Summary 89
    Chapter Review Problems 89
    Chapter Test 91
    Suggested Laboratory Exercises 92

CHAPTER 3 — GRAPHING — 95

- 3-1 **Rectangular Coordinate System** 96
- 3-2 **Graphing Linear Equations** 101
- 3-3 **Slope** 105
- 3-4 **Writing Equations of Lines** 116
- 3-5 **Applications and Uses of Graphs** 121

    Chapter Summary 128
    Chapter Review Problems 129
    Chapter Test 130
    Suggested Laboratory Exercises 131

CHAPTER 4 — FUNCTIONS — 137

- 4-1 **Functions** 138
- 4-2 **Using Function Notation** 146
- 4-3 **Linear Functions as Models** 153
- 4-4 **Direct and Inverse Variation** 160
- 4-5 **Quadratic Functions and Power Functions as Models** 166
- 4-6 **Exponential Functions as Models** 178

    Chapter Summary 183
    Chapter Review Problems 183
    Chapter Test 186
    Suggested Laboratory Exercises 187

CHAPTER 5 — MATHEMATICAL MODELS IN CONSUMER MATH — 195

- 5-1 **Mathematical Models in the Business World** 196
- 5-2 **Mathematical Models in Banking** 202
- 5-3 **Mathematical Models in Consumer Credit** 210
- 5-4 **Mathematical Models in Purchasing an Automobile** 217
- 5-5 **Mathematical Models in Purchasing a Home** 223
- 5-6 **Mathematical Models in Insurance Options and Rates** 230
- 5-7 **Mathematical Models in Stocks, Mutual Funds, and Bonds** 236
- 5-8 **Mathematical Models in Personal Income** 241

    Chapter Summary 250
    Chapter Review Problems 251
    Chapter Test 253
    Suggested Laboratory Exercises 254

Contents **vii**

CHAPTER 6 — MODELING WITH SYSTEMS OF EQUATIONS 259

- 6-1 **Solving Systems by Graphing** 260
- 6-2 **Solving Systems Algebraically** 266
- 6-3 **Applications of Linear Systems** 274
- 6-4 **Systems of Nonlinear Equations** 283

   Chapter Summary 286
   Chapter Review Problems 287
   Chapter Test 288
   Suggested Laboratory Exercises 289

CHAPTER 7 — PROBABILITY MODELS 293

- 7-1 **Sets and Set Theory** 294
- 7-2 **What Is Probability?** 298
- 7-3 **Theoretical Probability** 306
- 7-4 **Odds** 312
- 7-5 **Tree Diagrams** 316
- 7-6 ***Or* Problems** 321
- 7-7 ***And* Problems** 327
- 7-8 **The Counting Principle, Permutations, and Combinations** 330

   Chapter Summary 338
   Chapter Review Problems 339
   Chapter Test 340
   Suggested Laboratory Exercises 342

CHAPTER 8 — MODELING WITH STATISTICS 345

- 8-1 **Introduction to Statistics** 346
- 8-2 **Frequency Tables and Histograms** 352
- 8-3 **Reading and Interpreting Graphical Information** 358
- 8-4 **Descriptive Statistics** 363
- 8-5 **Variation** 369
- 8-6 **Normal Curve** 374
- 8-7 **Scatter Diagrams and Linear Regression** 384

   Chapter Summary 391
   Chapter Review Problems 392
   Chapter Test 393
   Suggested Laboratory Exercises 395

APPENDIX 1	**COMMONLY USED CALCULATOR KEYS**	401
APPENDIX 2	**FORMULAS USED IN THIS TEXT**	405
APPENDIX 3	**LEVELS OF DATA IN STATISTICS**	407

Answer Key A-1

Index I-1

Preface

> *"The longer mathematics lives the more abstract— and therefore, possibly also the more practical—it becomes."*
>
> —Eric Temple Bell (1883–1960)

TO THE INSTRUCTOR

There is no doubt that mathematics has become increasingly more important in our ever-changing world. For most people, the usefulness of mathematics lies in its applications to practical situations. Our goal in writing this book is to get students to think of mathematics as a useful tool in their chosen occupations and in their everyday lives.

This book was written and designed for students in a two-year associate in arts curriculum who are not planning additional course work in mathematics. We have written the book in "nonthreatening" mathematical language so that students who have been previously fearful of or intimidated by mathematics will be able to comprehend the concepts presented in the text. We have tried to write it in such a manner that students with backgrounds in fundamental algebra can understand and learn the ideas we have presented.

Various types of problems are included throughout the book in order to attempt to make students aware of their own thought processes. Our intent is to teach students how to approach a variety of problems with some basic skills and a plan for success. We hope that students will learn to use, or develop and then test, mathematical models against reality. Further, we have tried to be sensitive to various student learning styles. This was done by including problems in many formats (graphical, numerical, and symbolic) in order to give students many different opportunities to "see" the mathematics.

IN THE SIXTH EDITION

The elements that proved successful in previous editions remain in this edition. However, we have reordered several sections and added some new topics. Many of the problem sets have had extensive revision with expanded problem sets including many new problems. The answers to the odd exercises are included in the answer key in the back of the book, with answers to all problems in the Chapter Reviews and Chapter Tests in the key. Additional lab activities have been included in many of the chapters. Here is a list of the major changes included in the sixth edition.

- A brand new Chapter R, A Review of Algebra Fundamentals, has been added, giving students an opportunity to review the algebra skills needed to be successful in a modeling course.
- A section on scientific notation has been added to introduce students to this important method of handling large numbers.
- Chapter 1, Mathematical Model Fundamentals, has been reorganized.
- Geometric models has been moved from Chapter 5 to Chapter 2.
- Chapter 7, Probability Models, has been reorganized, creating additional sections.
- Chapter 8, Modeling with Statistics, now includes a section on reading and interpreting graphical information.

FEATURES OF THE BOOK

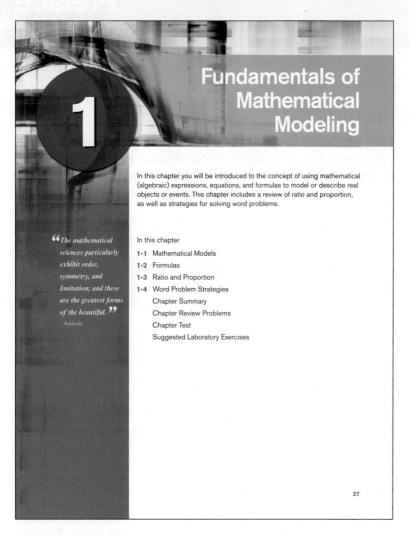

Each chapter opens with an introduction to the topics in the chapter, a chapter outline, and a quote about math.

We have included clear step-by-step *Examples* in each section to illustrate the concepts and skills being introduced. More examples have been added to the material in several sections, based on suggestions from other instructors.

Using a Function in a Word Problem

Forensic anthropologists (specialists who work for the police and use scientific techniques to solve crimes) can determine a woman's height in life by using the skeletal remains of her humerus bone. (This is the long bone that forms the upper arm.) The formula is $3.08x + 64.47$, where x represents the bone length in centimeters. Write this formula in function notation and find the estimated height in centimeters of a woman whose humerus measures 31.0 cm. Then convert this measurement to inches.

Because the formula is designed to measure a woman's height, we will use the variable h to represent the name of the function and x, the length of the humerus, will be the independent variable.

(a) Function notation:
$$h(x) = 3.08x + 64.47$$

(b) Evaluate the function if the length of the humerus is 31.0 cm.
$$h(31.0) = 3.08(31.0) + 64.47 = 159.95$$

Therefore, the predicted height of this woman, based on the length of her humerus, is 159.95 cm.

(c) There are approximately 2.54 cm in 1 in., so we can convert this answer to inches by dividing.
$$159.95 \text{ cm} \div 2.54 \text{ cm/in.} \approx 62.97 \text{ in. or } 63 \text{ in. (5 ft 3 in.)}$$

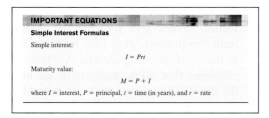

Important Equations are called out and featured in boxes that are easy to spot, perfect for quick reference and for review.

The inclusion of easy to find *Theorems, Rules, and Definitions* within the text allows students to focus on key topics within each section for comprehension and review.

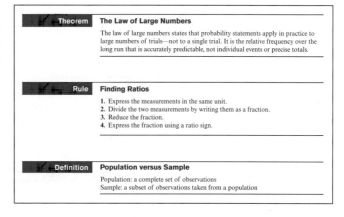

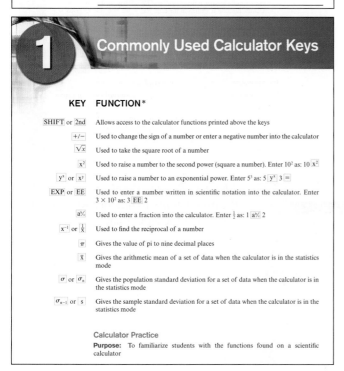

The American Mathematical Association of Two-Year Colleges (AMATYC) also recommends in *Crossroads* the routine use of calculators in the classroom. We have created *Calculator Mini-Lessons* throughout the book to aid students who are unfamiliar with calculators in becoming more proficient. Instructions for both a standard scientific calculator and a graphing calculator are provided. The use of graphing calculators is recommended in several areas, such as graphing nonlinear functions, science and technology applications, and linear regression. An *Appendix* that outlines the meaning and use of calculator keys is also included in the back of the book. However, the use of a graphing calculator is not essential to the successful completion of a course using this book.

Practice Sets appear at the end of each section. These give students the opportunity to practice what they have learned in the corresponding section. Application problems that allow students to see how the practiced concepts are used in real life are also included in the Practice Sets.

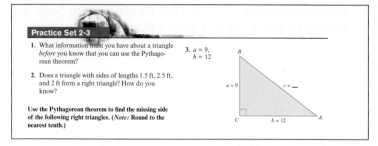

At the end of each chapter is a *Chapter Summary* listing key terms and formulas and points to remember.

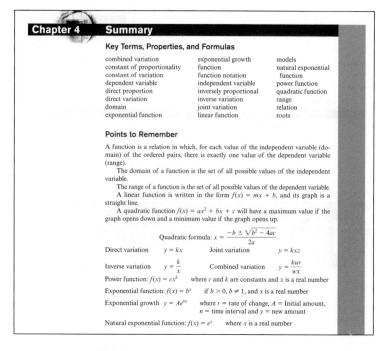

Chapter Review Problems give students the opportunity to practice what they have learned at the end of each chapter.

Chapter Tests are included at the end of each chapter so that students can test their understanding of all topics presented. All of the answers to these questions are included in the back of the book. Students are encouraged to practice test-taking skills by completing these tests outside of class in a timed environment without consulting the answer key until all problems have been completed.

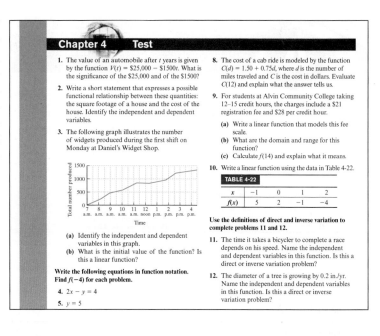

AMATYC, in its publication *Crossroads in Mathematics: Standards for Introductory College Mathematics before Calculus*, recommends that mathematics be taught as a laboratory discipline where students are involved in guided hands-on activities. We have included *Laboratory Exercises* at the end of each chapter and a wide variety of other activities in the ancillary materials available with this text. Some are designed to be completed as individual assignments and others require group work. There are assignments that require access to a computer lab and several that require online work. We have tried to make the labs versatile so the instructor can use any technology available. However, instructors are not required to have technology in order to teach a successful course using this book.

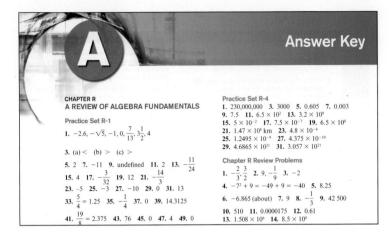

The *Answer Key* provides odd-numbered answers to the practice sets. The focus of the book is understanding the modeling process involved in solving a problem. Having answers available helps students know if their thinking process has been correct. All answers are provided for the chapter reviews and chapter tests.

Supplements

For the student	For the instructor
	Instructor Edition (ISBN: 978-1-133-36561-7) The *Instructor Edition* features an appendix containing the answers to all problems in the book.
Student Solutions Manual (ISBN: 978-1-285-42042-4) Go beyond the answers—see what it takes to get there and improve your grade! This manual provides step-by-step solutions to selected problems in the text, giving you the information you need to truly understand how these problems are solved.	**Instructor's Resource Manual** (ISBN: 978-1-285-42049-3) The *Instructor's Resource Manual* includes additional labs and group activities for each chapter of the text, lab notes for the instructor, a test bank with five tests per chapter as well as three final exams, and solutions to all problems in the text. The manual also includes worksheets that can be given to students for additional practice and a graphing calculator quick reference guide.
	PowerLecture with ExamView (ISBN: 978-1-285-42047-9) This DVD provides you with dynamic media tools for teaching. Create, deliver, and customize tests (both print and online) in minutes with ExamView Computerized Testing Featuring Algorithmic Equations. Easily build solution sets for homework or exams using Solution Builder's online solutions manual. Microsoft PowerPoint lecture slides and figures from the book are included on this DVD.

TO THE STUDENT

Why take math courses? You may have asked yourself or your advisor this very question. Perhaps you asked because you don't see a need for math in your primary area of study. Perhaps you asked because you have always feared mathematics. Well, no matter what your math history has been, your math future can be better. As you start this new course, try to cultivate a positive attitude and look at the tips we offer for success.

Math can be thought of as a tool. It does have a practical value in your daily life as well as in most professions. In some fields such as engineering, accounting, business, drafting, welding, carpentry, and nursing, the connection to mathematics is obvious. In others such as music, art, history, criminal justice, and early childhood education, the connection is not as clear. But, we assure you, there is one. The logic developed by solving mathematical problems can be useful in all professions. For example, those in the criminal justice field must put together facts in a logical way and come to a solution for the crimes they investigate. This involves mathematical processes.

Overcoming anxiety about math is not easy for most students. However, developing a positive attitude, improving your study habits, and making a commitment to yourself to succeed can all help. Enroll in any study-skills courses offered and take advantage of any tutoring services provided by your college. Do this on the first day of class, not after you've done poorly on two or three tests. To reduce your math anxiety, try these tips.

- Be well prepared for tests. Practice "taking tests" at home with a timer.
- Write down memory cues before beginning a test.
- Begin a test by first doing the problems with which you have the least trouble.
- Take advantage of all available help (tutoring services, skills lab, instructor office hours).
- Learn from your mistakes by reworking all problems missed on a test or homework assignment.
- Take math courses in the fall or spring semester—not during a short summer term.
- If you must take several math courses, take them in consecutive semesters.
- Form study groups to study outside of class time.

You *can* choose to be successful by using these tips and giving this course the time necessary to master the material.

ACKNOWLEDGMENTS

There have been many whose efforts have brought us to this point, and we are grateful to them for their suggestions and encouragement. Students at Alamance Community College who used the original manuscript offered many helpful ideas, and we are grateful to them for their assistance.

We particularly wish to thank several students in the Mechanical Drafting Technology program at Alamance Community College who provided original drawings for Section 5-3: Modeling and Patterns in Architecture: Perspective and Symmetry. They are Joseph F. Vaughn, Charles P. Hanbert, and Michael Wood.

Tim Beaver	Isothermal Community College
Jean Bevis	Georgia State University
Michael Bradshaw	Caldwell Community College and Technical Institute

Marie Cash	Fayetteville Technical Community College
Erica M. Farelli	Rowan Cabarrus Community College
Steven Felzer, Ph.D.	Lenoir Community College
John D. Gieringer	Alvernia College
Chuckie Hairston	Halifax Community College
Don Hancock	Pepperdine University
Sandee House	Georgia Perimeter College
Sally Jackman	Richland College
Connie Kiehn	Elsik High School, Alief, Texas
Lori Kiel	Fayetteville Technical Community College
Jeff Lewis	Johnson County Community College
Diana Ochoa	Conroe High School, Conroe, Texas
Libbie Reeves	Coastal Carolina Community College
John Robertson	Georgia Military College
Michelle Robinson	Fayetteville Technical Community College
Joy Sammons	Andrews College
Paul Shaklovitz	Pasadena High School, Pasadena, Texas
Lara Smith	Pitt Community College
Lee Ann Spahr	Durham Technical Community College
Carolyn Spillman	Georgia Perimeter College
Sr. Barbara Vano, OSF	Lourdes College
Amy O. Van Pelt	State University of New York at Paltz
David Wainaina	Coastal Carolina Community College
Sandra Kay Westbrook	Keller High School, Keller, Texas
Janet Yates	Forsyth Technical Community College

We would also like to acknowledge the support of several important people who have encouraged and assisted us throughout the development of this book as well as previous editions: Ray Harclerode, Wendy Weisner, Valerie Rectin, John-Paul Ramin, and our spouses, Lynn, Everett, and Rob.

Dan Timmons
Sonya McCook
Cathy Johnson

Keys to Success

BEFORE CLASS STARTS

- ✓ Find a quiet, comfortable place to work outside of class.
- ✓ Make both short-term and long-term study schedules.
- ✓ Discourage interruptions.
- ✓ Take short breaks occasionally.
- ✓ Do not procrastinate.
- ✓ Give yourself some warm-up time by working simple problems.

IN CLASS

- ✓ Attend class regularly.
- ✓ Ask questions early in the term.
- ✓ Listen for critical points.
- ✓ Lost? Mark your location in your notes and see your instructor during the next available office hour or at the end of class, if time permits.
- ✓ Review your notes from the previous class before going to class again.
- ✓ Be sure to read the text sections that correspond to your lecture notes.
- ✓ Form a study group with some of your classmates.

ABOUT THOSE CLASS NOTES

- ✓ Remember, your notes are your links between your class and your textbook.
- ✓ Never write at the expense of listening.
- ✓ Forget about correct grammar while taking notes.
- ✓ Use a lot of abbreviations.
- ✓ Be sure to copy down class examples.
- ✓ Rewrite, if you wish, your class notes to make them clearer and neater for future reference.
- ✓ Compare your notes with those of some of your classmates; they may have gotten some points that you missed and vice versa.

PROPERLY USE THIS TEXTBOOK

- ✓ Read the section due for lecture before class.
- ✓ Read each section twice, first quickly, and then slowly while referring to your class notes.
- ✓ Write things down as they occur to you in your reading because writing down a concept or idea is definitely linked to your thinking processes.
- ✓ Write notes to yourself in your textbook margins.
- ✓ Look up the definitions of unfamiliar terms.
- ✓ Highlight sparingly.

LOVE THOSE WORD PROBLEMS

- ✓ Think about the problem before jumping into a solution.
- ✓ Be sure to clearly delineate the questions to be answered.
- ✓ Break long, complicated problems into parts.
- ✓ Work with your study group but remember that you must be able to solve problems on your own at test time.
- ✓ Ask for help (from your instructor, tutor, classmates) when you need it.

TEST PREPARATION

- ✓ Be sure you know what topics are to be covered.
- ✓ Make up reasonable questions with your study group members for each other to practice.
- ✓ Review old quizzes or tests if you can.
- ✓ Reread those marginal notes you made in your textbook, particularly those that indicate weaknesses.
- ✓ Honestly admit your weaknesses and work on strengthening them.

R: A Review of Algebra Fundamentals

This chapter is intended as a brief summary of some of the major topics of an introductory algebra course. It is not intended to replace such a course. A review of algebraic properties, rules for solving equations in one variable, percents, and scientific notation are included in this chapter. Students should be familiar with these topics because these skills will be necessary for success in subsequent chapters.

> *"Don't lose the substance by grasping at the shadow."*
> – Aesop, "The Dog and the Shadow"

In this chapter

R-1 Real Numbers and Mathematical Operations

R-2 Solving Linear Equations

R-3 Percents

R-4 Scientific Notation

Chapter Summary

Chapter Review Problems

Chapter Test

Suggested Laboratory Exercises

Chapter R A Review of Algebra Fundamentals

Section R-1 Real Numbers and Mathematical Operations

Real Numbers

In algebra, the set of numbers commonly encountered in solving various equations and formulas and in graphing is called the **real numbers**. The set of real numbers can be broken down into several subsets:

1. The **rational numbers**
   (a) The **natural** or **counting numbers** = $\{1, 2, 3, 4, 5, \ldots\}$
   (b) The **whole numbers** = $\{0, 1, 2, 3, 4, 5, \ldots\}$
   (c) The **integers** = $\{\ldots, -2, -1, 0, 1, 2, \ldots\}$
   (d) The **rational numbers** = numbers that can be expressed as a ratio of two integers, such as $\frac{1}{2}, -\frac{5}{3}, 0, 6 \left(\text{as } \frac{6}{1}\right), \frac{-8}{-5}$, and so forth
2. The **irrational numbers** = numbers that cannot be expressed as the ratio of two integers, such as $\sqrt{2}$ or π (pi), or similar numbers that are nonterminating, nonrepeating decimal numbers

The set of real numbers is an infinite set of numbers, so it is impossible to write them all down. Thus, a **number line** is often used to picture the real numbers. The number line shown in Figure R-1 is a *picture* of all the real numbers.

Negative real numbers | Positive real numbers

$-7\ -6\ -5\ -4\ -3\ -2\ -1\ \ 0\ \ 1\ \ 2\ \ 3\ \ 4\ \ 5\ \ 6\ \ 7$

← Smaller numbers   Larger numbers →

FIGURE R-1
Number line.

The symbol $\sqrt{}$ is called a radical sign and indicates the **square root** of the number under the symbol. Finding a square root is the inverse of squaring a number. The square root of every number is either positive or 0, so negative numbers do not have square roots in the set of real numbers. Some square roots result in whole-number answers. For example, $\sqrt{25} = 5$, because $5^2 = 25$. Other square roots have inexact decimal equivalents and are classified as irrational numbers. The number $\sqrt{3}$ is an irrational number. If you use your calculator to derive a decimal representation for this number, it must be rounded. If we round it to the hundredths place, $\sqrt{3} = 1.73$.

All numbers, whether rational or irrational, can be located at some position on the number line. In comparing two real numbers, the number located farther to the right on the number line is the larger of the two numbers. Finding the decimal equivalent of square roots will help us locate them easily on the number line. (See Figure R-2.)

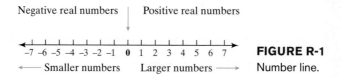

FIGURE R-2
Locating real numbers on a number line.

Operations with Real Numbers

When performing computations using algebra, both positive and negative numbers are used. In this section, we will *briefly* review addition, subtraction, multiplication, and division of real numbers.

Absolute Value Look at the number line in Figure R-3. Notice that the line *begins* at 0, has positive numbers to the right of 0, and negative numbers to the left. The **absolute value** of a number is its distance from 0. Distances are always measured with positive numbers, so the absolute value of a number is never negative. For example, the numbers 3 and −3 are both the same distance from 0. The +3 is three units to the right of 0 and the −3 is three units to the left of 0.

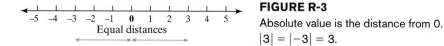

FIGURE R-3
Absolute value is the distance from 0.
$|3| = |-3| = 3$.

Because they are both the same distance from 0, they both have the same absolute value, 3. The symbol for the absolute value of 3, or any other number, is a pair of vertical lines with the number in between them: |3|. Using this symbol, we would write the absolute values discussed previously as

$$|3| = 3$$
$$|-3| = 3$$

Here are a few more examples:

$$|0| = 0$$
$$-|-8| = -8$$
$$|-0.45| = 0.45$$
$$\left|\frac{1}{2}\right| = \frac{1}{2}$$

Additive Inverses or Opposites Two numbers that are the same distance from 0 but on opposite sides of 0 on the number line are said to be **additive inverses** of each other. For example, both 6 and −6 have the same absolute value, so they are the same distance from 0. However, 6 is to the right of 0 and −6 is to the left of 0. Thus, 6 and −6 are additive inverses. (See Figure R-4.)

Here are some other additive inverses or opposites:

$$10, -10$$
$$0.34, -0.34$$
$$\frac{1}{2}, -\frac{1}{2}$$

FIGURE R-4
Additive inverses.

An important fact about additive inverses is that the sum of any number and its additive inverse is always equal to 0 [i.e., $2 + (-2) = 0$]. Because $0 + 0 = 0$, 0 is its own additive inverse.

Rule: Addition of Real Numbers

1. If all numbers are positive, then add as usual. The answer is positive.
2. If all numbers are negative, then add as usual. The answer is negative.
3. If one number is positive and the other negative, then
   (a) Find the absolute values of both numbers.
   (b) Find the difference between the absolute values.
   (c) Give the answer the sign of the original number with the larger absolute value.

Example 1: Addition of Real Numbers

Use the rule for addition of real numbers to verify these results.

$$5 + 2 = 7 \qquad -2.6 + 6.8 = 4.2$$

$$-5 + (-2) = -7 \qquad -\frac{7}{8} + \frac{3}{4} = -\frac{1}{8}$$

$$-5 + 2 = -3 \qquad 5 + (-2) = 3$$

Rule: Subtraction of Real Numbers

1. Change the number to be subtracted to its additive inverse or opposite.
2. Change the sign indicating subtraction to an addition sign.
3. Now follow the rules given for addition.

This rule gives students more trouble than any other rule in fundamental algebra. Often this rule is written as follows: "Change the sign of the number to be subtracted and then add."

Example 2: Subtraction of Real Numbers

Use the rule for subtraction of real numbers to verify these results.

$$5 - 2 = 5 + (-2) = 3 \qquad -8 - (-6) = -8 + (+6) = -2$$
$$8 - 15 = 8 + (-15) = -7 \qquad 5.4 - 9.2 = 5.4 + (-9.2) = -3.8$$

Rule: Multiplication and Division of Real Numbers

1. Multiply or divide the numbers as usual.
2. If both numbers have the same sign, then the answer is positive.
3. If the signs of the numbers are different, then the answer is negative.

Example 3 — Multiplication of Real Numbers

Use the rule for multiplication of real numbers to verify these results.

$$2(4) = 8 \qquad -\frac{1}{2}\left(\frac{1}{4}\right) = -\frac{1}{8}$$
$$-2(-4) = 8 \qquad -4.5(2) = -9$$
$$2(-4) = -8 \qquad 0(2) = 0$$
$$0(-5) = 0$$

Remember that 0 is neither positive nor negative.

CALCULATOR MINI-LESSON

Operations with Real Numbers

Your calculator knows all of the rules for adding, subtracting, multiplying, and dividing real numbers. It even knows the rule about division by 0. An important thing to remember is that the calculator will speed up your calculations, but it cannot read the problem. If you properly enter the operations, then the calculator will give you a correct answer. If you enter the numbers and operations improperly, then all the calculator will do is give you a wrong answer quickly.

When doing operations with signed numbers on your calculator, be sure to distinguish between the subtraction sign and the negative sign. On most scientific calculators, you must enter the number and then the sign into the calculator. For example, to do the problem $-4 - (-5)$, use the following keystrokes: 4 $\boxed{+/-}$ $\boxed{-}$ 5 $\boxed{+/-}$ $\boxed{=}$.

The second sign in the keystroke list is the subtraction key. All others are negative signs. Some calculators are direct entry, so enter the signs in the order given in the problem. A graphing calculator is a direct-entry calculator and has the negative key labeled $(-)$. Read your manual to be sure about operations with real numbers.

Example 4 — Division of Real Numbers

Use the rule for division of real numbers to verify these results.

$$-8 \div 2 = -4 \qquad 0 \div 5 = 0$$
$$-8 \div (-2) = 4 \qquad 0 \div (-8) = 0$$
$$-\frac{1}{2} \div \left(-\frac{1}{2}\right) = 1 \qquad 6 \div 0 = \text{undefined}$$
$$8.2 \div (-2) = -4.1 \qquad -7.8 \div 0 = \text{undefined}$$

Remember that division by 0 cannot be done in the real number system.

Reciprocals or Multiplicative Inverses By definition, two numbers whose products equal 1 are **reciprocals** or **multiplicative inverses** of each other. The reciprocal of $-\frac{2}{3}$ is found by "flipping it," giving $-\frac{3}{2}$. This number fulfills the definition because $-\frac{2}{3} \cdot -\frac{3}{2} = 1$. The number 0 does not have a reciprocal since $\frac{1}{0}$ is undefined. When dividing **rational numbers** (fractions), the quotient is the product of the first number and the reciprocal or multiplicative inverse of the second number or divisor. Example 5 demonstrates this calculation.

Example 5

Division of Rational Numbers

Give the reciprocal of $-\frac{1}{2}$, and then use it to complete the problem $\frac{3}{5} \div -\frac{1}{2}$.

First find the reciprocal of $-\frac{1}{2}$ by "flipping it" to give $-\frac{2}{1}$ or –2.

Then, complete the given problem by reciprocating the divisor and multiplying.

$$\frac{3}{5} \div -\frac{1}{2} = \frac{3}{5} \cdot -\frac{2}{1} = -\frac{6}{5}$$

Evaluation of Expressions Suppose that you had to find the numerical value of the following expression:

$$5 + 2(3)$$

There are two possible results, depending on which operation, addition or multiplication, is performed first. If the addition is done first, then the result will be as follows:

$$5 + 2(3) = 7(3) = 21$$

If the multiplication is done first, the result will be different:

$$5 + 2(3) = 5 + 6 = 11$$

The order in which operations are performed does make a difference. So, which is correct? We can easily find out if we first study the rules for **order of operations**.

Rule

The Order of Operations

1. If any operations are enclosed in parentheses, do those operations first.
2. If any numbers have exponents (or are raised to some power), do those next.
3. Perform all multiplication and division in order, from left to right.
4. Perform all addition and subtraction in order, from left to right.

Referring to the problem $5 + 2(3)$, you see that it contains two operations, addition and multiplication. If we follow the order of operations, the multiplication would be performed first, followed by the addition. Therefore, the correct result is 11.

With a little practice, this rule becomes very easy to remember. It will be an important part of many algebra problems, so it is worth your time to practice it. As an aid to remembering the rule, a "silly" statement is often used as a mnemonic device.

For example:

"Please Entertain My Dear Aunt Sally"

The first letter of each word (PEMDAS) corresponds to part of the rule for the order of operations. **P** stands for parentheses, **E** for exponents, **MD** for multiplication/division, and **AS** for addition/subtraction.

Example 6 — Order of Operations

Find the value of each of the following expressions by following the order of operations.

1. $2(3 + 5) =$ (parentheses)
 $2(8) = 16$ (multiply)
2. $3^2 + 5 - 7 =$ (exponents)
 $9 + 5 - 7 =$
 $14 - 7 = 7$ (add and subtract, from left to right)
3. $-2[4 + (-5 + 2)] =$ (parentheses)
 $-2[4 + (-3)] =$ (brackets)
 $-2[1] = -2$ (multiply)
4. $14 \div 2 \cdot 7 + 6 =$
 $7 \cdot 7 + 6 =$ (divide/multiply, left to right)
 $49 + 6 = 55$ (add)

If the expression contains a complicated fraction with several operations in the numerator, the denominator, or both, then you must evaluate the numerator and denominator as if they were two separate little expressions and then divide last. For example:

$$\frac{2(4 - 5)}{2(2) + 1}$$

CALCULATOR MINI-LESSON

The Order of Operations

Your calculator will perform several operations at once and is able to follow the order of operations automatically. Just be careful to enter all the operations, numbers, and symbols in the same order as they are written in the problem.

Some helpful hints:

1. If you use a graphing calculator, you must use the parentheses key when raising a negative number to an exponent. For example, if you enter $(-3)^2$ into a graphing calculator without the parentheses, the answer will be -9. The calculator will not assume that you are including the sign in the squaring process if you do not use grouping symbols. On most regular scientific calculators, entering 3 [+/−] [x^2] will tell the calculator to square the sign also.
2. When entering a problem like $\frac{25 + 31}{2}$ into your calculator, be sure to enter 25 [+] 31 [=] [÷] 2. If you do not enter the equal sign, the calculator will only divide 31 by 2, and then add 25. It is following the order of operations!

Chapter R A Review of Algebra Fundamentals

To evaluate this expression, start with the numerator and follow the order of operations until you arrive at one number. Next, do the same for the denominator. Finally, divide the two numbers to arrive at the correct answer.

$$\frac{2(-1)}{2(2)+1} = \frac{-2}{4+1} = \frac{-2}{5} = -0.4$$

Practice Set R-1

1. List the following set of numbers in numerical order from lowest to highest.

   $$\left\{-\sqrt{5},\ 4,\ 3\frac{1}{2},\ -1,\ 0,\ -2.6,\ \frac{7}{13}\right\}$$

2. List the following set of numbers in numerical order from lowest to highest.

   $$\left\{-3.5,\ |-2|,\ \sqrt{5},\ -1,\ \frac{1}{2},\ 1\frac{3}{8},\ -1.625\right\}$$

3. Use $<$, $>$, or $=$ to complete each of the following.
   (a) $\sqrt{7}\ ?\ 3$ (b) $-5\ ?\ -5\frac{1}{4}$ (c) $\pi\ ?\ 0.3$

4. Use $<$, $>$, or $=$ to complete the following.
   (a) $\frac{8}{5}\ ?\ 2.3$ (b) $-3.1\ ?\ -3.8$ (c) $4.5\ ?\ \sqrt{15}$

Evaluate each of the following by applying the appropriate rule for basic operations with signed numbers.

5. $5 + (-3)$
6. $-11 + 8$
7. $-6 + (-5)$
8. $-13 + (-4)$
9. $60 \div 0$
10. $-567 \div 0$
11. $|6 - 8|$
12. $-|-2 - 10|$
13. $-1\frac{1}{3} + \frac{7}{8}$
14. $\frac{3}{4} - 2\frac{4}{5}$
15. $(-2.5)(-1.6)$
16. $(-1.1)(3.8)$
17. $\left(\frac{1}{4}\right)\left(-\frac{3}{8}\right)$
18. $\left(-\frac{1}{3}\right)\left(-\frac{6}{7}\right)$
19. $-(-2) + 10$
20. $7 - (-2) + 6$
21. $-\frac{8}{9} \div \frac{4}{21}$
22. $-\frac{5}{8} \div \frac{15}{24}$

Evaluate each of the following expressions, being careful to follow the order of operations.

23. $7 \cdot 3 - 2 \cdot 13$
24. $(-3)(5) - (-8)(2)$
25. $\dfrac{15}{[-16 - (-11)]}$
26. $\dfrac{27}{(-24 + 21)}$
27. $8 - (-6)(4 - 7)$
28. $40 + (-2)(8 - 3)$
29. $\dfrac{2(-6 + 6)}{23 - 97}$
30. $\dfrac{2(5) - 10}{-7 - (-2)}$
31. $(-8)^2 - 7(8) + 5$
32. $(-3)^3 - 5(-2) + 3$
33. $\dfrac{2^2 + 4^2}{5^2 - 3^2}$
34. $\dfrac{12^2 - 10^2}{5^2 + 1^2}$
35. $\dfrac{3(-5 + 1)}{12(3) + (-5 + 2)(-3 - 1)}$
36. $\dfrac{5(-8 + 3)}{13(-2) + (-6 - 1)(-4 + 1)}$
37. $\dfrac{5}{8} - 5\left(\dfrac{1}{8}\right)$
38. $\dfrac{5}{6} - 3\left(\dfrac{1}{6}\right)$
39. $1.25 + \dfrac{6.5}{0.5} + (0.25)^2$
40. $2.5 + \dfrac{7.5}{0.3} + (0.5)^2$
41. $\dfrac{3^3 - 2^3}{-4(-3 + 1)}$
42. $\dfrac{2^3 - 4^3}{-8(-3 + 2)}$
43. $-4[(-2)(6) - 7]$
44. $-6[(7)(-2) - 3]$
45. $(3 - 8)(-2) - 10$
46. $-20 - (-1)(-7 - 11)$
47. $\dfrac{-|-8|}{-2}$
48. $3\sqrt{-4}$
49. What is the sum of $|-4|$ and its opposite?
50. What is the product of $-\dfrac{2}{3}$ and its reciprocal?

Section R-2 Solving Linear Equations

In our review of equation-solving procedures, extensive use of the properties of real numbers will be required. The principles used are designed to result in what are called **equivalent equations**. For example, are $x + 3 = 5$ and $x = 2$ equivalent equations? If x is replaced with the number 2 in both equations, then both equations can be seen to be true mathematical statements. Because of this, they are equivalent equations.

The properties of equality that we study in algebra are very important in the process of solving equations. These properties are briefly reviewed in this section.

Rule: The Addition Property of Equality

For any real numbers a, b, and c,

$$\text{if } a = b, \text{ then } a + c = b + c$$

This property simply says that the same number may be added to (or subtracted from) both sides of an equation and the new equation will be equivalent to the original equation.

Example 1: Solving a Linear Equation Using the Addition Property of Equality

Solve the equation $x + 6 = 13$ for the value of x.

To solve this, or any, equation you must isolate the variable on one side of the equal sign. To do this, the 6 must be removed. Because it is added, we can remove it by adding its additive inverse. The additive inverse of 6 is -6, so add -6 to both sides.

$$x + 6 + (-6) = 13 + (-6)$$
$$x + 0 = 7$$
$$x = 7$$

Check this answer by substituting 7 for x in the original equation.

$$\text{Check: } x + 6 = 13$$
$$7 + 6 = 13$$
$$13 = 13$$

In solving equations, you will often be adding additive inverses to both sides.

Rule: The Multiplication Property of Equality

For any real numbers a, b, and c ($c \neq 0$), if $a = b$, then $ac = bc$.

This property says that if we multiply both sides of an equation by the same number, the new equation will be equivalent to the original one. One use of this property is to change the coefficient of the variable to a 1 when solving an equation. Another use might be the elimination of fractions from an equation by multiplying the entire equation by the least common denominator of the fractions in the equation. Look at the next two examples illustrating these concepts.

Example 2

Solving a Linear Equation Using the Multiplication Property of Equality

Solve $2h = 26$ for the value of h.

The coefficient of the variable in this equation is 2, not 1. If we multiply by the reciprocal of 2, then by the inverse property of multiplication the result will equal 1. So, using the multiplication property of equality, we multiply both sides of the equation as follows:

$$\frac{1}{2}(2h) = \frac{1}{2}(26)$$

$$h = 13$$

Check your answer by substituting 13 for h in the original equation.

$$2h = 26$$
$$2(13) = 26$$
$$26 = 26$$

Example 3

Solving a Linear Equation Containing Fractions

Solve the equation $\frac{1}{2}x - 3 = \frac{2}{3} + 3x$.

The easiest way to solve an algebraic equation containing fractions is to first multiply the entire equation by the least common denominator (LCD) of the fractions in the equation. This will result in an equivalent equation with only integer coefficients. The LCD of this equation is 6, so we will multiply all terms by 6.

$$\frac{1}{2}x - 3 = \frac{2}{3} + 3x$$

$6\left(\frac{1}{2}x\right) - 6(3) = 6\left(\frac{2}{3}\right) + 6(3x)$	(multiply each term by 6)
$3x - 18 = 4 + 18x$	(simplify)
$3x - 18 + (-18x) = 4 + 18x + (-18x)$	(add $-18x$ to each side)
$-15x - 18 = 4$	(simplify)
$-15x - 18 + 18 = 4 + 18$	(add 18 to each side)
$-15x = 22$	(simplify)
$x = -\dfrac{22}{15}$	(divide both sides by -15)

Often several steps will be required to solve a particular equation. Several properties may be used. Remember that parentheses must be removed first in the solving process if at all possible.

One of the properties of real numbers that is quite useful in algebra is the **distributive property**. This property allows us to remove parentheses in an equation so the properties of equality can be applied to find the solution. The distributive property states that $a(b + c) = ab + ac$ for all real numbers a, b, and c.

Example 4 Solving a Linear Equation Using the Distributive Property

Solve $2(x + 4) = 18$ for the value of x.

$$2(x + 4) = 18$$

Remove the parentheses with the distributive property.

$$2x + 8 = 18$$

Remove the 8 by adding its inverse, -8.

$$2x + 8 + (-8) = 18 + (-8)$$
$$2x + 0 = 10$$
$$2x = 10$$

Make the coefficient of the variable x a 1 by multiplying by the reciprocal of 2 (this is equivalent to dividing both sides by 2).

$$\frac{1}{2}(2x) = \frac{1}{2}(10)$$
$$x = 5$$

Check your answer by substituting 5 for x in the original equation.

$$2(x + 4) = 18$$
$$2(5 + 4) = 18$$

Now follow the order of operations to evaluate this expression.

$$2(9) = 18$$
$$18 = 18$$

So, x does equal 5.

A summary of the steps needed to solve any linear equation is given in Table R-1.

TABLE R-1 Solving Linear Equations

To Solve Linear Equations: **Solve:** $\dfrac{1}{2}(x + 5) = 2(x - 1) + 5$

Step	Description	Work
Step 1	Remove any grouping symbols, such as parentheses, using the distributive property.	$\dfrac{1}{2}x + \dfrac{5}{2} = 2x - 2 + 5$
Step 2	Clear the equation of all fractions by multiplying all terms on each side of the equation by the lowest common denominator of all the fractions present.	$2\left(\dfrac{1}{2}x\right) + 2\left(\dfrac{5}{2}\right) = 2(2x) - 2(2) + 2(5)$ $x + 5 = 4x - 4 + 10$
Step 3	Simplify each side of the equation by combining any like terms.	$x + 5 = 4x + 6$
Step 4	Using the addition property of equality, gather all terms containing the specified variable on one side of the equation and place all other terms on the other side of the equation.	$x - x + 5 = 4x - x + 6$ $5 = 3x + 6$ $5 - 6 = 3x + 6 - 6$ $-1 = 3x$
Step 5	Using the multiplication property of equality, find the value of the variable in question.	$-\dfrac{1}{3} = \dfrac{3x}{3}$ $-\dfrac{1}{3} = x$
Step 6	Check your answer for correctness.	$\dfrac{1}{2}\left(-\dfrac{1}{3} + 5\right) = 2\left(-\dfrac{1}{3} - 1\right) + 5$ $\dfrac{1}{2}\left(-\dfrac{1}{3} + \dfrac{15}{3}\right) = 2\left(-\dfrac{1}{3} - \dfrac{3}{3}\right) + \dfrac{15}{3}$ $\dfrac{1}{2}\left(\dfrac{14}{3}\right) = 2\left(-\dfrac{4}{3}\right) + \dfrac{15}{3}$ $\dfrac{7}{3} = -\dfrac{8}{3} + \dfrac{15}{3}$ $\dfrac{7}{3} = \dfrac{7}{3}$ So, $x = -\dfrac{1}{3}$ is correct.

Practice Set R-2

Solve each of the following linear equations.

1. $x + 6 = -4$
2. $x - 5 = -2$
3. $3x + 7 = x$
4. $2x + 3 = 5x$
5. $2(x - 8) = 4$
6. $-3(x + 3) = 12$
7. $35 - 3x = 5$
8. $24 + 2x = -8$
9. $-4x + 5 = -54 + 41$
10. $3x + 2 = -20 + 31$
11. $7 - 3x = 2 - 5x$
12. $10 - 2x = 3 + 4x$
13. $3x + 9 = 42$
14. $5x - 8 = 17$
15. $-7(2x + 3) = -7$
16. $-3(5x + 1) = -6$
17. $3(x - 2) + 2 = 11$
18. $4(x + 1) + 8 = 16$
19. $3x + 5 = -5x - 8 + 1$
20. $2x - 6 = -3x + 6 - 2$
21. $-\dfrac{3x}{8} = -\dfrac{15}{32}$
22. $\dfrac{2x}{5} = -\dfrac{7}{8}$
23. $\dfrac{x}{5} - 12 = 7$
24. $2 - f = \dfrac{f}{5}$
25. $w + \dfrac{3}{4} = \dfrac{5}{8}$
26. $2x + \dfrac{1}{3} = -\dfrac{2}{5}$
27. $\dfrac{2}{3}x + 8 = \dfrac{1}{2}$
28. $\dfrac{5}{8}x - \dfrac{1}{2} = 4$
29. $\dfrac{1}{3} + \dfrac{1}{6}x - 2 = \dfrac{2}{3}x + \dfrac{1}{2}$
30. $\dfrac{3}{4} + \dfrac{1}{2}x + 6 = -\dfrac{5}{8}x - \dfrac{1}{4}$
31. $\dfrac{3}{4}(4 - x) = 3 - x$
32. $3(x - 5) + 1 = \dfrac{1}{4}(2x - 8)$
33. $k - 76.98 = 43.56$
34. $m + 36.28 = 1.5$
35. $x - 0.5x = 12$
36. $x + 0.07x = 2.14$
37. $3.5x + 3 = x - 1.75$
38. $6 - 1.75x = 4.25 + x$
39. $-4y + 3 = 12 - 4y$
40. $2x + 1 = 6x - 5 - 4x$
41. $7a - (a - 5) = -10$
42. $3s + 5(s - 3) + 8 = 0$
43. $3(x - 5) - 5x = 2x + 9$
44. $5x - 3(1 - 2x) = 4(2x - 1)$
45. $6x - 2(3 - 4x) = 6(3x + 2)$
46. $2p + 4(p - 3) = 5p - 1$
47. $4(3 - x) + 2x = -12$
48. $3 - 2(x - 4) = 3(3 - x)$
49. $3(x - 2) + 7 = 5 - 2(x + 3)$
50. $4(2x - 8) + 1 = 3 - 4(x - 1)$

Section R-3 Percents

A **percent** is a ratio whose second term is 100. Percent means parts per hundred. The word percent comes from the Latin phrase *per centum*, which means per hundred. In mathematics, we use the symbol % for percent. For example, 5% means 5 parts per 100 parts, which can also be written 5/100, or 0.05. In working with problems involving percents, some conversions of percents to decimals or fractions may be needed. Review the following rules.

Changing a Percent to a Fraction Remove the percent sign and multiply the number by 1/100. Reduce the fraction.

$$6\% = 6 \cdot \frac{1}{100} = \frac{6}{100} = \frac{3}{50}$$

$$150\% = 150 \cdot \frac{1}{100} = \frac{150}{100} = \frac{3}{2}$$

Changing a Percent to a Decimal Remove the percent sign and multiply the number by 0.01. (This is the same as moving the decimal two places to the left.)

$$6\% = 6 \cdot 0.01 = 0.06$$

$$150\% = 150 \cdot 0.01 = 1.5$$

Changing a Fraction or Decimal to a Percent Multiply the fraction or decimal by 100 and add a percent sign.

Changing a fraction to a percent: $\quad \frac{3}{50} \cdot 100 = \frac{300}{50} = 6\%$

Changing a decimal to a percent: $\quad 0.06 \cdot 100 = 6\%$

Note: It is often easier to change a fraction to a percent by first changing it to a decimal number and then multiplying it by 100. This is especially true when using a calculator for these conversions.

Changing a fraction to a percent on the calculator: $\quad \frac{3}{4} = 3 \div 4 = 0.75 \quad 0.75 \cdot 100 = 75\%$

Remember that the percent sign is very important in determining the value of the number. The percent symbol means "out of one hundred."

$$0.05 = 5\% = \frac{5}{100} \qquad \text{(5 out of 100)}$$

$$0.05\% = \frac{0.05}{100} = \frac{5}{10,000} \qquad \text{(0.05 out of 100 or 5 out of 10,000)}$$

The basic model for problems involving percents is shown next. A percent is a ratio with a denominator of 100. Expressing the given percent as a ratio and using the cross-multiplication property to complete the problem can easily solve many percent problems. The basic model for a percent problem is as follows.

IMPORTANT EQUATIONS

$$\frac{P}{B} = \frac{r}{100}$$

where r is the rate (percent written as a ratio), P is the percentage, and B is the base.

Example 1

Finding a Percent Using a Proportion

Frances Cosgrove receives $1850 per month as salary. Her rent is $555 per month. What percent of her salary is spent for rent?

Percent = r = ?   Salary = B = $1850   Rent = P = $555

$$\frac{P}{B} = \frac{r}{100} \quad \text{(formula)}$$

$$\frac{\$555}{\$1850} = \frac{r}{100} \quad \text{(substitute values)}$$

$(\$555)(100) = \$1850r$ (cross-multiplication property)

$\$55{,}500 = \$1850r$ (simplify)

$30 = r$ (divide both sides by $1850)

Therefore, her rent is 30% of her monthly income.

Example 2

Using a Proportion to Solve a Percent Problem

When the Band Boosters finished their fall fundraiser, they had $819 in the bank. If this is 65% of their goal for the year, what is the fundraising goal for this year?

Percent = r = 65   Money raised = P = $819   Goal = B = ?

$$\frac{P}{B} = \frac{r}{100} \quad \text{(formula)}$$

$$\frac{\$819}{B} = \frac{65}{100} \quad \text{(substitute values)}$$

$(\$819)(100) = 65B$ (cross-multiplication property)

$\$81{,}900 = 65B$ (simplify)

$\$1260 = B$ (divide both sides by 65)

Therefore, the fundraising goal for the year is $1260.

Another method of solving percent problems is to translate the problem into an algebraic equation and solve it using your algebra skills. These problems can be fairly easy to solve if you remember a few key words in mathematics. Two commonly used words that you may encounter in percent problems are

of, which means "multiply"

is, which means "equal to"

To find the percent of a number using an equation, the percent is changed to a decimal and then multiplied times the original quantity, which is called the *base*. The answer to this problem is called a *percentage*. For example, 20% of 55 is solved by the computation $0.20 \times 55 = 11$. In this problem, the 0.20 is the rate (the percent), the 55 is the base (the original total quantity), and the 11 is the percentage (the answer to the percent problem). The basic model is

Rate (percent as a decimal) · Base (original total amount) = Percentage (answer)

If you know any two of the three parts of this equation, you can easily solve for the third variable. Look at the following example demonstrating this process.

Example 3 — Finding a Number if the Percent Is Known

12% *of* what number *is equal to* 6?

Our goal is to "translate" this sentence into an algebraic equation. To do this, we convert the 12% to its decimal equivalent, 0.12, locate the words "of" and "is equal to," and rewrite the sentence algebraically.

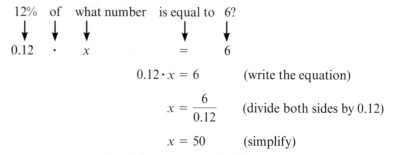

$$0.12 \cdot x = 6 \quad \text{(write the equation)}$$
$$x = \frac{6}{0.12} \quad \text{(divide both sides by 0.12)}$$
$$x = 50 \quad \text{(simplify)}$$

Check: To check your answer, calculate 12% of 50 to see if the answer is 6.

$$12\% \text{ of } 50 = 0.12 \cdot 50 = 6$$

Example 4 — Finding a Percentage

75% of $300 is how many dollars?

Translate the problem into an equation and solve. Remember to change the percent to the equivalent decimal number.

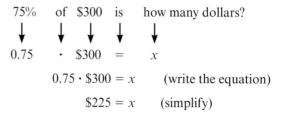

$$0.75 \cdot \$300 = x \quad \text{(write the equation)}$$
$$\$225 = x \quad \text{(simplify)}$$

The method demonstrated in Example 4 is very useful to us in our daily lives. It is used to calculate sales tax (tax rate times cost of items purchased), tips (15%–20% times the cost of the meal), amount saved at a sale (percent savings times purchase amount), and simple interest earned on investments (interest rate times investment times number of years).

Example 5 — Tipping the Waiter

At the bottom of many restaurant bills today, there is a list of the amount of money you should leave your waiter if you plan to tip 15%, 18%, or 20%. Calculate these three amounts on a dinner bill of $54.50.

$$\text{Model: } x\% \cdot \text{cost of dinner} = \text{tip}$$

(a) 15% of $54.50 = 0.15 \cdot $54.50 = $8.175 = $8.18
(b) 18% of $54.50 = 0.18 \cdot $54.50 = $9.81
(c) 20% of $54.50 = 0.20 \cdot $54.50 = $10.90

Section R-3 **Percents** 17

Example 6

Finding a Missing Percent
What percent of 200 is 13?

Translate the problem into an equation and solve. Since the final answer is supposed to be a percent, remember to express the final answer in that form.

$$x \cdot 200 = 13 \qquad \text{(write the equation)}$$

$$x = \frac{13}{200} \qquad \text{(divide both sides by 200)}$$

$$x = 0.065 \qquad \text{(simplify)}$$

Now multiply the decimal number by 100 to change it into the equivalent percent.

$$0.065 \cdot 100 = 6.5\%$$

We are often interested in the percent increase or decrease of two quantities that have changed over time. For example, if I make $6.50 an hour and receive a $0.13 per hour increase, I can calculate the percent raise that I have received using the following formula.

IMPORTANT EQUATIONS

Percent Increase/Decrease

$$\text{Percent increase/decrease} = \frac{\text{new value} - \text{original value}}{\text{original value}} \cdot 100$$

If my new salary is $6.63 per hour and my original salary is $6.50 per hour, I can calculate the percent increase of my raise as follows:

$$\text{Percent increase/decrease} = \frac{\text{new value} - \text{original value}}{\text{original value}} \cdot 100$$

$$\text{Percent increase/decrease} = \frac{6.63 - 6.50}{6.50} \cdot 100 = \frac{0.13}{6.5} \cdot 100 = 2\%$$

Therefore, I have received a 2% increase in my salary. This formula can be used to calculate percent gains and losses in the stock market, percent increases and decreases in the prices of goods and services, and so forth.

Example 7

Calculating Percent Loss
Joanna loves doughnuts. In June 2000, she bought stock in a doughnut company for $61 per share. In January 2008, she went on a diet and decided to sell her doughnut stock. The selling price at that time was $6.05 per share. Calculate her percent loss on this stock purchase.

$$\text{Percent loss} = \frac{\text{new value} - \text{original value}}{\text{original value}} \cdot 100$$

$$\text{Percent loss} = \frac{6.05 - 61}{61} \cdot 100 = \frac{-54.95}{61} \cdot 100$$

$$\approx -0.9008 \cdot 100 = -90.08\% \approx -90\%$$

Therefore, Joanna has lost 90% of her investment in the doughnut stock.

Practice Set R-3

Change each percent to a fraction. Reduce it to its lowest terms.

1. 8%
2. 15%
3. 180%
4. 150%
5. $33\frac{1}{3}\%$
6. $64\frac{1}{2}\%$

Change each percent to a decimal.

7. 5%
8. 8%
9. 0.05%
10. 0.125%
11. 1.5%
12. 4.9%
13. 125%
14. 200%

Change each fraction or decimal to a percent.

15. 0.005
16. 0.25
17. $\frac{2}{3}$
18. $\frac{7}{2}$
19. 1.5
20. 2.55

Solve the following word problems.

21. Find 15% of 75.
22. Find 27% of 250.
23. Find 1.5% of 32.
24. Find 0.75% of 36.5.
25. 2.5% of what number is 33.5?
26. 42% of what number is 25.2?
27. 15% of what number is 7.5?
28. 1.25% of what number is 0.15?
29. 306 is what percent of 450?
30. 33.5 is what percent of 150?
31. 0.375 is what percent of 25?
32. 10.2 is what percent of 85?
33. A basketball player's free throw average is the ratio of the number of free throws made to the number of attempted free throws. This number is usually reported as a percent. Calculate the free throw average for a player who makes 18 free throws out of 24 attempts.
34. Calculate the free throw average in percent form for a basketball player who makes 20 free throws out of 25 attempts.
35. Elijah's free throw average this season is 65%. Based on this statistic, if he shoots 12 free throws during a basketball game, how many do you expect him to complete?
36. Tameka is a leading scorer on her team with a free throw average of 82%. If she shoots 15 free throws in the next game, how many do you expect her to complete?
37. In many restaurants, an 18% gratuity will be automatically added to a bill for a table of eight people or more. If the total bill for a party of 10 people is $325, how much will an 18% gratuity be for this bill?
38. The standard tip for service in a restaurant is 15% of the total bill. If your meal costs $52.75, how much should you leave for the standard tip? For exceptional service, some customers tip 20% of the total bill. How much would a 20% tip be for this meal?
39. By weight, the average adult is composed of 26% skin. If a person weighs 180 lb, how many pounds of skin does this person have?
40. By weight, the average adult is composed of 17.5% bone. If a person weighs 110 lb, how many pounds of bone does this person have?
41. The local sales tax at the beach is 9%. If you purchase a bathing suit for $42.50, how much sales tax will you be charged?
42. In addition to the state sales tax rate of 5%, Wright County assesses a local tax of 1.5% on

sales in the county. How much tax money will the local government receive if I purchase a television for $395.99?

43. According to the American Medical Association, 6 out of 200 Americans sleep more than 8 hours each night. Give this ratio as a percent. If 700 people are surveyed, based on this statistic how many do you expect to sleep more than 8 hours per night?

44. According to the American Medical Association, 30% of Americans sleep 7 hours each night. If 500 people are surveyed, based on this statistic how many would you expect to sleep 7 hours per night?

45. According to the U.S. Census Bureau statistics from the 2000 census, Georgia has 5,185,965 residents ages 25 and older. Of this number, 8.3% have graduate or professional degrees. Based on these figures, how many Georgian citizens have professional or graduate degrees?

46. The U.S. Census Bureau report from 2000 states that 14.5% of Georgian workers ages 16 and older carpool to work. If there were approximately 3,832,803 workers counted in the 2000 census, how many of them carpooled to work?

47. Your hourly rate of pay is $7.75 per hour. If you receive a 4% increase in your hourly rate, what is the new rate?

48. Bernie's Burgers just increased the price of its Mega Burger by 5%. If the Mega Burger cost $2.29 before the increase, how much does it cost now?

49. The advertisement for a deskjet printer gives a sale price of $151.80. The regular price of the printer is $172.50. What is the percent discount on this printer?

50. The population of Swan Quarter was 550 in 2001. Ten years later, it had decreased to 536. By what percent did the population decrease?

51. Marisol's annual salary as an engineer is $75,000. After receiving a promotion to project supervisor, her new annual salary is $80,000. What percent increase in her salary resulted from her promotion?

52. Florida teachers received a $1250 raise this year. If Will's salary last year was $32,000, what percent raise did he receive?

Section R-4 Scientific Notation

Very large or very small numbers occur frequently in applications involving science. For example, the mass of an atom of hydrogen is 0.00000000000000000000000167 grams. On the other hand, the mass of the Earth is 6,000,000,000,000,000,000,000 tons. You can easily see how difficult it is to write these numbers as decimals or whole numbers, and then try to carry out any calculations. Trying to keep up with the number of zeros needed would drive you crazy! **Scientific notation** allows us to write these large numbers in a shorthand notation using powers of 10. The following definition gives the basic format for writing numbers in scientific notation.

Definition Scientific Notation

A positive number is written in scientific notation if it is written in the form

$$a \times 10^n$$

where $1 \leq a < 10$ and n is an integer.

If $n < 0$, then the value of the number is between 0 and 1.
If $n = 0$, then the value of the number is greater than or equal to 1 but less than 10.
If $n > 0$, then the value of the number is greater than or equal to 10.

The value of n is important in helping us determine the value of the number expressed in scientific notation. If $n > 0$, then we move the decimal point in a to the right n places denoting a large number greater than or equal to 10. If $n < 0$, then we move the decimal point in a to the left n places creating a decimal number between 0 and 1. If $n = 0$, then we do not move the decimal at all.

Example 1 — Converting a Number from Scientific Notation to Decimal Notation

Convert the following numbers to decimal notation:

(a) 6.03×10^{-5} (b) 1.3×10^6 (c) 3.102×10^0

(a) Because the exponent is negative, move the decimal in the number 6.03 to the left 5 places using 4 zeros as placeholders.

$$6.03 \times 10^{-5} = 0.0000603$$

(b) Because the exponent is positive, move the decimal in the number 1.3 to the right 6 places, adding 5 zeros.

$$1.3 \times 10^6 = 1{,}300{,}000$$

(c) Because the exponent is zero, don't move the decimal at all.

$$3.102 \times 10^0 = 3.102$$

When you are solving problems such as an exponential growth problem, your calculator may automatically switch the format of the answer to scientific notation. Many calculators have a display that reads $\boxed{1.3\ E\ 8}$ or $\boxed{1.3^{\ 08}}$. This strange notation is "calculator talk" for scientific notation and you should record the answer properly as 1.3×10^8.

To convert a positive number into scientific notation, reverse the previous process. Relocate the decimal in the number so that the value of a will be greater than or equal to 1 and less than 10. If the value of the original number is greater than or equal to 10, the exponent n will be positive and represent the number of places that you moved the decimal. If the value of the original number is between 0 and 1, the exponent n will be negative.

Example 2 — Converting a Number from Decimal Notation to Scientific Notation

Convert the mass of a hydrogen atom, 0.00000000000000000000000167 grams, and the mass of the Earth, 6,000,000,000,000,000,000,000 tons, to scientific notation.

Mass of a hydrogen atom = 0.00000000000000000000000167 grams

Relocate the decimal to the right of the number 1 giving a value for a of 1.67. The value of the number is between 0 and 1 and the decimal moved 24 places so the value of n is -24.

$$0.00000000000000000000000167 \text{ grams} = 1.67 \times 10^{-24} \text{ grams}$$

Mass of the Earth = 6,000,000,000,000,000,000,000 tons

Relocate the decimal to the right of the number 6 giving a value for a of 6. The value of this number is greater than 10, and the decimal moved 21 places so the value of n is $+21$.

$$6{,}000{,}000{,}000{,}000{,}000{,}000{,}000 \text{ tons} = 6 \times 10^{21} \text{ tons}$$

CALCULATOR MINI-LESSON

Scientific Notation

When entering numbers into a calculator in scientific notation, you generally enter the value of a and the value of n. The base 10 is understood by the calculator if you use the scientific notation keys. Look for a key on your calculator that is labeled $\boxed{\text{EE}}$ or $\boxed{\text{EXP}}$. Those are the keys used to enter numbers in scientific notation into a calculator. Some calculators may require that you put the calculator into scientific notation mode. Look for an $\boxed{\text{SCI}}$ key or check your calculator instruction booklet to be sure.

Example 3 — Calculations Using Scientific Notation

Use a calculator to find the answer to the problem $(2.3 \times 10^4)(1.5 \times 10^8)$ and express the final answer in correct scientific notation.

To enter the problem $(2.3 \times 10^4)(1.5 \times 10^8)$, type 2.3 $\boxed{\text{EE}}$ 4 $\boxed{\times}$ 1.5 $\boxed{\text{EE}}$ 8. The answer display will look similar to $\boxed{3.45 \text{ E } 12}$ or $\boxed{3.45^{12}}$.
This represents the answer 3.45×10^{12}.

If you take a chemistry course, you will learn about **Avagadro's number**, named in honor of the Italian physicist Amedeo Avagadro. Avagadro's number is absolutely huge. It is 602,200,000,000,000,000,000,000. In scientific notation the is 6.022×10^{23}. The word "mole" is used to stand for this number of atoms just as the word "dozen" stands for 12 items.

In chemistry calculations, a **mole** of atoms of any element is the number of atoms of that element that, if all piled up in one place, would have a mass equal to the atomic mass of that element, interpreted as grams. So lead (Pb) has an atomic mass of 207.2 (see any periodic table), and a mole (Avagadro's number) of Pb atoms would have a mass of 207.2 grams. Similarly, a mole of carbon atoms (atomic mass 12.0) would have a mass of 12 grams.

Example 4 — Avagadro's Number

Use the definition of Avagadro's number and the information above to answer the following questions

(a) How many atoms are in 10 grams of carbon?
(b) What would be the mass of 1.5055×10^{24} atoms of lead?

(a) Ten grams of carbon are equivalent to $\frac{10}{12} = \frac{5}{6}$ of the atomic mass of carbon. Since 12.0 grams of carbon = a mole of carbon, 10 grams are equivalent to five-sixths of a mole, or five-sixths of Avagadro's number of atoms.

$$\frac{5}{6}(6.022 \times 10^{23} \text{ atoms}) \approx 5.018 \times 10^{23} \text{ atoms}$$

So, 10 grams of pure carbon contains about 501,800,000,000,000,000,000,000 atoms.

(b) Dividing the number of atoms by Avagadro's number will give the number of moles of lead in 1.5055×10^{24} atoms of lead.

$$\frac{1.5055 \times 10^{24} \text{ atoms}}{6.022 \times 10^{23} \text{ atoms/mole}} = 2.5 \text{ moles}$$

Now each mole of lead has a mass of 207.2 grams.

$$(2.5 \text{ moles})(207.2 \text{ grams/mole}) = 518 \text{ grams of lead.}$$

Practice Set R-4

Express each number in decimal notation.

1. 2.3×10^8
2. 6.5×10^4
3. 3×10^3
4. 4×10^5
5. 6.05×10^{-1}
6. 4.15×10^{-3}
7. 3×10^{-3}
8. 4×10^{-5}
9. 7.5×10^0
10. 8.2×10^0

Express the number in scientific notation.

11. 650
12. 724
13. 3,200,000,000
14. 14,800,000
15. 0.05
16. 0.0025
17. 0.00000075
18. 0.00001203
19. 6.5
20. 8
21. distance from the Sun to the Earth: 147,000,000 km
22. mass of an oxygen molecule: 0.000000000000000000000053 g

Multiply and write the answers in scientific notation.

23. $(0.00000008)(6000)$
24. $(750,000)(23,000,000)$
25. $(2.45 \times 10^{-2})(5.1 \times 10^{-8})$
26. $(4.05 \times 10^{-6})(8.001 \times 10^8)$
27. $(3.5 \times 10^0)(1.25 \times 10^{-10})$
28. $(1.415 \times 10^{-2})(6.1 \times 10^0)$
29. $(5.15 \times 10^{12})(9.1 \times 10^8)$
30. $(6.25 \times 10^2)(3.215 \times 10^9)$
31. Approximately how many atoms are in one gram of pure gold? (See Example 4 for help with this problem.)
32. Approximately how much would an 8.0×10^{24} atom of silver weigh, in grams? (See Example 4 for help with this problem.)

Chapter R Summary

Key Terms, Properties, and Formulas

absolute value $|x|$
additive inverse (opposite)
Avagadro's number
distributive property
equivalent equations
integers
irrational numbers
mole
multiplicative inverse
natural numbers
 (counting numbers)
number line
order of operations
percent
rational numbers
real numbers
reciprocal
scientific notation
square root
whole numbers

Points to Remember

Rules for addition of signed numbers: If all numbers are positive or negative, then add as usual and keep the same sign. If one number is positive and the other negative, then find the difference of the absolute values of both numbers and give the answer the sign of the original number with the larger absolute value.

Rules for subtraction of signed numbers: Change the number to be subtracted to its additive inverse or opposite and follow the rules for addition.

Rules for multiplication and division of signed numbers: Multiply or divide the numbers as usual. If both numbers have the same sign, then the answer is positive. If the signs of the numbers are different, then the answer is negative.

Order of operations:

1. Parentheses
2. Exponents
3. Multiplication and division in order from left to right
4. Addition and subtraction in order from left to right

Addition property of equality:

$$\text{If } a = b, \text{ then } a + c = b + c$$

Multiplication property of equality:

$$\text{If } a = b, \text{ then } ac = bc \quad (c \neq 0)$$

Distributive property:

$$a(b + c) = ab + ac$$

$$\text{Percent increase/decrease} = \frac{\text{new value} - \text{original value}}{\text{original value}} \cdot 100$$

Chapter R Review Problems

1. Give the additive inverse and multiplicative inverse of $\frac{2}{3}$.
2. Give the opposite and the reciprocal of -9.

Evaluate each of the following (remember the order of operations).

3. $-|-3 + 5|$
4. $-7^2 + 9$
5. $6 - 3(4 - 7) \div 4$
6. $23.89 \div (-3.48)$

7. $3^2 + |-28| - (6^2 - 8)$

8. $-\dfrac{2}{3} \div \dfrac{4}{5} - \left(-\dfrac{1}{2}\right)$

Express each number in decimal notation.

9. 4.25×10^4

10. 5.1×10^2

11. 1.75×10^{-5}

12. 6.1×10^{-1}

Write each number in scientific notation.

13. 1,508,000

14. 8.5

15. 0.278

16. 0.000108

Solve each of the following equations.

17. $2x - 9 = -5x + 7$

18. $2x - 7x = 15$

19. $3 = \dfrac{1}{3} - 2x$

20. $-5(n - 2) = 8 - 4n$

21. $\dfrac{2x}{3} - 5 = 7$

22. $2x + 3(x - 5) = 15$

23. $-2(x - 1) - 3x = 8 - 4x$

24. $7(x - 2) - 6(x + 1) = -20$

Solve the following percent problems.

25. What is 12.5% of 150?

26. What percent of 80 is 10?

27. 4% of what number is 64?

Solve each of the following word problems.

28. Juan is buying a 2-year certificate of deposit that pays simple interest at a rate of 8% per year. He is investing $2000. How much interest will he earn at the end of 2 years? ($I = Prt$)

29. If 160 lb of an alloy contains 76.8 lb of copper, what percent of the alloy is copper?

30. Jerry's monthly salary is $1800. After he received a 1.5% raise, what was his new monthly salary?

31. A copper alloy contains 80% copper by weight. How much copper is in 2.00 kg of this alloy?

32. A brass rod has a length of 1.5671 m when measured at room temperature. If this rod expands 0.45% due to heating, what will its length be then?

33. A 50-L volume of oxygen is placed under pressure so that the volume is decreased to 46.1 L. By what percent was the volume decreased?

34. Chocolate malt balls used to sell for seventy-five cents a pound, but you see that the new price is eighty-one cents a pound. What is the percent increase?

35. McPherson's Cleaners raised the price of dry cleaning a winter coat from $4.00 to $5.00. The same percent increase was applied to all items cleaned. If the old cost of cleaning a suit was $9.00, what is the new cost?

36. Engagement rings and sets at Frank's Jewelry Store are 40% off the regular price. If the sales price of a bridal set is $599.99, find the original price of these rings.

37. Marguerite's new hourly wage as a culinary assistant is $8.25/hour. If this rate represents a 10% increase from her former hourly wage, calculate her past hourly rate.

38. Your best friend has been dieting and going to the gym 3 times a week to work out. She has just reported that her weight today was 110 lb which represents a 12% loss from her weight when she began her fitness program. Calculate your friend's former weight.

39. If the average height of male students at Alamance Community College is determined to be 70 inches ±5%, what are the heights of the tallest and shortest male students who were measured? Give your answer in feet and inches.

40. You and two of your friends go to a restaurant together. You decide to split the bill equally, including a 15% tip on the price of the meal and 7.5% sales tax. If the total cost of the food alone is $62.40, how much would your part of the bill be?

Chapter R   Test

1. $|8| = |-8| = 8$. Explain what this means.
2. What is the "additive inverse" of a number?

Simplify each of the following:

3. $2(5 - 7)^3$
4. $|3^2 - 2(6 - 5)|$
5. $\dfrac{6(-5)}{2} + 3(6 - 9) - 7$
6. Write the number 2.08×10^{-4} in decimal form.
7. Write the number 6,120,000,000 in scientific notation.

Solve each of the following equations.

8. $3(y + 7) = 2y - 5$
9. $\dfrac{5x}{6} = 2x - 7$
10. $6x + x - 0.9 = 6x + 0.9$
11. $4y - 6(y + 4) = 1 - y$
12. $7 - (3x - 1) = 5x$

Solve the following percent problems.

13. Find 15% of 85.
14. What percent of 150 is 33.5?
15. 2.5% of what number is 16.25?

Solve the following word problems.

16. If the sales tax rate in the county is 6%, find the sales tax on a purchase of $29.95.
17. It is customary to leave a waitress a 15% tip. If the total cost of a meal is $45.80, how much should you leave as a tip?
18. Maria invested $15,000 in a technology company in early 2011. By January of 2012, her investment was only worth $6500. What is the percent decrease in Maria's investment?
19. A voltmeter consistently reads 6% too high. What is the actual voltage of a certain source if this meter reads 120 volts?
20. The distance between two plates is measured to be 0.453 cm. If the measuring instrument has an error range (tolerance) of ±0.15%, what are the largest and smallest possible actual distances between these two plates?

Suggested Laboratory Exercises

These are some problems that are good for group discussion and collaborative work. Work with your chosen group to set up and solve these "fun" problems.

Lab Exercise 1   What Money?

You approach a co-worker, Alice, to get some change for the drink machine. You ask Alice to change a dollar for you. She replies that she does have some change in her pocket but can't change a dollar. Then you ask if she can change a half-dollar (I guess you have one of those old Kennedy halves on you). Alice says that she can't change a half-dollar either. At this point, you ask how much change she has. Alice replies that she has $1.15 in change. What coins (any valid U.S. coins less than $1 are possible) does Alice actually have in her pocket? How did you arrive at this answer?

Lab Exercise 2

Sudoku Puzzles

Sudoku is a logic-based, combinatorial number-placement puzzle. It is easy to learn but can be very challenging to master. The objective is to fill a 9×9 grid so that each column, each row, and each of the nine 3×3 boxes (also called blocks or regions) contains the digits from 1 to 9, only one time each. The numbers can appear in any order, and diagonals are not part of the game. Each grid will have some numbers already assigned to a location. Your goal is to fill in the others so that all numbers 1–9 are in a column, block, or row one time only. Puzzles range in difficulty levels from Easy (for beginners) to Hard (for masters). For more information and additional puzzles, go to www.sudoku.com. Try your skills with this puzzle, which is rated Easy.

		2				8	5	
8		9		3				4
	6				1			
		2			6	1	8	5
1	7	4	3		8	2	6	9
6	5	8	9			7		
			8				3	
5				9		4		2
	3	6			5			

Lab Exercise 3

Our Growing Population

The most recent census was taken in the year 2010. The goal of the census is to gather statistical data about the population of the United States as a whole and statistics about each individual state in the union. The data concerning population numbers for each state and many other interesting statistics can be obtained at www.census.gov.

For this lab, go to the website www.census.gov/dmd/www/schoolessons.html. Choose "the state of your state" worksheet, and use the data available at this website to complete the worksheet. You will learn how your state has changed in population, its percent increase or decrease in population, and what percentage of the entire United States your state's population represents. You will also be able to compare data about your state with that of other states in the United States.

1 Fundamentals of Mathematical Modeling

In this chapter you will be introduced to the concept of using mathematical (algebraic) expressions, equations, and formulas to model or describe real objects or events. This chapter includes a review of ratio and proportion, as well as strategies for solving word problems.

> *The mathematical sciences particularly exhibit order, symmetry, and limitation; and these are the greatest forms of the beautiful.*
> – Aristotle

In this chapter

- **1-1** Mathematical Models
- **1-2** Formulas
- **1-3** Ratio and Proportion
- **1-4** Word Problem Strategies
- Chapter Summary
- Chapter Review Problems
- Chapter Test
- Suggested Laboratory Exercises

Section 1-1 Mathematical Models

A model of an object is not the object itself but is a scaled-down version of the actual object. We are all familiar with model airplanes and cars. Architects and engineers build model buildings or bridges before constructing the actual structure. Machine parts are modeled by draftspersons, and nurses learn anatomy from models of the human body before working on the real thing. All models have two important features. The first is that a model will contain many features of the real object. The second is that a model can be manipulated fairly easily and studied so that we can better understand the real object. In a similar way, a **mathematical model** is a mathematical *structure* that approximates the important aspects of a given situation. A mathematical model may be an equation or a set of equations, a graph, table, chart, or any of several other similar mathematical structures.

The process of examining a given situation or "real-world" problem and then developing an equation, formula, table, or graph that correctly represents the main features of the situation is called *mathematical modeling*. The thing that makes "real-life" problems so difficult for most people to solve is that they appear to be simple on the surface but are often complicated with many possible variables. You have to study the problem and then try to connect the information given in the problem to your mathematical knowledge and to your problem-solving skills. To do this, you have to build a mental picture of just what is going on in a given situation. This mental picture is your *model* of the problem. In the real world, construction and interpretation of mathematical models are two of the more important uses of mathematics. As you work through this book, you will have many chances to construct your own mathematical models and then work with them to solve problems, make predictions, and carry out any number of other tasks.

Definition Mathematical Model

A mathematical model is a mathematical structure that approximates the important features of a situation. This model may be in the form of an equation, graph, table, or any other mathematical tool applicable to the situation.

When people attempt to construct a model that duplicates what is observed in the real world, the results may not be perfect. The more complicated the system or situation, the greater the amount of information that must be collected and analyzed. It can be very difficult to account for all possible variables or causes in real situations. For example, when you turn on your television set tonight to get the weather forecast for tomorrow, the forecaster uses a very complicated model to help predict future events in the atmosphere. The model being used is a *causal model*. No causal model is ever a perfect representation of a physical situation because it is very difficult, if not impossible, to account for all possible causes of the results observed.

Causal models are based on the best information and theory currently available. Many models used in business, industry, and laboratory environments are not able to give definitive answers. Causal models allow predictions or "educated guesses" to be made that are often close to the actual results observed. As more information is gained, the model can be refined to give better results. If this were a meteorology class, a psychology class, or similar study, then many models would be of this type.

Some things can be described precisely if there are just a few simple variables that can be easily measured. The models used to describe these situations are called *descriptive models*. The formula for the area of a rectangle, $A = lw$, models the area of a rectangle and can be used to calculate the area of a rectangular figure precisely. It may be of some comfort to you to know that most of the models that you study in this book will be of the descriptive type.

It is important to note that even the most commonly used formulas were not always so certain. The various formulas, procedures, or concepts that we use in our workplaces and around our homes were all developed over a period of time by trial-and-error methods. As *experiments* were done and the results gathered, the formulas have been refined until they can be used to predict events with a high degree of accuracy. In this textbook, you will learn to do some of the same kinds of things to develop models and procedures to solve specific problems.

Example 1

A Price-Demand Model

Suppose a grocery store sells small bunches of flowers to its customers. The owner of the store gathers data over the course of a month comparing the demand for the flowers (based on the number of bunches sold) to the prices being charged for the flowers. If he sells 15 bunches of flowers when the price is $2 per bunch but only 10 bunches when the price is $4 a bunch, he can graph this information to model the price-demand relationship.

The graph in Figure 1-1 shows us that, as the price increases, the demand for flowers decreases. This model can help the store owner set a reasonable price so that he will make a profit but also sell his flowers before they wilt!

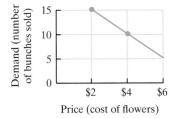

FIGURE 1-1
Price-demand model.

Example 2

Formulas as Models

Financial planning involves putting money into sound investments that will pay dividends and interest over time. Compound interest helps our investments grow more quickly because interest is paid on interest plus the original principal.

The formula $M = P(1 + i)^n$, where P = principal, n = years, and i = interest rate per period, can be used to calculate the value of an investment at a given interest rate after a designated number of years. Table 1-1 shows the value of a $1000 investment compounded yearly at 10% for 50 years. As you can see, the growth is phenomenal between the 20th and 50th year of the investment. This is an example of exponential growth calculated using a formula.

TABLE 1-1

Example 2

Number of Years	Compounded Yearly ($)
0	1,000.00
1	1,100.00
2	1,210.00
5	1,610.51
10	2,593.74
20	6,727.50
30	17,449.40
40	45,259.26
50	117,390.85

These examples illustrate two of the types of problems that we will examine throughout this book. We will use graphs, charts, and formulas to model our problems. It may be difficult to predict the future, but analyzing trends through mathematical modeling can give us insight into our world and how it works. Through modeling we can make educated guesses about the future, and, it is hoped, use these models to give us some control over our fates.

Section 1-2 Formulas

A formula is an equation that contains more than one variable. It can be thought of as a "recipe" for solving a particular type of problem. There are many standard formulas that we can access to help us solve problems. A formula can be used as a model for problems in the areas of geometry, banking and finance, and science.

The formula $d = rt$ relates distance (d) to rate (r) and time (t). This is the formula that you use every day when you decide what time to leave to arrive at a particular destination on time (or how fast to drive to get there). It is also the formula that patrol officers use to catch you speeding on the highway.

Example 1 Distance Formula

1. The Shearin family is traveling across the country to California for their summer vacation. If they plan to drive about 8 hours per day and can average 60 mph, how far can they travel per day?

$$d = rt \quad \text{(distance formula)}$$
$$d = (60\,\text{mph})(8\,\text{hr}) \quad \text{(substitute given values)}$$
$$d = 480\,\text{mi}$$

2. If they decide to travel at least 600 miles per day, how long will they be traveling each day if they can average 60 mph on the highway?

$$d = rt \quad \text{(distance formula)}$$
$$600\,\text{mi} = (60\,\text{mph})(t) \quad \text{(substitute given values)}$$
$$\frac{600}{60} = \frac{60t}{60} \quad \text{(divide both sides by 60)}$$
$$10\,\text{hr} = t$$

Example 2 The Perimeter of a Dog Pen

The perimeter of a rectangular dog pen is 40 feet. If the width is 6 feet, find the length of the pen.

$$2L + 2W = P \quad \text{(perimeter formula)}$$
$$2L + 2(6\,\text{ft}) = 40\,\text{ft} \quad \text{(substitute given values)}$$
$$2L + 12 = 40 \quad \text{(simplify)}$$
$$2L = 28 \quad \text{(add } -12 \text{ to both sides)}$$
$$L = 14\,\text{ft} \quad \text{(divide both sides by 2)}$$

Therefore, the length of the dog pen is 14 feet.

Some formulas are complicated and require the use of a scientific calculator or computer to simplify computations. Therefore, it is important that you become proficient at using your calculator. Look at the following example and use your calculator to follow the steps in calculating the correct answer.

Example 3 — The Compound-Interest Formula

Lucero deposits $2500 into a savings account that pays 4.5% interest compounded monthly. Find the value of her account after 5 years.

Let M = the maturity value of the account, P = $2500, r = 4.5% or 0.045, n = 12 months, and t = 5 years and substitute these values into the compound-interest formula.

$$M = P\left(1 + \frac{r}{n}\right)^{nt} \quad \text{(compound-interest formula)}$$

$$M = 2500\left(1 + \frac{0.045}{12}\right)^{(12)(5)} \quad \text{(substitute values)}$$

$$M = 2500(1 + 0.00375)^{60} \quad \text{(simplify fraction and exponent)}$$

$$M = \$3129.49 \quad \text{(round answer to the nearest penny)}$$

Note: For exact results, *do not* round any part of the answer until you have completed the problem. Then round the final answer to the nearest penny.

Example 4 — Calculating Your BMI

Your body mass index, or BMI, is often calculated by your doctor or other medical professional to help determine your weight status. Table 1-2 shows the categories determined by your BMI according to the Center for Disease Control (CDC).

TABLE 1-2 BMI and Weight Status

BMI	Weight Status
Below 18.5	Underweight
18.5–24.9	Normal
25.0–29.9	Overweight
30.0 and Above	Obese

Source: CDC.gov/healthyweight

To calculate a BMI, the CDC uses the following formula, where W is the person's weight in pounds and H is the person's height in inches.

$$BMI = \frac{W}{H^2} \cdot 703$$

(a) Find the weight status of a person who weighs 155 pounds and who is 5 feet 10 inches tall.

$$(5 \text{ ft})(12 \text{ in/ft}) + 10 \text{ in} = 70 \text{ in} \quad \text{(determine the height in inches)}$$

$$BMI = \frac{W}{H^2} \cdot 703 = \frac{155}{70^2} \cdot 703 \approx 22.2$$

This person's weight status is normal, according to Table 1-2.

(b) Find the weight status of a person who weights 155 pounds and is 5 feet 2 inches (62 inches) tall.

$$BMI = \frac{W}{H^2} \cdot 703 = \frac{155}{62^2} \cdot 703 \approx 28.3$$

This person's weight status is overweight, according to Table 1-2.

If you wish to look at other medical applications of formulas, you may want to do Lab Exercise 4 at the end of this chapter.

As you can see from some of these examples, we may often need to rearrange the problem to answer the question asked. One way to do this is to rewrite the given formula by solving for a specified variable. For example, in the formula $d = rt$, we can rewrite this formula solving for r ($r = \frac{d}{t}$) or solving for t ($t = \frac{d}{r}$).

To solve a formula for a given variable, we must isolate that variable on either side of the equal sign. All other variables will be on the other side. To do this, we follow the same algebraic rules and principles that we have used to solve equations in the past. Look at the following example.

Example 5

Rewriting a Formula

Solve the formula for the area of a triangle for the variable h.

$$A = \frac{1}{2}bh \qquad \text{(formula for the area of a triangle)}$$

$$2(A) = 2\left(\frac{1}{2}bh\right) \qquad \text{(multiply both sides by 2)}$$

$$2A = bh$$

$$\frac{2A}{b} = \frac{bh}{b} \qquad \text{(divide both sides by } b\text{)}$$

$$\frac{2A}{b} = h$$

The same relationship is maintained among the variables, but the formula has been rewritten to solve for the value of h.

Example 6

Rewriting an Equation

Solve $h = vt + 16t^2$ for v.

$$h = vt + 16t^2 \qquad \text{(given equation)}$$

$$h - 16t^2 = vt + 16t^2 - 16t^2 \qquad \text{(subtract } 16t^2 \text{ from both sides)}$$

$$h - 16t^2 = vt$$

$$\frac{h - 16t^2}{t} = \frac{vt}{t} \qquad \text{(divide both sides by } t\text{)}$$

$$\frac{h - 16t^2}{t} = v \qquad \text{(rewritten equation solved for } v\text{)}$$

Practice Set 1-2

Evaluate each formula using the given values of the variables.

1. Distance formula: $d = rt$, if $r = 50$ mph and $d = 125$ mi
2. Distance formula: $d = rt$, if $d = 170$ mi and $t = 2.5$ hr
3. Simple interest: $I = Prt$, if $P = \$5000$, $r = 5\%$, and $t = 2$ years
4. Simple interest: $I = Prt$, if $I = \$616$, $P = \$2800$, and $t = 4$ years
5. Area of a circle: $A = \pi r^2$, if $r = 3$ in. (use 3.14 for π)
6. Area of a circle: $A = \pi r^2$, if $A = 12.56$ ft^2 (use 3.14 for π)
7. Area of a triangle: $A = \frac{1}{2}bh$, if $A = 36$ in.^{2} and $b = 8$ in.
8. Area of a triangle: $A = \frac{1}{2}bh$, if $A = 150$ cm^2 and $h = 40$ cm
9. Hero's formula for the area of a triangle: $A = \sqrt{s(s-a)(s-b)(s-c)}$, if $a = 5$ in., $b = 12$ in., $c = 13$ in., and $s = 15$ in.
10. Hero's formula for the area of a triangle: $A = \sqrt{s(s-a)(s-b)(s-c)}$, if $a = 3$ in., $b = 4$ in., $c = 5$ in., and $s = 6$ in.
11. Conversion of Fahrenheit to Celsius: $C = \frac{5}{9}(F - 32)$, if $F = 68°$
12. Conversion of Fahrenheit to Celsius: $C = \frac{5}{9}(F - 32)$, if $F = -4°$
13. Conversion of Celsius to Fahrenheit: $F = \frac{9}{5}C + 32$, if $C = -10°$
14. Conversion of Celsius to Fahrenheit: $F = \frac{9}{5}C + 32$, if $C = 100°$
15. Slope of a line: $m = \frac{y_2 - y_1}{x_2 - x_1}$, if $(x_1, y_1) = (2, -4)$ and $(x_2, y_2) = (-1, -3)$
16. Slope of a line: $m = \frac{y_2 - y_1}{x_2 - x_1}$, if $(x_1, y_1) = (-1, 0)$ and $(x_2, y_2) = (-3, -2)$
17. z-scores: $z = \frac{x - \bar{x}}{s}$, if $\bar{x} = 100$, $x = 95$, and $s = 15$
18. z-scores: $z = \frac{x - \bar{x}}{s}$, if $\bar{x} = 17.3$, $x = 25.2$, and $s = 7.9$
19. Pythagorean theorem: $a^2 + b^2 = c^2$, if $a = 3$, and $b = 4$
20. Pythagorean theorem: $a^2 + b^2 = c^2$, if $a = 12$ and $b = 5$
21. Compound interest: $M = P\left(1 + \frac{r}{n}\right)^{nt}$, if $P = \$5000$, $r = 4.5\%$, $n = 12$, and $t = 10$
22. Compound interest: $M = P\left(1 + \frac{r}{n}\right)^{nt}$, if $P = \$12,000$, $r = 3.25\%$, $n = 12$, and $t = 5$
23. Quadratic formula: $x = \frac{-b \pm \sqrt{b^2 - 4ac}}{2a}$, if $a = 1$, $b = 5$, and $c = -6$
24. Quadratic formula: $x = \frac{-b \pm \sqrt{b^2 - 4ac}}{2a}$, if $a = 2$, $b = -9$, and $c = -5$
25. Exponential growth: $y = Ae^{rn}$, if $A = 1,500,000$ bacteria, $r = 5.5\%$ per hour, $n = 7$ hr
26. Exponential growth: $y = Ae^{rn}$, if $A = 45,000$ people, $r = 1.5\%$ per year, $n = 10$ years

Solve each formula or equation for the designated letter.

27. $I = Prt$ for r (simple interest)
28. $V = lwh$ for w (volume of a rectangular solid)
29. $A = \frac{1}{2}bh$ for b (area of a triangle)
30. $V = \frac{1}{3}Bh$ for h (volume of a pyramid)
31. $P = 2L + 2W$ for L (perimeter of a rectangle)
32. $P = 2L + 2W$ for W (perimeter of a rectangle)

33. $A = \frac{1}{2}(B + b)h$ for B (area of a trapezoid)

34. $A = \frac{1}{2}(B + b)h$ for h (area of a trapezoid)

35. $2x + 3y = 6$ for y (linear equation)

36. $2x - y = 10$ for y (linear equation)

37. $A = \frac{x + y}{2}$ for x (average of two numbers)

38. $A = \frac{x + y}{2}$ for y (average of two numbers)

39. $F = \frac{9}{5}C + 32$ for C (Celsius to Fahrenheit)

40. $C = \frac{5}{9}(F - 32)$ for F (Fahrenheit to Celsius)

41. Calculate the body mass index of a person who is 6 feet 1 inch tall with a weight of 170 pounds. What is this person's weight status? (See Example 4 in Section 1-2.)

42. What is the BMI and weight status of a person who is 5 feet tall and weighs 160 pounds? (See Example 4 in Section 1-2.)

Section 1-3   Ratio and Proportion

The distance from the Earth to the Moon is approximately 240,000 miles. The distance from Jupiter to one of its moons is approximately 260,000 miles. Although these distances differ by 20,000 miles, they are approximately the same. This relationship is indicated by the quotient, or ratio, of the distances.

$$\frac{240{,}000}{260{,}000} = \frac{12}{13}$$

The **ratio** of one number to another number is the quotient of the first number divided by the second. We can express the ratio of 7 to 8 in the following ways:

1. Use a ratio sign   7 : 8   (7 to 8)
2. Use a division symbol   7 ÷ 8   (7 divided by 8)
3. Write as a fraction $\frac{7}{8}$ (seven-eighths)
4. Write as a decimal   0.875   (eight hundred seventy-five thousandths)

When working with ratios, be sure that all distances, masses, and other measurements are expressed in the same units. For example, to find the ratio of the height of a tree that is 4 feet high to the height of a tree that is 50 inches high, you must first express the heights in the same unit. Remember that 1 foot = 12 inches, so 4 feet = 4(12) = 48 inches. Therefore, the ratio will be $\frac{48 \text{ inches}}{50 \text{ inches}}$, which reduces to $\frac{24 \text{ inches}}{25 \text{ inches}}$. Because the units are alike, they will cancel leaving the ratio as a pure number without units, 24 : 25.

Rule   Finding Ratios

1. Express the measurements in the same unit.
2. Divide the two measurements by writing them as a fraction.
3. Reduce the fraction.
4. Express the fraction using a ratio sign.

Example 1 — Simplifying Ratios

Express the following ratios in simplest form.

(a) $5:15$
(b) $18:24$
(c) $20x:35x$

(a) Remember that another way to write $5:15$ is as a fraction, $\frac{5}{15}$. Because 5 is the greatest number that will divide evenly into the numerator and denominator, the fraction can be reduced to $\frac{1}{3}$. So the ratio $5:15$ simplifies to $1:3$.
(b) The greatest common factor for 18 and 24 is 6. $18:24$ will simplify to $3:4$.
(c) The greatest common factor for $20x$ and $35x$ is $5x$. $20x:35x$ will simplify to $4:7$.

Example 2 — Simplifying the Ratios of Two Measured Amounts

State each ratio in simplest form.

(a) 20 min : 2 hr
(b) 6 ft : 120 in.
(c) 6 weeks : 3 days

(a) Remember that to find the ratio, the measurements must be in the same unit. Because 1 hr = 60 min, then 2 hr = 2(60) = 120 min. Divide each by 20 min.

$$20 \text{ min} : 120 \text{ min} = 1:6$$

(b) 1 ft = 12 in., so 6 ft = 6(12) = 72 in. Divide each by 24 in.

$$72 \text{ in.} : 120 \text{ in.} = 3:5$$

(c) 1 week = 7 days, so 6 weeks = 6(7) = 42 days. Divide each by 3 days.

$$42 \text{ days} : 3 \text{ days} = 14:1$$

A ratio is a comparison of two numbers having the same units of measure. These like units will cancel, leaving all ratios as pure fractions. A **rate** is similar to a ratio, but the units of measure are different. Because the units are different, they will not cancel but are usually reduced to a **unit rate** having a denominator of 1. For example, 55 mph is a rate that can be written as 55 mi : 1 hr.

Example 3 — Calculating Unit Rates

Find each unit rate.

(a) $4.35 for 3 lb of cheese
(b) 544 mi in 8 hr
(c) 337.5 mi on 13.5 gal of gas

(a) Write the rate as a fraction.

$$\frac{\$4.35 \div 3}{3 \text{ lb} \div 3} = \frac{\$1.45}{1 \text{ lb}} = \$1.45/\text{lb}$$

The unit rate is $1.45/pound.

(b) Write the rate as a fraction.

$$\frac{544 \text{ mi} \div 8}{8 \text{ hr} \div 8} = \frac{68 \text{ mi}}{1 \text{ hr}} = 68 \text{ mi/hr}$$

The unit rate is 68 miles/hour.

(c) Write the rate as a fraction.

$$\frac{337.5 \text{ mi} \div 13.5}{13.5 \text{ gal} \div 13.5} = \frac{25 \text{ mi}}{1 \text{ gal}} = 25 \text{ mi/gal}$$

The unit rate is 25 miles/gallon.

Example 4 Calculating Batting Averages

The rate comparing the number of hits made by a baseball player to the number of times at bat results in an important baseball statistic known as a batting average. Ty Cobb had one of the best batting averages of all time, 0.367. Calculate the batting average for a player who has 57 hits out of 210 times at bat.

Write the given numbers in fraction form, and then change it to a decimal.

$$\text{Batting average} = \frac{\text{hits}}{\text{at bats}} = \frac{57}{210} = 57 \div 210 = 0.271$$

An equation that states that two ratios or rates are equal is called a **proportion**. Proportions can be written in different ways. For instance,

$$\frac{4}{6} = \frac{2}{3} \text{ (equal ratios)} \qquad 18:20 :: 9:10 \text{ (equal ratios)}$$

The first of these proportions is read "4 is to 6 as 2 is to 3." The next proportion is read "18 is to 20 as 9 is to 10." As you can see, another way to express a proportion is to write it using colons. For example, the proportion $\frac{a}{b} = \frac{c}{d}$ can be written as $a:b :: c:d$. Both forms of the proportions are read "a is to b as c is to d."

The four quantities of a proportion are called its *terms*. So the terms of $a:b :: c:d$ are a, b, c, and d. The first and last terms of a proportion are called the **extremes**, and the second and third terms are called the **means**. Therefore, in the proportion $a:b :: c:d$, a and d are the extremes and b and c are the means.

We can use the multiplication property of equality to show that in any proportion the product of the extremes is equal to the product of the means. This property is known as the cross-multiplication property. By applying this property to the proportion $a:b :: c:d$, we get the equation $ad = bc$. This fact can be used to solve proportions.

Rule Cross-Multiplication Property

1. $a:b :: c:d$ is equivalent to $ad = bc$.
2. $\dfrac{a}{b} = \dfrac{c}{d}$ is equivalent to $ad = bc$.

Example 5 Using the Cross-Multiplication Property to Solve a Proportion

Solve the following equation for x: $\dfrac{5}{4} = \dfrac{3}{x}$

$$\frac{5}{4} = \frac{3}{x} \qquad \text{(original equation)}$$

$$5(x) = 4(3) \qquad \text{(cross-multiplication property)}$$

$$5x = 12 \quad \text{(simplify)}$$

$$x = \frac{12}{5} = 2.4 \quad \text{(divide both sides by 5)}$$

Example 6 — Using the Cross-Multiplication Property to Solve a Proportion

Solve the following equation for x: $\dfrac{2x - 3}{5} = \dfrac{x + 2}{6}$

$$\frac{2x - 3}{5} = \frac{x + 2}{6} \quad \text{(original equation)}$$

$$6(2x - 3) = 5(x + 2) \quad \text{(cross-multiplication property)}$$

$$12x - 18 = 5x + 10 \quad \text{(distributive property)}$$

$$7x = 28 \quad \text{(subtract } 5x \text{ and add 18)}$$

$$x = 4 \quad \text{(divide both sides by 7)}$$

Proportions can also be used to solve application problems that involve ratios or rates. Begin by setting up a ratio or rate using the related parts given in the problem. Then, set up a similar ratio or rate containing a variable representing the quantity that you are asked to find. The variable may be in the numerator or in the denominator. Just be sure that the units in the numerators and denominators of both fractions are alike. Then, use the cross-multiplication property to solve the problem. Look at the following examples that use these steps.

Example 7 — Solving a Word Problem Using a Proportion

John Cook is an Iowan farmer participating in a field test using a new type of hybrid corn. His first crop yielded 5691 bushels of corn last year from a 35-acre field. If he plants the same type of corn in a 42-acre field this year, assuming adequate growing conditions, how many bushels should he expect to harvest?

Let x = the number of bushels of corn.

The given rate is 35 acres yields 5691 bushels. Set up a proportion using this rate to solve for the unknown.

$$\frac{35 \text{ acres}}{5691 \text{ bushels}} = \frac{42 \text{ acres}}{x} \quad \text{(proportion)}$$

$$35x = 42(5691) \quad \text{(cross multiply)}$$

$$35x = 239{,}022 \quad \text{(simplify)}$$

$$x = 6829.2 \text{ bushels} \quad \text{(divide both sides by 35)}$$

Therefore, he should expect to harvest approximately 6829 bushels of corn from a 42-acre field.

Example 8 — Solving a Word Problem Using a Proportion

On average it takes a small construction crew 8 hours to lay 1500 square feet of plywood sub-flooring. At this rate, how long will it take this crew to lay 3750 square feet of sub-flooring?

Let x = the number of hours it will take to lay 3750 ft² of sub-flooring.
The given rate is 8 hours:1500 square feet. Set up a proportion using this rate to solve for the unknown.

$$\frac{8 \text{ hours}}{1500 \text{ ft}^2} = \frac{x}{3750 \text{ ft}^2} \quad \text{(proportion)}$$

$$1500x = 8(3750) \quad \text{(cross multiply)}$$

$$1500x = 30000 \quad \text{(simplify)}$$

$$x = 20 \text{ hours} \quad \text{(divide both sides by 1500)}$$

Therefore, this crew should be able to lay 3750 ft² of sub-flooring in about 20 hours.

Practice Set 1-3

Express the following ratios in simplest form.

1. 45 min : 2 hr
2. 1 hr : 35 min
3. 4 in. : 4 ft
4. 3 ft : 60 in.
5. 6 ft : 3 yd
6. 2 yd : 12 ft
7. 8 weeks : 16 days
8. 5 days : 1 week
9. 25 mL : 1 L
10. 2400 mL : 2 L

Find the unit rate for each of the following.

11. 304 mi on 9.5 gal of gas
12. 450 mi on 20 gal of gas
13. $3.50 for 10-min phone call
14. $86.40 for 720 kWh of electricity
15. $48 for 10 days of a classified ad
16. $340 for 40 hr of work
17. 24 lb turkey for 15 people
18. 5.5 lb hamburger for 12 people

Solve the following proportions for x.

19. $\dfrac{x}{5} = \dfrac{3}{4}$
20. $\dfrac{9}{2x} = \dfrac{6}{4}$
21. $\dfrac{30}{126} = \dfrac{5}{3x}$
22. $\dfrac{2x}{7} = \dfrac{8}{14}$
23. $\dfrac{3x + 6}{35} = \dfrac{2x - 18}{5}$
24. $\dfrac{x - 2}{7} = \dfrac{2x + 2}{28}$
25. $\dfrac{15}{18} = \dfrac{x - 1}{x}$
26. $\dfrac{3}{x + 4} = \dfrac{5}{2x + 3}$
27. $\dfrac{3}{x + 1} = \dfrac{18}{9x - 3}$
28. $\dfrac{x + 8}{6} = \dfrac{2x - 8}{3}$

Solve the following word problems using proportions.

29. Carpets Plus is having a special this week. They will install 1500 square feet of carpet for $2235. Find the unit rate for this special.

30. Mill Outlet Village is having a sale on silk this week. The cost of 4 yd of silk is $47.80. Find the unit rate for this purchase.

31. Steak is on special this week for $8.99 per pound. If there are 16 oz in a pound, find the cost per ounce.

32. Six cans of baked beans are on special for $7.14. What is the cost of one can of beans?

33. Which is the better buy: Brand X green beans costing $1.49 for 8 oz or Brand Z that costs $2.12 for 12 oz?

34. Which is the better deal: five 2-L colas for $4.95 or three 2-L colas for $2.99?

35. There are 120 calories in $\frac{3}{4}$ cup of a fruit and yo-gurt cereal. How many calories are in 1 cup of this cereal?

36. An 8 oz glass of skim milk has 90 calories. If your child drinks 28 oz of milk in a day, how many calories has he or she consumed?

37. Jewell's recipe for $2\frac{1}{2}$ dozen blueberry muffins calls for $1\frac{1}{4}$ cups of flour. How many muffins can be made with 3 cups of flour?

38. A sugar cookie recipe that makes 3 dozen cookies requires $2\frac{1}{2}$ cups of powdered sugar. If you want to make enough cookies so that each of 30 students in your child's class gets exactly two cookies, how much powdered sugar will you need?

39. Olivia needs to determine the approximate height of a tree in her yard. She measures the length of the shadow of the tree to be 20 ft. At the same time, standing beside the tree, her shadow is 10.5 ft long and her height is 5 ft 4 in. Approximately how tall is the tree?

40. Everett is 5 ft 10 in. tall and casts a shadow 8 ft 9 in. long. A flag pole next to him casts a shadow of 14 ft. How tall is the flag pole?

41. In an electric circuit, the current increases with the voltage. If the current is 1.8 A when the voltage is 18 V, what is the voltage when the current is 5.4 A?

42. When an electric current is 10 A, the electromotive force is 50 V. Find the force when the current is 25 A if the force is proportional to the current.

43. A person who weighs 110 lb on Earth would weigh approximately 19.4-lb on the Moon. How much would a 200-lb man weigh on the Moon?

44. A person who weighs 110 lb on Earth would weigh approximately 19.4 lb on the Moon. How much would a 155-lb woman weigh on the Moon?

45. Last year, a gardener used 10 lb of fertilizer to prepare his garden plot, which was 400 square feet. If he increases the size of his garden to 500 square feet this year, how much fertilizer will he need to purchase?

46. A two-cycle engine such as an outboard motor uses a gas-oil mixture that must be mixed in a ratio of 20:1. How many ounces of oil will be needed to mix with 2 gallons of gasoline in order to provide fuel for a two-cycle engine? (Note: 1 gallon = 128 fluid ounces.)

47. The amount of money paid by *JTM Magazine* for an article is based on the number of words in the article. If the magazine pays $240 for a 1200-word article, how much will it pay for an article of 1500 words?

48. An automatic doughnut machine can produce 1500 doughnuts in 9 hours. How long would it take the machine to make 300 doughnuts?

49. Licensing requirements for a day care center in Alabama require that a center have 1 adult for every 15 children ages 1 to $2\frac{1}{2}$ years old. If Tiny Tot Day Care Center has 3 adults on staff to care for toddlers in this age group, what is the maximum number of toddlers they can accept?

50. Licensing requirements for a day care center in Alabama require that a center hire 1 adult for every 6 infants from ages 3 weeks to 6 months. How many adult workers must be hired if the school enrolls 15 babies in this age bracket?

51. House plans are drawn to a scale which is stated as a rate on a blueprint. If a house plan is drawn to a scale of 1 in. = 8 ft, give the true dimensions of a room if the blueprint shows a length of $2\frac{3}{4}$ in. and a width of $1\frac{15}{16}$ in.

52. Road maps are drawn to scale. Using a scale of 1 in. = 10 mi, how many inches should a distance of 185 mi measure with a ruler on this road map?

53. The speed of military jets is often measured in Mach units. A speed of Mach 1 = 761.2 mph in the Earth's atmosphere. If a jet travels at a speed of Mach 3.1, what is the equivalent speed in mph?

54. If a jet travels at a speed of 1903 mph, what is its speed in Mach units?

55. A fencing crew can install a chain-link fence at the rate of 400 linear feet per day. How long will it take them to install 1000 linear feet of fence?

56. A single roofer can lay 8 squares of shingles in one 8-hour day. How long would it take two roofers to lay 32 squares of shingles?

57. You are asked to administer 60 mg of medication orally to a patient. In the drug cabinet you find a stock solution of this medication, which states the concentration on the label as 40 mg/mL. How many milliliters of the stock supply should you give to the patient?

58. A doctor prescribed ampicillin to a child that weighs 9.3 kg. The label on the box recommends a daily dosage of 100 mg/kg. How much medication should this child receive daily?

59. Marisol walks 1.8 miles every weekday in approximately 30 minutes. If she walks at the same rate for 45 minutes on Saturday morning, how many miles will she cover?

60. My car's speedometer has malfunctioned and now displays only my speed in kilometers per hour (kph). If 1 mph = 1.609 kph and I am traveling at a speed of 98 kph, approximately how fast am I traveling in miles per hour?

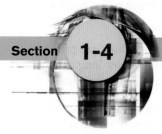

Section 1-4 Word Problem Strategies

To be successful in solving word problems, you must have a firm knowledge of the fundamental rules of algebra and be able to exercise some common sense. *Common sense* is a way of saying that it takes experience to become adept at solving word problems. How do you acquire that needed experience? There's only one way. Practice solving word problems. There will be lots of practice problems in this book. You should look at the homework and practice assignments as a chance to improve your skills. Use the five steps to solving word problems listed in Table 1-3.

Table 1-4 provides a list of key words to look for in solving word problems. These words indicate mathematical operations that are part of the problem. This is not a complete list, so space is left for you to write in any other words you discover while solving problems in this book.

Another word commonly found in word problems is the word *and*. The word *and* is not one of those listed in the table because it generally serves another purpose. It usually shows the *location of an operation*. In each of the following examples, note that the indicated operation is placed where the word *and* is located.

The sum of 7 *and* 5 is 7 + 5

The difference between 7 *and* 5 is 7 − 5

There are many other key words in problems that will tell you what the problem expects you to do. For example, *compute*, *draw*, *write*, *show*, *identify*, *graph*, and *state*. Can you think of any others?

TABLE 1-3 Five Steps to Solving Word Problems

Step 1 Read the problem. Put down your pencil and read the problem all the way through. Decide what quantity the problem is asking you to find. Remember that mathematics has a special language of its own. It may be a good idea to circle the key words that you find in the problem. Use Table 1-4 as a reference.

Step 2 Select a variable to represent one of the quantities in the problem. Then, write expressions for other unknown quantities in terms of the selected variable. The statement immediately preceding the question mark is usually the key statement. Read it carefully.

Step 3 Translate the word problem into a mathematical statement (equation/model) that describes the conditions of the problem. It may be helpful to start with a "word" equation, and then substitute the variables and given values in place of the words they represent. This is the tough step. Look for patterns or try to match a new problem to a similar one that you have already worked. Guessing is allowed here.

Step 4 Using algebra, solve the equation you have written for the value(s) that is (are) required.

Step 5 Check your solution in the *original wording* of the problem, not in your derived equation. Is your answer reasonable?

TABLE 1-4 Indicator Words

Addition	Subtraction	Multiplication	Division	Equals
plus	less	product	divided by	gives
more	minus	double	goes into	are
more than	take away	triple	quotient	is
sum	subtract	of	divide	the same as
total	subtracted from	twice	per	equals
increased by	difference	times		yields
added to	less than			
sum of	fewer			
increase of	decreased by			
	loss of			

Example 1

Solving a Simple Word Problem

Step 1 Read the problem.

One number is 12 more than another number. The sum of the two numbers is 156. What are the two numbers?

Step 2 List quantities and assign variables.

Let one number be represented by "x." If this is the first number, then the other number will be $x + 12$. Also note that the word *sum* means that the two numbers are to be added.

Steps 3 and 4 Translate into an algebraic equation and solve.

One number plus another number = 156.

$$x + (x + 12) = 156$$
$$x + x + 12 = 156 \quad \text{(remove parentheses)}$$
$$2x + 12 = 156 \quad \text{(combine like terms)}$$
$$2x = 144 \quad \text{(add } -12 \text{ to both sides)}$$
$$x = 72 \quad \text{(divide both sides by 2)}$$
$$x + 12 = 84 \quad \text{(add 12 to 72 to calculate the second number)}$$

So, the two numbers are 72 and 84.

Step 5 Check the answer.

Do our two numbers fulfill the requirements of the problem? Is one number 12 more than the other?

$$84 - 72 = 12$$

So, the answer is "yes."
  Do the two numbers add up to 156?

$$72 + 84 = 156$$

Yes, they do.

Example 2 — Problem Solving

Step 1 Read the problem.

The number of students enrolled in English 101 classes is four more than twice as many as are enrolled in Psychology 101 classes. The total enrollment for the classes combined is 160 students. How many are enrolled in each subject?

Step 2 List quantities and assign variables.

  Number of students in Psychology 101 = x

  Number in English 101 = $2x + 4$ (four more than twice Psychology 101)

Steps 3 and 4 Translate into an algebraic equation and solve.

Now we write the equation for the total number of students.

  Number of psychology students + English students = 160

$$x + 2x + 4 = 160$$
$$3x + 4 = 160 \quad \text{(combine like terms)}$$
$$3x = 156 \quad \text{(add } -4 \text{ to both sides)}$$
$$x = 52 \quad \text{(divide both sides by 3)}$$

Therefore, there are 52 students in Psychology 101 classes and

$$2x + 4 = 2(52) + 4 = 108 \text{ students in English 101 classes.}$$

Step 5 Check the answer.

Does the number of students match the requirements of the problem? Yes.

$$108 + 52 = 160 \quad \text{(total students enrolled)}$$

Example 3 Calculating Averages

Mary scored 88, 98, 82, and 75 on her first four of five math tests. What must she score on the fifth test to average a score of 85 for all five tests?

The desired average of 85 is obtained by dividing the sum of the scores by 5. So, we can multiply 85 by 5 to get the total that all five grades must add up to.

$$5(85) = 425$$

The sum of all five grades, including the unknown one, must be 425.

$$88 + 98 + 82 + 75 + x = 425$$
$$343 + x = 425 \quad \text{(combine like terms)}$$
$$x = 82 \quad \text{(subtract 343 from both sides)}$$

Mary must score an 82 on the fifth test to have an average score of 85.

Example 4 Consecutive Integers

The sum of four consecutive integers is -2. What are the integers?

First, be sure you understand what *consecutive integers* are. *Consecutive* means one directly after the other. For example, 5, 6, and 7 are consecutive integers and so are $-11, -10,$ and -9. From this brief example, you should see that to get from one integer to the next you just add 1.

Let x be the first of the four integers. The second would be $x + 1$, and the third $x + 2$, and the fourth $x + 3$.

The problem states that the sum of the four integers is -2, so we write this equation to represent the relationship

$$x + (x + 1) + (x + 2) + (x + 3) = -2$$
$$4x + 6 = -2 \quad \text{(combine like terms)}$$
$$4x = -8 \quad \text{(add } -6 \text{ to both sides)}$$
$$x = -2 \quad \text{(divide both sides by 4)}$$

Answer:

First: $x = -2$

Second: $x + 1 = -2 + 1 = -1$

Third: $x + 2 = -2 + 2 = 0$

Fourth: $x + 3 = -2 + 3 = 1$

Check:
$$-2 + (-1) + 0 + 1 = -2$$

Consider a problem involving odd or even integers. If we assign x to the first number, the second must be designated $x + 2$. This allows us to skip the number immediately next to our number. For example, if the first number is odd, the next odd number is 2 steps away ($x + 2$). If the first number is even, the next even number is also 2 steps away ($x + 2$). If a problem asks you to find three consecutive odd integers or three consecutive even integers, in order to work the problem, you would designate them as x, $x + 2$, and $x + 4$.

Example 5 — Determining an Original Value after a Percent Increase

In January, Mariana's car insurance premium was $66.98. This was a 1% price increase from her monthly cost last year. What was the monthly cost of her car insurance last year?

Let $x =$ the amount of her premium last year. The new monthly premium represents the old premium plus 1% of the old premium. Write a model to represent this situation and solve.

$$x + 0.01x = \$66.98$$
$$1.01x = \$66.98$$
$$\frac{1.01x}{1.01} = \frac{\$66.98}{1.01}$$
$$x = \$66.32$$

Therefore, Mariana's monthly car insurance premium last year was $66.32. You can check your answer by checking to see if it works in the original wording of the problem.

Check: monthly cost last year + increase = $66.98

monthly cost last year = $66.32

increase = (1%)($66.32) = $0.66

new price = $66.32 + $0.66 = $66.98

This satisfies the original wording in the problem.

Practice Set 1-4

Write an algebraic expression for each statement given below. Let x represent the unknown quantity in each one.

1. Twice a number increased by 6
2. Three less than twice a number
3. The product of a number and 7 decreased by 2
4. The sum of 3 and 4 times a number
5. Triple the sum of 4 and a number
6. Twice the difference of a number and 6
7. Five less than one-third of a number
8. A number decreased by three-fourths of itself

Different quantities are being compared in each of the following problems. Write an algebraic expression for each quantity.

9. Ann earns $5000 more per year than Bill.
   Ann's salary =
   Bill's salary =

10. Friday's class attendance was 3 more than half of Monday's attendance.
    Friday's attendance =
    Monday's attendance =

11. The length of a rectangle is 5 m more than twice the width of the rectangle.
    Length =
    Width =

12. A rod is cut into three segments and each is 1 in. longer than the previous one.
    Short rod =
    Middle-sized rod =
    Long rod =

Write an equation that properly models each problem below. Then solve the problem.

13. Five times a number plus 5 has the same value as twice the number minus 10. What is the number?

14. Sixteen less than 8 times a number is 80. Find the number.

15. Elena and Emily are twins and together they have saved a total of $72. Elena has saved twice as much as Emily. How much has each girl saved?

16. On an algebra test, the highest grade was 56 points more than the lowest grade. (In statistics, the difference between the largest number and the smallest number is known as the range.) The sum of the two grades was 128. What were the highest and lowest grades on the test?

17. A piece of rope 60 m long is cut into two pieces so that the longer piece is twice the length of the shorter. How long are the two pieces?

18. Mario is 18 years older than Joshua. If the sum of their ages is 46 years, how old is each?

19. Jeff wants to earn an A in his math course this semester. To do so, he must have an average of 90 or above. If his test scores this semester are 88, 91, and 95, what is the score that he would need on his final test to attain an average of exactly 90?

20. In an introductory algebra course, the lowest average a student can have to earn a B is 84. If Michelle hopes to earn a B in the course and her grades are 60, 92, 78, and 89, what must she score on the last test to attain this average? Will she earn a B in the course?

21. Frank's electric bill for the month of March was $85.78. The electric company charges a flat monthly fee of $20.00 for service plus $0.14 per kilowatt-hour of electricity used. Approximately how many kilowatt-hours of electricity did Frank use in March?

22. Michelle's cell phone bill last month was $69.90. Her basic monthly charge for unlimited minutes in her calling area is $29.95. Miscellaneous fees and taxes this month totaled $9.95. She is charged $0.40 per minute when she is out of her service area. How many "out of area" minutes was she charged for this month?

23. In the Las Vegas area, a taxi charges $3.20 as an initial charge, and then $0.20 for each $\frac{1}{11}$ of a mile traveled. There is also an additional $1.20 airport fee if the cab picks you up there. Calculate the fare from the airport to Lake Las Vegas, a distance of 18 miles.

24. The metered rate of fare for a New York City taxi is $2.50 upon entry and $0.40 per $\frac{1}{5}$ of a mile when the taxi is traveling 6 miles or more. Using these figures, calculate the cost of a taxi ride from the Empire State Building to JFK Airport, a distance of 16 miles.

25. The Mighty Mites scored 39 points in their basketball game on Saturday. This was one point less than twice what their opponents scored. Find the number of points scored by their opponents.

26. The Perry brothers, Jim and Gaylord, were natives of North Carolina and were outstanding pitchers in the major leagues during the 1960s and 1970s. Together they won 529 games. Gaylord won 99 more games than Jim. How many games did each brother win?

27. The sum of three consecutive integers is 87. What are the three integers?

28. Are there three consecutive integers whose sum is 100? Explain your answer.

29. The sum of three consecutive odd integers is −273. What are the numbers?

30. The sum of three consecutive odd integers is 1503. What are the integers?

31. A real estate agent estimates the value of a certain house and lot to be $175,000. If the house is worth approximately 6.5 times the value of the lot, how much is each worth separately?

32. A house costs 7 times as much as the lot it stands on, and together the house and lot cost $164,000. What is the cost of the lot?

33. You are vacationing in Japan and want to buy a steak for dinner. The steak costs 949 yen. If the exchange rate is 77 yen to the dollar, how many dollars will the steak cost you?

34. The exchange rate for the British pound is approximately 0.634 pounds = 1 U.S. dollar. If you return from Great Britain with 65 pounds, at that exchange rate how many U.S. dollars will you receive when you exchange your pounds for dollars?

35. The Indianapolis 500 is a 200-lap race covering 500 miles. In 2011, it was won by Dan Wheldon with an average speed of 170.265 miles per hour. Calculate the total time of the race.

36. In 1911, Ray Harroun won the first Indianapolis 500 in a Marmon Wasp with an average speed of 74.602 mph. Calculate the total time of this race.

37. On the first day of the semester, a history class contains an equal number of male and female students. Eight male students drop the course before the end of the semester leaving twice as many females as males in the class. What was the original number of male and female students on the roll?

38. After the 2006 election, there were two independents in the U.S. Senate and an equal number of Democrats and Republicans. Of the 100 senators, how many Republicans and Democrats were in the Senate?

39. A farmer delivers 252 lb of apples to the local grocer. The grocer divides the apples into an equal number of 5-lb and 2-lb bags. How many bags of apples did he have to sell?

40. The new movie theater complex has a total capacity of 800 occupants. There are three theaters. There are 150 more seats in Theater A than in Theater B. Theater C has 270 seats. How many seats are in Theater A?

41. Ford Motor Company had a total second quarter pretax profit of $483 million in 2007. The profit from financial services was $273 million less than the profit from the automotive division. How much profit did each sector make during this quarter?

42. In 2010, Leslie Moonves, CEO of CBS, Inc., earned approximately $3.91 million dollars less than Samuel J. Palmisano, CEO of IBM, Inc. (This comparison does not include stock options or other incentives.) Together, their annual salaries totaled $56.73 million. Find the salary of each of these CEOs.

43. During the 2010 football season, the average home attendance at Penn State games was 1044 fewer than the average home attendance at Ohio State games. If the sum of the average attendances for these two colleges is 209,512 persons, find the average home attendance for both colleges.

44. A piece of pipe is 40 in. long. It is cut into three pieces. The longest piece is 3 times as long as the middle-sized piece, and the shortest piece is 23 in. shorter than the longest piece. Find the lengths of the three pieces of pipe.

45. A radio is marked down 15% resulting in a sale price of $127.46. What was the original price of the radio?

46. Wade's Jewelers is advertising a man's watch for $168.75. The ad states that this is a 25% savings off the regular price. Find the regular price of the watch.

47. The retail price of a pair of shoes is determined by adding a 65% markup amount to the wholesale price. (Markup amount equals the markup % times the wholesale cost.) If the retail price of a pair of shoes is $125.40, calculate the wholesale price.

48. Another shoe store uses a 40% markup on the wholesale price. Find the wholesale price of a pair of shoes that sells for $63.

49. In order to encourage employees to join the company 401(k) plan, on December 31, a local company adds to each employee's account an amount equal to 20% of that employee's annual contribution. Drema's account contained $1200 on January 1, which included both her annual contributions and her employer's contribution.

How much money did Drema actually deposit into the account that year?

50. Growing up, you lived in a small town. When you returned home after attending college, the mayor proudly announced that the population had grown by 5% in the last four years. If the current population is 882, what was the population four years ago?

51. Should a young woman who weighs 120 pounds and is 5 feet tall be concerned about the status of her weight? (Use the BMI formula to help you answer this question.)

52. Should a young man who is 5 feet 7 inches tall with a weight of 195 pounds be concerned about the status of his weight? (Use the BMI formula to help you answer this question.)

Chapter 1 Summary

Key Terms, Properties, and Formulas

extremes
mathematical model
means

proportion
rate
ratio

unit rate

Points to Remember

Cross-multiplication property:

$$\text{If } \frac{a}{b} = \frac{c}{d}, \text{ then } ad = bc$$

To calculate a BMI, the CDC uses the following formula, where W is the person's weight in pounds and H is the person's height in inches.

$$BMI = \frac{W}{H^2} \cdot 703$$

Chapter 1 Review Problems

Solve each formula or equation for the indicated variable.

1. $I = Prt$ for t (simple interest)
2. $2x + 3y = 9$ for y (linear equation)
3. $C = \pi d$ for d (circumference of a circle)
4. $a + b + c = P$ for c (perimeter of a triangle)

Express the following ratios in simplest form.

5. 27 min : 3 hr
6. 4 weeks : 21 days
7. 6 in. : 5 ft

Find each unit rate.

8. $60 in 5 hr
9. 44 bushels from 8 trees
10. $12.80 for 3.5 lb fish

Solve the following proportions for x.

11. $\dfrac{x}{3} = \dfrac{4}{7}$

12. $\dfrac{2}{3} = \dfrac{8}{2x}$

13. $\dfrac{x-3}{8} = \dfrac{3}{4}$

14. $\dfrac{4x-3}{7} = \dfrac{2x-1}{3}$

Solve each of the following word problems.

15. The area of a triangle can be calculated with the formula $A = \frac{1}{2}bh$. Find the area of a triangle whose base is 3 in. and whose height is 4 in.

16. A pump can fill a 2400-L water tank in 50 min. How much water can it pump in half an hour?

17. An advertisement claims that, in a recent poll, three out of four dentists recommended brushing with Wenderoth toothpaste. If there were 92 dentists polled, how many favored Wenderoth toothpaste?

18. A telephone company charges for long-distance calls as follows: $1.24 for the first 4 minutes plus $0.28 for each additional minute or fraction thereof. If you make a long-distance call and the charges are $3.76, how many minutes long was your call?

19. A local cellular phone company charges Elaine $50 per month and $0.36 per minute of phone use in her usage category. If Elaine was charged $99.68 for a month's cellular phone use, determine the number of whole minutes of phone use.

20. A plumber charged $115 to install a water heater. This charge included $40 for materials and $30 per hour labor charge. How long did it take for the plumber to make this installation?

21. Dave has a board that is 33 in. long. He wishes to cut it into two pieces so that one is 3 in. longer than the other. How long must each piece be?

22. A piece of rope is 79 cm long. It is cut into three pieces. The longest piece is 3 times as long as the shortest piece, and the middle-sized piece is 14 cm longer than the shortest piece. Find the lengths of all three pieces.

23. Fred bought an MP3 player and a calculator for a total cost of $208. If the MP3 player cost $140 more than the calculator, how much did the calculator cost?

24. A child's piggy bank contains only dimes and nickels. There are 52 coins in the bank, and the number of dimes is 2 less than twice the number of nickels. How much money is in the bank?

25. It is 182 mi over the river and through the woods to Grandma's house. If you were to travel country roads and average only 52 mph, how long would it take you to get to Grandma's house?

26. To find the total cost of production of an item, an economist uses the formula $T = UN + F$, where T is the total cost, U is the cost per unit, N is the number of units produced, and F is fixed costs (e.g., heat, electricity, etc.). If it costs $15 per unit to make the product, fixed costs are $2500, and the total cost of production for the last week was $16,750, how many units were produced?

27. Explain why the following simplification is incorrect. Then correctly simplify the initial expression.

$$2 + 3(2x + 4) = 5(2x + 4) = 10x + 20$$

28. Write an equation that says "5 times a number plus 3 is the same as the number times 6." What is the number?

29. The second angle in a triangle is 3 times the first, and the third angle is two-thirds of the second. What size are the three angles?

30. Suppose that you are 5 feet 3 inches tall and cast a shadow 10 feet 6 inches long. If you are standing next to a tree that casts a shadow that is 52 feet long, about how tall is the tree?

Chapter 1 Test

Solve each formula for the indicated variable.

1. $V = lwh$ for w (volume of a rectangular solid)

2. $h = vt - 16t^2$ for v (height above ground of projectile)

Express the following ratios in simplest form.

3. $4 \text{ hr} : 1 \text{ day}$

4. $10 \text{ ft} : 160 \text{ in.}$

Find each unit rate.

5. $413.20 for 4 days

6. 7.5 lb for six people

7. 50 eggs from 10 chickens

Solve the following proportions for x.

8. $\dfrac{7}{12} = \dfrac{3x}{10}$

9. $\dfrac{x-4}{8} = \dfrac{2x+3}{9}$

Solve the following word problems.

10. In a sample of 85 fluorescent lightbulbs, three were found to be defective. At this rate, how many defective bulbs should be found in 510 bulbs?

11. A machine can process 300 parts in 20 min. How many parts can be processed in 45 min?

12. To purchase a wallpaper border for the living room, José needs to find the perimeter of the room to know how long the border should be. If the room is rectangular, with a length of 20 ft and width of 12 ft, how many feet of border will he need to complete the job? ($P = 2L + 2W$)

13. Car rental company A rents cars for $20 plus $0.10 per mile driven. Company B rents cars for $10 plus $0.30 per mile. For what number of miles driven would the cost of renting a car from company A equal that of company B?

14. You are scheduling passengers for cruise ships. One ship holds twice as many passengers as a second ship. Together they can hold 2250 passengers. How many passengers will the smaller ship accommodate?

15. Michelle's age is 10 years less than 5 times Sarah's age. If the sum of their ages is 44 years, find their ages.

16. A 21-ft board is cut into a long piece and a short piece. The long piece is 1 ft more than 3 times the short piece. Find the length of each piece.

17. In recent months, the price of a gallon of milk has risen by 12.3% to a new high of $2.83. What was the price per gallon before the increase?

18. If the population of a small town has decreased by 22% over a period of 8 years to a current population of 28,000 people, what was the population 8 years ago? At what yearly rate is the population dropping?

19. You wish to have an average grade of at least 84 for a math course so that you will earn a grade of B. If four tests will be given, what must you make on the fourth test if you made grades of 75, 82, and 80 on the first three? Is it mathematically possible for you to get a grade of B?

20. A house and the lot that it is built on have a combined value of $152,000. If the value of the house is 7.5 times the value of the lot, what is the value of the house and the lot separately?

Suggested Laboratory Exercises

These are some problems that are good for group discussion and collaborative work. Work with your chosen group to set up and solve these "fun" problems.

Lab Exercise 1 — A Hen and a Half?

If a hen and a half lay an egg and a half in a day and a half, how many eggs will six hens lay in seven days?

Lab Exercise 2 — Max It Out

Your instructor will give you 12 equal-length items (straws, toothpicks, etc.). Assume that each item represents a length of fence 10 ft long. You are to use these pieces of fence to construct a rectangle. You may not bend or break the individual pieces of fence. Start forming rectangles with a width of 10 ft, then a width of 20 ft, and so forth (fill in the following table). What is the largest area that may be enclosed using all of the pieces of fencing? What happens to the perimeter as the areas of the various rectangles change? Did you have any trouble completing the chart? If so, what was the problem?

Width (ft)	Length (ft)	Area (ft²)	Perimeter (ft)
10			
20			
30			
40			
50			
60			

Lab Exercise 3 — Batter Up

Baseball fanatics love to keep all kinds of statistics about players and teams. The sports pages of the newspaper are full of them during every baseball season. One odd statistic is called the *slugging percentage*. It is calculated by taking the total number of bases that the player got per hit and then dividing by the total number of times the player batted, whether he got a hit or not.

1. Let S = a single (one base per hit), D = a double (two bases per), T = a triple (three bases per), H = a home run (four bases per), and A = the total number of at bats. Write an equation for calculating a player's slugging percentage.
2. Get a baseball statistics or record book from the school library or gather statistics from a website such as www.mlb.com and calculate the slugging percentage for a current player and a retired player (e.g., Ty Cobb).
3. Compare your results with published statistics (either from a book or perhaps from the previous website) to see if your model is a good one.

Lab Exercise 4 — BMR

Your basal metabolic rate, or BMR, is the rate at which your body consumes calories for basic functions such as pumping blood, breathing, and maintaining an appropriate body temperature. In other words, it is the rate at which your body consumes calories while it is at rest. The calculation of BMR takes into consideration your gender, weight, height, and age.

The BMR equations for women and men are

Women: BMR = 655 + (4.35 × weight in pounds) + (4.7 × height in inches) − (4.7 × age in years)
BMR = 655 +   4.35w +   4.7h −   4.7a

Men: BMR = 66 + (6.23 × weight in pounds) + (12.7 × height in inches) − (6.8 × age in years)
BMR = 66 +   6.23w +   12.7h −   6.8a

The Harris Benedict equation uses the BMR to calculate the number of calories that a person expends daily based on the level of exercise. To determine your total daily calorie needs, multiply your BMR by the appropriate activity factor, as follows:

1. If you are sedentary (little or no exercise): Calorie-Calculation = BMR × 1.2
2. If you are lightly active (light exercise/sports 1–3 days/week): Calorie-Calculation = BMR × 1.375
3. If you are moderately active (moderate exercise/sports 3–5 days/week): Calorie-Calculation = BMR × 1.55
4. If you are very active (hard exercise/sports 6–7 days a week): Calorie-Calculation = BMR × 1.725
5. If you are extra active (very hard exercise/sports & physical job or 2 × training): Calorie-Calculation = BMR × 1.9

Source: www.cdc.gov/healthyweight or www.bmi-calculator.net/bmr-calculator/bmr-formula.php

If you anticipate beginning a weight-loss plan, you can use these equations to help you determine the number of calories that you should consume each day in order to lose weight. Consuming fewer calories than you burn will result in weight loss. Be sure to consult a doctor prior to beginning any weight-loss plan.

1. Maria is a moderately active 25-year-old who is 5 ft 6 in. tall and who weighs 160 pounds.
   (a) Calculate Maria's BMI to determine her weight status. (See Example 4, Section 1.2.)
   (b) Calculate her BMR.
   (c) Use her BMR to calculate the number of calories she expends daily. Round to the nearest whole number.
   (d) Maria plans to reduce her daily caloric intake by 15% in order to lose weight. How many calories should she consume daily on this diet plan? Round to the nearest whole number.

2. Go to a professional sports website (mlb.com, nba.com, usta.com nhl.com, nfl.com, or fifa.com) and look up the statistics for a famous athlete. Their height and weight will be given as part of their vital statistics. Use these numbers and assume that they are extremely active. How many calories should they be consuming to maintain their weight?

3. *Privately* calculate your own BMI and BMR numbers.
   (a) Do you think that you need to lose a little weight? Remember that you should not start on a major calorie-reduced diet without first seeking the advice of your doctor, as more than just your weight is affected by reduced caloric intake. If you have other medical conditions, these may be affected as well.
   (b) Calculate the number of calories that you would need to consume daily to maintain your current weight using the level of your usual physical activity as listed in the Harris Benedict equation.

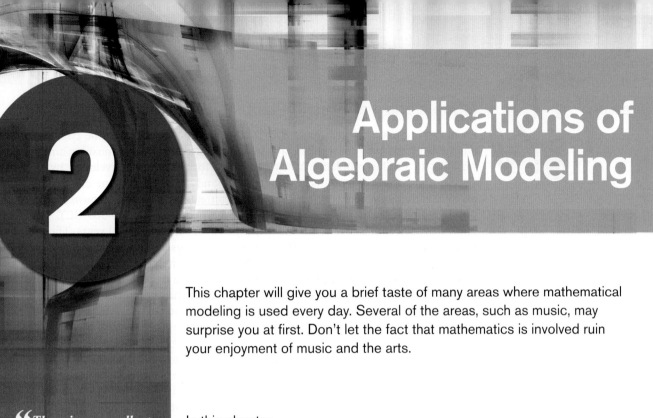

Applications of Algebraic Modeling

This chapter will give you a brief taste of many areas where mathematical modeling is used every day. Several of the areas, such as music, may surprise you at first. Don't let the fact that mathematics is involved ruin your enjoyment of music and the arts.

"There is no excellent beauty that hath not some strangeness in the proportion."
– Sir Francis Bacon

In this chapter

2-1 Models and Patterns in Plane Geometry
2-2 Models and Patterns in Triangles
2-3 Models and Patterns in Right Triangles
2-4 Models and Patterns in Art, Architecture, and Nature
2-5 Models and Patterns in Music
　　　Chapter Summary
　　　Chapter Review Problems
　　　Chapter Test
　　　Suggested Laboratory Exercises

54 Chapter 2 Applications of Algebraic Modeling

Section 2-1 Models and Patterns in Plane Geometry

Formulas are mathematical models that have been developed over the course of hundreds of years by mathematicians who noticed constant relationships among variables in similar problems. Many geometric formulas were developed by the Greeks and ancient Egyptians after much trial and error. Euclid, one of the most prominent mathematicians of antiquity, wrote a treatise called *The Elements*, which was a compilation of the geometric knowledge of the time (365–300 BC). This book was used as the core of geometry textbooks for the next 2000 years. The use of geometric formulas has made the design of buildings and construction projects much simpler. If there were no formula to calculate the area of a rectangle, for example, a person wanting to carpet a floor shaped like a rectangle would need to roll out a large sheet of paper to cover the floor, trace the outline of the floor, cut it out, take it to the store, trace it onto a piece of carpet, and cut out a suitable piece to finish the job. How much simpler it is to measure the length and width of the room and use the formula to calculate the square footage required for the job!

Perimeter and area are two basic geometric concepts that have many real-life applications. The **perimeter** of any plane geometric figure is the distance around that figure, or the sum of the lengths of its sides. Perimeter is measured in linear units, such as inches, feet, yards, meters, or kilometers. Calculating the length of a wallpaper border needed for all four kitchen walls in a house would require finding the perimeter of the room. Buying the correct amount of fencing needed to enclose a dog lot would also require the calculation of the perimeter of the lot to be surrounded.

A rectangle is a four-sided geometric figure with four right angles and opposite sides the same length. Because perimeter is the sum of the lengths of the sides of any figure, the perimeter of a rectangle such as the one in Figure 2-1 would be $l + w + l + w$ or $P = 2l + 2w$. There are other formulas for the perimeter of geometric figures, such as the square ($P = 4s$) and triangle ($P = a + b + c$), but many real-life shapes are not perfectly geometric. Therefore, it is usually best to think of the "formula" for perimeter as "add all the sides together."

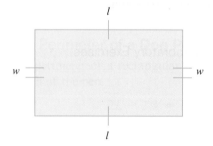

FIGURE 2-1

Example 1 Finding the Perimeter of a Rectangle

Christopher is building a $7\frac{1}{2}$-ft × 13-ft rectangular dog pen in his back yard. If fencing costs $4.37 per ft, find the cost of enclosing the dog pen.

Since the dog pen is rectangular, we can use the formula $P = 2l + 2w$ to find the perimeter and, therefore, the amount of fencing he needs to buy.

$$P = 2l + 2w$$

$$P = 2(13 \text{ ft}) + 2\left(7\frac{1}{2} \text{ ft}\right)$$

$$P = 26 \text{ ft} + 15 \text{ ft}$$
$$P = 41 \text{ ft}$$

Christopher needs 41 ft of fencing to enclose the lot, and the cost is $4.37 per foot. To find his total cost, we multiply these numbers.

$$\text{Cost} = (41 \text{ ft})(\$4.37/\text{ft}) = \$179.17$$

The total cost for enclosing the dog pen will be $179.17.

The distance around the outside of a circle is called its **circumference.** The formulas that are used to calculate the area and circumference of circles involve the use of the irrational number *pi* (π), which is derived from the ratio of the circumference of a circle to its diameter. The value of pi is approximately 3.14. The symbol π appears on your calculator and will display a value for pi that contains more than two decimal places. To calculate the distance around any circle, use the formula $C = \pi d$ or $C = 2\pi r$, where C, d, and r are the circumference, diameter, and radius, respectively, of the circle. The two formulas are equivalent, since the length of the diameter is twice that of the radius of the circle. (See Figure 2-2.)

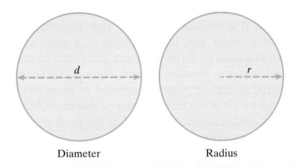

Diameter   Radius

Diameter = 2(radius) or $d = 2r$

FIGURE 2-2

Example 2 Finding the Circumference of a Circle

The diameter of a bicycle tire is 60 cm (see Figure 2-3). Through what distance does the tire go when it makes one revolution? Give the answer to the nearest centimeter.

Because the diameter of the tire is given, we will use the formula $C = \pi d$ and the rounded value 3.14 for π.

$$C = \pi d$$
$$C = (3.14)(60 \text{ cm})$$
$$C = 188.4 \text{ cm}$$
$$C \approx 188 \text{ cm}$$

FIGURE 2-3

Example 3 — Window Molding

At $0.30 per foot, how much will it cost to put molding around the window pictured in Figure 2-4?

We need to calculate the distance around the outside of the window. The window is 5 ft tall and 4 ft 6 in. wide.

(a) Adding the three straight sides together gives a total of 14 ft 6 in., or 14.5 ft.
(b) To calculate the length of the curved part of the window, we need to use the circumference formula ($C = \pi d$). Because the curve is only half a circle, we will halve the circumference after it is calculated. To make our calculations easier, we will convert 4 ft 6 in. to 4.5 ft.

$$C = \pi d$$
$$C = \pi(4.5 \text{ ft}) \approx 14.14 \text{ ft.}$$
$$\text{Length of arc} = 14.14 \text{ ft} \div 2 = 7.07 \text{ ft.}$$

(c) Total distance around the window would be:
$$14.5 \text{ ft} + 7.07 \text{ ft} = 21.57 \text{ ft}$$

(d) Cost of molding is (21.57 ft)($0.30 per foot) = $6.47

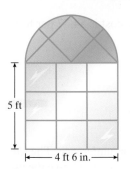

FIGURE 2-4

The **area** of any flat surface is the amount of surface enclosed by the sides of the figure. It is measured in square units (square feet, square meters, etc.) because we want to know how many squares of a given size will be needed to cover the surface in question. Buying enough grass seed to cover a lawn involves the use of area calculations. Painting a room requires that the square footage of the walls be calculated so that an adequate amount of paint can be purchased. There are area formulas for all of the basic geometric shapes, including squares, rectangles, circles, triangles, and many more. Several of the most commonly used formulas are probably familiar to you.

IMPORTANT EQUATIONS
Common Area Formulas

Circle	Square
$A = \pi r^2$	$A = s^2$

Rectangle
$A = lw$

When trying to find the answer to a word problem involving perimeter or area, first decide on the formulas needed for the problem. Then organize the data, apply the formula, and calculate the answer. Be sure the answer makes sense for your problem. Look at the following examples.

Example 4 — Wallpapering a Bedroom

Juan needs to purchase wallpaper to cover the four walls of his rectangular bedroom and a border to use at the top of his walls as a decorative accent. The dimensions of the room are 12 ft 6 in. by 15 ft 3 in., and his ceilings are 10 ft high. Calculate the amount of wallpaper needed and the length of the border he needs to buy. (For this problem, do not subtract the area of windows or doors that may be in the walls.)

(a) *Wallpaper:* This calculation will require the use of the area formula for rectangles:

$$A = lw$$

First, note that 15 ft 3 in. = 15.25 ft, and 12 ft 6 in. = 12.5 ft. Using these forms of the measurements will make the calculation easier. (See Table 2-1.)
Therefore, Juan will need 555 ft² of wallpaper to cover the bedroom walls.

TABLE 2-1

	Length × Width = Area		
	(ft)	(ft)	(ft²)
Front wall	15.25	10	152.5
Back wall	15.25	10	152.5
Side wall	12.5	10	125
Side wall	12.5	10	125
		Total area	555

(b) *Wallpaper border:* Calculation of the perimeter will be required in this instance. We need to add the lengths of the walls to calculate the total length of border that Juan needs to purchase.

$$15.25 \text{ ft} + 12.5 \text{ ft} + 15.25 \text{ ft} + 12.5 \text{ ft} = 55.5 \text{ ft of border}$$

Or, using the perimeter formula for a rectangle:

$$2l + 2w = 2(15.25 \text{ ft}) + 2(12.5 \text{ ft}) = 55.5 \text{ ft of border}$$

FIGURE 2-5

Many shapes are not standard shapes. A window, for example, might be a combination of a rectangle and a half-circle (or semicircle). (See Figure 2-5.) A roller rink is a rectangle with a half-circle on each end. Problems involving the area of a figure that is a combination of several shapes must be worked in parts. The areas of all the parts are then totaled to get the final answer.

Example 5

Area of the Floor at the Roller Rink

The hardwood floor needs to be replaced at the local roller rink. The measurements are given in the diagram found in Figure 2-6. Calculate the amount of flooring needed for this project.

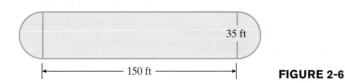

FIGURE 2-6

(a) Calculate the area of the center rectangle.

$$A = lw = (150 \text{ ft})(35 \text{ ft}) = 5250 \text{ ft}^2$$

(b) Calculate the area of the circle (two half circles). Because the diameter of the circle is given as 35 ft, we must divide it by 2 to determine the radius needed for the area formula.

$$r = 35 \text{ ft} \div 2 = 17.5 \text{ ft}$$
$$A = \pi r^2 = \pi(17.5)^2 \approx 962.113 \text{ ft}^2$$

(c) Find the total area.

$$5250 \text{ ft}^2 + 962.113 \text{ ft}^2 = 6212.113 \approx 6212 \text{ ft}^2$$

Practice Set 2-1

Use the appropriate formula to find the answers to problems 1 to 4. Round answers to the nearest tenth, if necessary.

1. Calculate the perimeter of a professional football field that is 360 feet long and 160 feet wide.

2. Calculate the circumference of a rug having a diameter of 8 feet.

3. Calculate the area of new carpet needed in a rectangular room that measures 11 feet by 15 feet. Give your answer in square feet and in square yards.

4. Calculate the area of the lawn covered by an oscillating sprinkler that sprays water in a half-circle with a radius of 10 feet.

5. A square has a perimeter of 54 in. What is the area of this square?

6. A rectangle has an area of 65 cm² and a length of 13 cm. Find the perimeter of this rectangle.

7. Julianne is planting a ground cover in one area of her yard. The area she wants to cover is rectangular shaped and measures 8.75 ft by 4.5 ft. The manager of the garden shop tells her that she will need to purchase four plants for every square foot of area she wishes to cover. How many plants will she need to buy to complete this project?

8. How many plants spaced every 6 in. are needed to surround a circular walkway with a 25-ft radius?

9. Mike is buying a rectangular building lot measuring 220 ft by 158 ft in a new golfing community. He plans to sow new grass on this lot, and each 40-lb. bag of turf grass blend will cover 1000 ft² of lawn. How many bags does he need to buy?

10. A builder is completing the flooring in the living room shown in the drawing. He needs to purchase an adequate amount of baseboard molding and enough carpet to cover the floor. The openings for the two doors are 3 ft wide. Calculate the number of feet of molding he must purchase. Then calculate the number of *square yards* of carpet needed to complete the job. (*Remember:* 1 yd = 3 ft, so 1 yd² = 9 ft²)

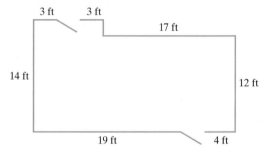

11. A soup can is in the shape of a cylinder. If it is cut and flattened into geometric shapes, it consists of two circles and a rectangle. If a can is 4.5 in. tall and measures 7.9 in. around, how many square inches of paper will be needed for the label on the can? Round to the nearest tenth.

12. If the soup can in problem 11 has a diameter of 2.5 in., how many square inches of metal will be needed to make the can? (Ignore the overlap of metal needed when actually producing the can.) Round to the nearest tenth.

13. Will a rectangular piece of wrapping paper that is 20 in. by 15 in. cover a shirt box that is 12 in. long, 8 in. wide, and 3 in. deep? Explain.

14. The poster you are giving your friend is in a cardboard tube with a 2-in. diameter and 26-in. length. Will a rectangular piece of wrapping paper that is 8 in. by 28 in. completely cover the tube? Explain.

15. PIZZAZZ Pizza Parlor sells an 8-in. pepperoni pizza for $6.99. The price of its 16-in. pizza

is $15.95. Which pizza is the better buy per square inch?

16. A 12-in. by 24-in. rectangular pizza is the same price as a pizza with a 10-in. diameter. Which shape will give you more pizza?

17. Find the area of the shaded parts of the following figure. Round to the nearest tenth.

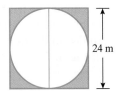

18. Find the area of the shaded parts of the following figure. Round your answer to the nearest tenth.

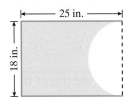

19. A rectangular lot with a 50-ft frontage and a 230-ft depth sold for $8625. Find the following:
    (a) the cost of the lot per square foot
    (b) the cost of the lot per front foot
    (c) the cost of the lot per acre (1 acre = 43,560 ft²)

20. A cement walk 3 ft wide is placed around the outside of a rectangular garden plot that is 20 ft by 30 ft. Find the cost of the walk at $12.25 a square yard.

21. What effect would each of the following have on the area and perimeter of a square?
    (a) doubling the length of the sides
    (b) halving the length of the sides
    (c) tripling the length of the sides
    (d) doubling the length of the diagonal distance between opposite corners

22. What effect would each of the following have on the area and circumference of a circle?
    (a) doubling the diameter
    (b) halving the diameter
    (c) doubling the radius

23. Eight circles of equal size are inscribed in a rectangle so that four are in each row. The circles touch each other and the sides of the rectangle. What percent of the area of the rectangle is covered by the circles?

24. Can two geometric figures have the same perimeter and area and yet not be identical? Explain how you arrived at your answer.

25. Let us revisit the window in Example 3. To determine the amount of molding (L) needed to go around the outside of this shape of window, you need to add one width (w), two heights (h), and the distance around the semicircular top (half of the circumference of a circle whose radius is half of the width: $C = 2\pi r$). So, a formula for calculating the perimeter would be written as:

$$w + 2(h) + \pi\left(\frac{w}{2}\right) = \text{amount of molding needed}$$

$$w + 2h + \pi \cdot \frac{w}{2} = L$$

Assume that you have 20 ft ($L = 20$ ft) of molding that you may use.

$$w + 2h + \pi \cdot \frac{w}{2} = 20 \text{ ft}$$

(a) Solve this equation for h.
(b) Construct a graph of height versus width.
(c) Make a table of values by choosing various widths in whole feet and calculating the value that the height would need to be to keep a constant perimeter of 20 ft.
(d) How many of such windows of different widths and heights can be constructed with a perimeter of 20 ft?

26. A rectangular backyard is roughly 70 ft by 90 ft. The back wall of the house extends for 40 ft along one of the longer sides of the backyard. How many feet of fencing would be needed to fence in the entire backyard, anchoring the fence to the back corners of the house on each end?

27. The owner of the house in problem 26 wants to put a small fish pond in the backyard. If the pond is to be 8 feet in diameter, what will its area be?

28. How many square feet of sod would be needed to cover the backyard in problems 26 and 27, after the pond is installed?

29. If brick pavers cost $0.59 each, how much would it cost to put a circular walkway 2 feet wide around the fish pond described in problem 27? For the purposes of this problem, assume that four pavers laid together cover approximately 1 square foot.

30. Referring back to problems 26 to 29, if sod costs $3.00 per square yard (plus installation labor costs, of course), how much will the sod cost to cover the backyard after the pond and walkway are installed?

Section 2-2 Models and Patterns in Triangles

A triangle is the simplest of the polygons, having only three sides and three angles. Angles are measured in degrees and classified according to the number of degrees in each. An **acute angle** measures between 0° and 90°; a **right angle** measures exactly 90°; an **obtuse angle** measures between 90° and 180°. Similarly, a triangle can be classified by the sizes of its angles. While the sum of the three angles in every triangle is 180°, if all angles are acute angles, the triangle is classified as an **acute triangle**. If one of the angles is an obtuse angle, the triangle is classified as an **obtuse triangle**. If one of the angles is a 90° angle or right angle, the triangle is classified as a **right triangle**. Right triangles have special properties and will be studied in depth in the next section. These classifications of triangles are illustrated in Figure 2-7.

Acute Triangle
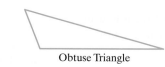
Obtuse Triangle

Right Triangle

FIGURE 2-7

Example 1 Angle Measurement

A triangle has two angles that measure 20° and 42°. Calculate the size of the third angle and classify the triangle as right, acute, or obtuse.

Since the total number of degrees in a triangle is 180°, we subtract the measures of the two given angles from 180° to find the measure of the third.

$$180° - 20° - 42° = 118°$$

The third angle measures 118° which is an obtuse angle. Therefore, this triangle is an obtuse triangle.

Triangles can also be classified by the lengths of their sides. A triangle having three equal sides is called an **equilateral triangle**. In addition to having three equal sides, an equilateral triangle has three equal angles each measuring 60°. An **isosceles triangle** has two sides that are the same length and two angles that are the same size. A **scalene triangle** has three sides that are all different lengths and three angles of different measurements. These classifications of triangles are illustrated in Figure 2-8.

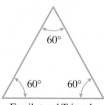
Equilateral Triangle
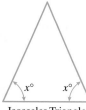
Isosceles Triangle

Scalene Triangle

FIGURE 2-8

Each corner of a triangle has a point called the vertex (plural: vertices). Every triangle has three vertices which are usually labeled with capital letters used to name the triangle. The base of a triangle can be any one of the three sides but is usually the one drawn at the bottom of the triangle. In an isosceles triangle, the base is usually the unequal side. The length of the base is used along with the length of the height for calculating the area of the triangle. The height of a triangle is a perpendicular

Section 2-2 **Models and Patterns in Triangles** 61

line segment from the base to the opposite vertex. Look at the labeled parts of $\triangle ABC$ shown in Figure 2-9.

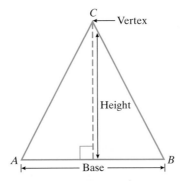

FIGURE 2-9

Recall from Section 2-1 that the **area** of any flat surface is the amount of surface enclosed by the sides of the figure. Remember that the units associated with area are square units such as cm², ft², or m². In this section, we will look at two area formulas that can be used to calculate the area of a triangle.

IMPORTANT EQUATIONS

Area of a Triangle

$A = 0.5bh$ where $b =$ the length of the base and $h =$ the length of the height

---

In order to calculate the area of a triangle using this formula, the lengths of the base b and the height h must be known. The height is a perpendicular line segment drawn from a vertex to the opposite side which we call the base. In a right triangle, one of the sides (or legs) forming the right angle serves as the height while the other side (or leg) forming the right angle serves as the base. Additionally, if we designate the unequal side of an isosceles triangle as the base, the height will divide the base into two equal lengths. Look at the following example illustrating the use of this formula to calculate the area of three triangles.

Example 2 — **Calculating the Area of a Triangle**

Use the area formula $A = 0.5bh$, where $b =$ base and $h =$ height, to calculate the area of the following triangles.

(a)

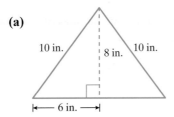

Since this triangle is an isosceles triangle, the height divides the base into two equal parts. Before we use the formula, we must calculate the length of the base.

$$\text{Base} = 2(6 \text{ in.}) = 12 \text{ in.}$$

$$h = 8 \text{ in.} \quad b = 12 \text{ in.}$$

$$A = 0.5(12 \text{ in.})(8 \text{ in.}) = 48 \text{ in.}^2$$

(b)

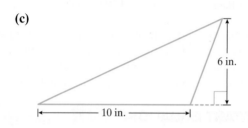

In a right triangle, the legs serve as the base and height of the triangle.

$$h = 9 \text{ cm} \quad b = 6 \text{ cm}$$
$$A = 0.5(6 \text{ cm})(9 \text{ cm}) = 27 \text{ cm}^2$$

(c)

$$h = 6 \text{ in.} \quad b = 10 \text{ in.}$$
$$A = 0.5(10 \text{ in.})(6 \text{ in.}) = 30 \text{ in.}^2$$

In many real-life applications where we are required to find the area of an object in the shape of a triangle, we may not be able to find the height of the triangle. Another formula, called Hero's formula, can also be used to find the area of a triangle using only the lengths of the sides of the triangle.

IMPORTANT EQUATIONS

Hero's Formula for the Area of a Triangle

$$A = \sqrt{s(s-a)(s-b)(s-c)}$$

where

$$s = \frac{a+b+c}{2} \quad \text{(called the semi-perimeter)}$$

and a, b, and c are the lengths of the sides of the triangle.

Hero's formula avoids the problem of trying to measure or calculate the height of a particular triangle.

Example 3 Making Pennants

Calculate the amount of material that will be needed to make 10 pennants in the shape of a triangle if the sides of each triangle measure 12 in., 20 in., and 20 in. (See Figure 2-10.)

Because we know only the lengths of the sides of the triangle, we can use Hero's formula to calculate the area of each flag.

Section 2-2 **Models and Patterns in Triangles** 63

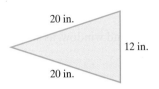

FIGURE 2-10

Step 1

$$s = \frac{a + b + c}{2}$$

$$s = \frac{12 + 20 + 20}{2}$$

$$s = 26 \text{ in.}$$

Step 2

$$A = \sqrt{s(s - a)(s - b)(s - c)}$$

$$A = \sqrt{26(26 - 12)(26 - 20)(26 - 20)}$$

$$A = \sqrt{26(14)(6)(6)} = \sqrt{13,104} = 114.5 \text{ in.}^2$$

Because 10 pennants are to be made, at least 10×114.5 in.$^2 \approx 1145$ in.^{2} of material will be needed.

We can convert this answer to square feet by dividing the total number of square inches by 144 in.^{2} (because 1 ft^2 = 144 in.^{2}).

$$\frac{1145 \text{ in.}^2}{144 \text{ in.}^2/\text{ft}^2} = 7.95 \text{ ft}^2 \quad \begin{array}{l}\text{(or approximately 8 ft}^2 \text{ of material} \\ \text{will be needed for these pennants)}\end{array}$$

When calculating the area of a triangular figure, determine the formula you will use based on the information that you are given. If you know the base and height of a triangle, the formula $A = 0.5bh$ is simple and easy to use. However, if you don't know the height of the triangle but can determine the lengths of the three sides, you can use Hero's formula to calculate the area.

Example 4

Area of an Irregular Figure

Calculate the area of the shaded part of Figure 2-11.

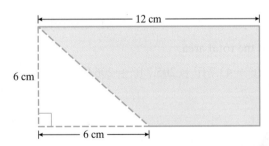

FIGURE 2-11

To solve this problem, we must first calculate the area of the rectangle ($A = lw$) and then subtract the area of the triangle. Because we know the length of the base and height of the triangle, we use the formula $A = 0.5bh$ to find its area.

In the rectangle, $l = 12$ cm and $w = 6$ cm

In the triangle, $b = 6$ cm and $h = 6$ cm

Rectangle: $A = lw = (12 \text{ cm})(6 \text{ cm}) = 72 \text{ cm}^2$
Triangle: $A = 0.5bh = 0.5(6 \text{ cm})(6 \text{ cm}) = \underline{-18 \text{ cm}^2}$
54 cm^2

The area of the shaded portion of the figure is 54 cm^2.

Example 5 — Area of the Wall of a Shed

Calculate the area of the side of the shed, including the door and window, as shown in Figure 2-12.

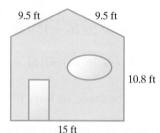

FIGURE 2-12

We will subdivide the figure into two parts, a rectangle and a triangle, calculate the areas of each, and total the areas. Because we are not given a height for the triangular portion, we must use Hero's formula to calculate the area.

(a) Calculate the areas of the rectangle having $l = 15$ feet and $w = 10.8$ feet.

$$A = lw = (15 \text{ ft})(10.8 \text{ ft}) = 162 \text{ ft}^2$$

(b) Calculate the area of the triangle using Hero's formula. Let $a = 9.5$ feet, $b = 9.5$ feet, and $c = 15$ feet. Round the answer to the nearest tenth.

Step 1
$$s = \frac{a + b + c}{2}$$
$$s = \frac{9.5 + 9.5 + 15}{2}$$
$$s = \frac{34}{2} = 17$$

Step 2
$$A = \sqrt{s(s - a)(s - b)(s - c)}$$
$$A = \sqrt{17(17 - 15)(17 - 9.5)(17 - 9.5)}$$
$$A = \sqrt{17(2)(7.5)(7.5)}$$
$$A = \sqrt{1912.5} \approx 43.7 \text{ ft}^2$$

(c) Find the total area.

$$162 \text{ ft}^2 + 43.7 \text{ ft}^2 = 205.7 \text{ ft}^2 \approx 206 \text{ ft}^2$$

Practice Set 2-2

Calculate the size of $\angle C$, the third angle in each triangle, and classify the triangle as acute, obtuse, or right in problems 1 to 6.

1. $\angle A = 35°$ and $\angle B = 55°$
2. $\angle A = 25°$ and $\angle B = 40°$
3. $\angle A = 101°$ and $\angle B = 36°$
4. $\angle A = 78°$ and $\angle B = 43°$
5. $\angle A = 90°$ and $\angle B = 29°$
6. $\angle A = 32°$ and $\angle B = 51°$

Find the area of the figures below. Remember to label your answer with the correct units.

7.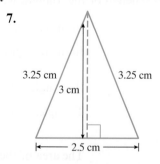

Section 2-2 **Models and Patterns in Triangles** 65

8.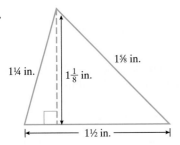

9.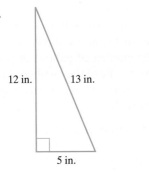

10.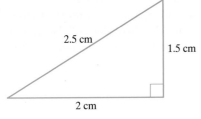

11.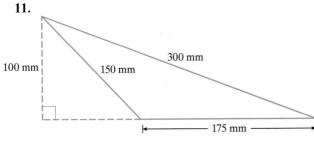

12.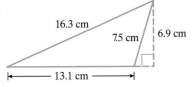

13.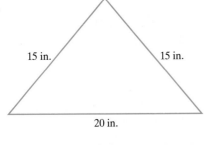

14.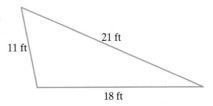

Find the area of the shaded parts of the figure below.

15.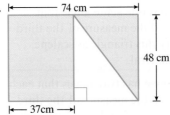

16. $r = 5$ in.

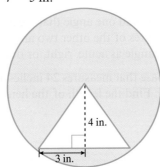

17. $\triangle ABC$ is an equilateral triangle with $AC = 10$ cm.

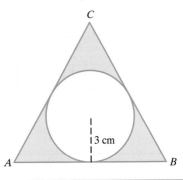

18.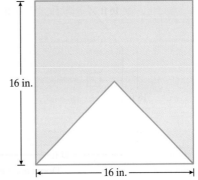

Complete the following word problems using the appropriate formulas based on the information given in the problem. Draw diagrams as needed to assist you with the solution.

19. A right triangle has one acute angle that measures 45°. Find the measure of the third angle and classify this triangle as scalene, isosceles, or equilateral.

20. A right triangle has one acute angle that measures 38°. Find the measure of the third angle and classify this triangle as scalene, isosceles, or equilateral.

21. An isosceles triangle has two angles that each measures 32°. Find the measure of the third angle and classify the triangle as acute, right, or obtuse.

22. An isosceles triangle had one angle that measures 98°. Find the measures of the other two angles and classify the triangle as acute, right, or obtuse.

23. A triangle has a base that measures 24 inches and an area of 132 in². Find the length of the height of this triangle.

24. A triangle has a height that measures 16 cm and an area of 104 cm². Find the length of the base of this triangle.

25. A sailor wishes to make a new sail for his boat. The sail is triangular shaped and measures 18 ft by 11 ft by 25 ft. Calculate the number of square feet of material that will be needed for this sail.

26. The Dodsons are planning to paint the outside of their house this spring. The paint can states that 1 gallon of paint will cover 350 ft² of surface. Using the dimensions of the house shown in the drawing, calculate the number of gallons of paint (rounded to the nearest gallon) they will need to buy to complete the project. (Don't include paint for the doors or windows!)

27. Tameka's back yard has three trees that form a small triangle. She is planning to plant a flower garden in this triangular area and must calculate the area in order to buy the correct amount of fertilizer and mulch. She measures the following distances for the three sides of the triangle: 10.5 ft, 8.25 ft, and 5.25 ft. Calculate the area of this garden.

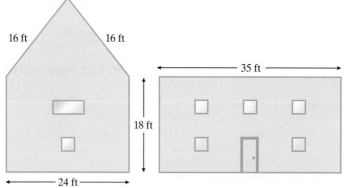

Square windows are 2.5 ft by 2.5 ft
Rectangular windows are 2.5 ft by 6 ft

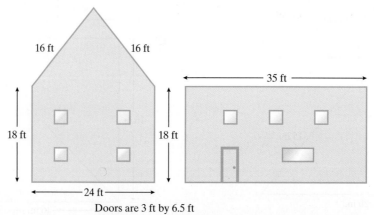

Doors are 3 ft by 6.5 ft

28. Miriam is sewing an appliqué of an ice cream cone on a flag which the local ice cream shop will use as a display for advertisement. Use the diagram to find the amount of material needed for the ice cream cone appliqué.

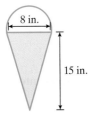

The Great Pyramid of Giza

29. The Great Pyramid of Giza is one of the Seven Wonders of the World and until the 19th century, was the tallest building in the world. The base of the pyramid measures 756 ft and the edges measure 612 ft. Each face is an isosceles triangle. Find the area of one of the triangular faces of the pyramid.

30. If each block used to build the Great Pyramid of Giza has a face measuring 10 ft by 10 ft, approximately how many blocks would be needed for one face of the triangle? To find your answer, use your results from problem 29. Round to the nearest whole number.

Section 2-3 Models and Patterns in Right Triangles

The triangle is structurally the strongest of the polygons. Triangles are evident in the construction of tall towers and bridges. Right triangles are especially useful in building projects. The framing of a house requires that the studs be braced with boards that form right triangles. Tall towers are often braced with guy wires to give support to the tower. The tower, the ground, and the guy wire form a right triangle.

In a right triangle, the sides of the triangle have lengths that demonstrate certain relationships. One of these relationships is stated by the **Pythagorean theorem.**

Theorem   Pythagorean Theorem

In a right triangle, the sum of the squares of the legs of the triangle equals the square of the hypotenuse, or as illustrated in Figure 2-13:

$$a^2 + b^2 = c^2$$

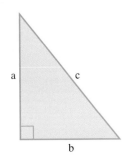

FIGURE 2-13

This theorem is only true for right triangles. It allows us to calculate the length of the third side of a right triangle if the lengths of the other two sides are known. Remember that the **hypotenuse** (c) is always the longest side of the triangle and is directly opposite the right angle. The other two sides of the triangle (a and b) are called the legs.

Example 1 — Using the Pythagorean Theorem

Find the length a in the right triangle shown in Figure 2-14.

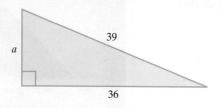

FIGURE 2-14

The base b is 36, and the hypotenuse c is 39. Substitute these values into the Pythagorean theorem.

$$a^2 + b^2 = c^2$$
$$a^2 + 36^2 = 39^2 \quad \text{(substitute the given values)}$$
$$a^2 + 1296 = 1521 \quad \text{(simplify by squaring)}$$
$$a^2 + 1296 - 1296 = 1521 - 1296 \quad \text{(subtract 1296 from both sides)}$$
$$a^2 = 225$$
$$a = \sqrt{225} = 15 \quad \text{(take the square root of 225)}$$

Example 2 — Guy Wires for the Communications Company

A 40-foot tower will be constructed for a cellular communications company. Guy wires will be needed for stability and will be attached 5 feet below the top of the tower. The wires will be anchored to the ground at a distance of 50 feet from the base of the tower. (Assume the tower makes a right angle with level ground.) What length of wire will be needed if the tower is to be stabilized with four guy wires? (See Figure 2-15.)

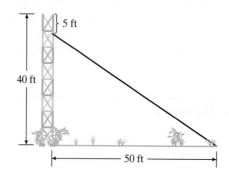

FIGURE 2-15

$$\text{Let side } a = 40 \text{ ft} - 5 \text{ ft} = 35 \text{ ft}$$
$$\text{side } b = 50 \text{ ft}$$
$$\text{side } c = ?$$
$$a^2 + b^2 = c^2$$
$$(35)^2 + (50)^2 = c^2$$
$$1225 + 2500 = c^2$$

$$3725 = c^2$$
$$\sqrt{3725} = c$$
$$61.03 = c$$

Because we need to have four guy wires, the total amount of wire needed will be:

$$4 \times 61.03 \text{ ft} = 244.12 \text{ ft},$$

or approximately 245 ft to complete the job.

In trying to solve some geometry problems, figures can be subdivided into right triangles in order to calculate certain lengths. An isosceles triangle, for example, can be divided into two congruent right triangles by drawing in a perpendicular segment called the height from the vertex angle of the triangle to the base. (See Figure 2-16.)

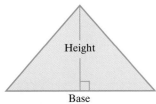

FIGURE 2-16

Example 3

Pitch of the Roof

Calculate the pitch (or slope) of the roof shown in Figure 2-17.

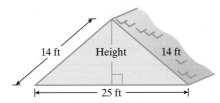

FIGURE 2-17

If a perpendicular line is drawn from the peak of the roof to a horizontal line connecting the ends of the house, two right triangles are formed. Remember that the height of an isosceles triangle will bisect the base.

Using the Pythagorean theorem, we can calculate the length of h.

$$a^2 + b^2 = c^2$$
$$12.5^2 + h^2 = 14^2$$
$$h = \sqrt{14^2 - 12.5^2} = 6.3 \text{ ft}.$$

Therefore, the slope of this roof line will be

$$\frac{6.3 \text{ ft}}{12.5 \text{ ft}} = 0.504 \approx \frac{1}{2}.$$

Example 4

The Area of a Composite Figure Using the Pythagorean Theorem

Find the painted area of the end of the house pictured in Figure 2-18. Do not use Hero's formula to find this area.

To find the area of the end of the house, we divide it into a rectangular portion and a triangle. We calculate the area of the rectangular portion using the formula $A = lw$, and then subtract the area of the two windows as calculated with the same formula.

Area = $(18)(24) = 432 \text{ ft}^2$ (total rectangular area)

Area = $(2.5)(2.5) = 6.25 \text{ ft}^2$ (area of the small window)

Area = $(2.5)(6) = 15 \text{ ft}^2$ (area of the rectangular window)

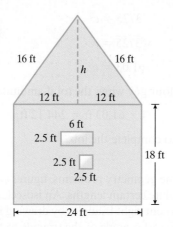

FIGURE 2-18

Therefore, $432 - 6.25 - 15 = 410.75$ ft² is the painted area of the rectangular portion of the house.

Next, we need to calculate the area of the triangular portion of the end of the house. The easiest area formula for a triangle is $A = 0.5bh$. In this problem, the base is 24 ft. We do not know the height, but we can calculate the height using the Pythagorean theorem.

$$a^2 + b^2 = c^2$$
$$h^2 + 12^2 = 16^2$$
$$h = \sqrt{16^2 - 12^2} = \sqrt{112} \approx 10.6 \text{ ft}$$

Therefore,

$$\text{Area of the triangular portion} = 0.5bh$$
$$\text{Area of the triangular portion} = 0.5(24)(10.6)$$
$$\text{Area of the triangular portion} = 127.2 \text{ ft}^2$$

Finally, to find the total painted area of the end of the house, we add our two results.

$$410.75 \text{ ft}^2 + 127.2 \text{ ft}^2 = 537.95 \text{ ft}^2 \text{ of painted surface}$$

Practice Set 2-3

1. What information must you have about a triangle *before* you know that you can use the Pythagorean theorem?

2. Does a triangle with sides of lengths 1.5 ft, 2.5 ft, and 2 ft form a right triangle? How do you know?

3. $a = 9$, $b = 12$

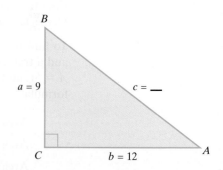

Use the Pythagorean theorem to find the missing side of the following right triangles. (*Note:* **Round to the nearest tenth.**)

4. $b = 25$, $c = 35$

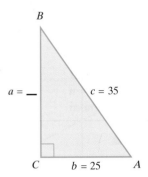

5. $b = 6$, $c = 12$

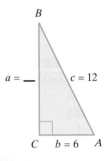

6. $a = 45$, $b = 45$

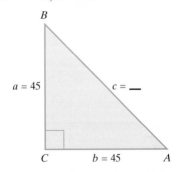

7. $a = 10$, $c = 26$

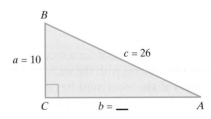

8. $a = 7$, $c = 25$

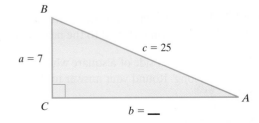

Complete the following word problems. Draw a sketch to help you solve each problem.

9. The bottom of a 25-ft ladder leaning against a house is 7 ft from the house. If the ground is level, how high on the house does the ladder reach?

10. A 6-ft ladder is placed against a wall with its base 2.5 ft from the wall. How high above the ground is the top of the ladder?

11. A rope 13 m long from the top of a flagpole will just reach a point on the ground 9 m from the foot of the pole. Find the height of the flagpole.

12. Clint wants to keep a young tree from blowing over during a high wind. He decides that he should tie a rope around the tree at a height of 180 cm, and then tie the rope to a stake 200 cm from the base. Allowing 30 cm of rope for doing the tying, will the 270-cm rope he has be long enough to do the job?

13. A flat computer monitor measures approximately 10.5 in. high and 13.5 in. wide. A monitor is advertised by giving the approximate length of the diagonal of its screen. How should this monitor be advertised?

14. A baseball diamond is really a square, and the distance between consecutive bases is 90 ft. How far does a catcher have to throw the ball to get it to second base if a runner tries to steal second?

15. Jackson is 64 miles east of Lazy Day Resort. Fairfield Heights is 25 miles south of Jackson. A land developer proposes building a shortcut road to directly connect Fairfield Heights and Lazy Day. Sketch a drawing and find the length of this new road.

16. A new housing development extends 4 miles in one direction, makes a right turn, and then continues for 3 miles. A new road runs between the beginning and ending points of the development. What is the perimeter of the triangle formed by the homes and the road? What is the area of the housing development?

17. An isosceles triangle has a base of 28 in. and two congruent sides of 25 in. each. Find the height of the triangle. (Round to the nearest tenth.)

18. For a "shed" roof, the height h is half the width w of the building. Find the length of the rafter r if a building is 42 ft wide.

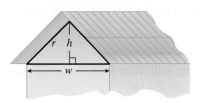

19. The slope, or pitch, of the roof on a new house should be in the ratio 11:20. If it is 30 ft from the front of the house to the back of the house, how far is it from the edge of the roof to its peak?

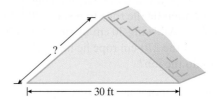

20. A yardstick held vertically on level ground casts a shadow 16 in. long. What would be the distance from the top of the yardstick to the edge of its shadow?

21. Find the area and perimeter of the triangle shown in the following figure.

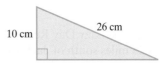

22. Find the area of the figure shown. $\angle C$ is a right angle.

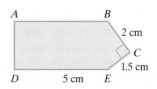

23. Find the distance from A to B on the graph. Round to the nearest tenth.

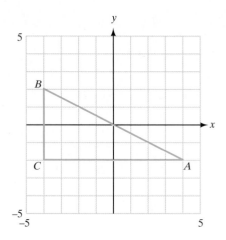

24. Find the distance from R to S on the graph. Round to the nearest tenth.

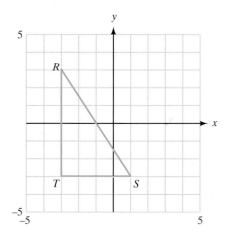

25. A rectangular doorway's inside dimensions are 200 cm by 75 cm. Could a circular mirror measuring 220 cm in diameter pass through the doorway?

26. A carpenter is building a floor for a rectangular room. To avoid problems with the walls and flooring, the corners of the room must be right angles. If he builds an 8-ft by 12-ft room and measures the diagonal from one corner to the other as 14 ft 8 in., is the floor "squared off" properly?

27. Find the length of a side of a square whose diagonal is 10 in. long. Round to the nearest tenth.

28. Find the length of a side of a square whose diagonal is 16 cm long. Round your answer to the nearest tenth.

Section 2-4 Models and Patterns in Art, Architecture, and Nature

Perspective and Symmetry

The visual arts such as sculpture, painting, drawing, and architecture are directly related to mathematics. The mathematical terms *perspective*, *proportion*, and *symmetry* are important elements of art and architecture. In fact, architects, industrial designers, illustrators of all kinds, and engineers will all, at one time or another, be involved in producing a pictorial representation of the objects they are building or designing.

In order to paint or draw with realism, an understanding of **perspective** is essential. Perspective is a geometric way of thinking that allows an artist or architect to give the sense of depth or three dimensions to a flat painting or drawing. (The original form of the word *perspective* means "to see through.") The process of producing a flat drawing that "looks" three-dimensional is based on the laws of optics. Because this section is part of a math book and not an art or architecture book, we will limit our presentation of topics to straight-line drawings only. Of course, no painting, drawing, or photograph (photographs are perspective "drawings") can really duplicate our visual perception of a scene or object.

The basis of perspective drawing is a principle that roughly says that all lines that are parallel in a real scene are not parallel on the perspective drawing but actually converge at a single point called the principal vanishing point. This vanishing point is located on a horizontal line called the horizon line or eye level in the drawing. The horizon line is at the level of the observer's eye when looking at the drawing. A simple example of this is shown in Figure 2-19. Imagine that you are standing near a long, straight road with a fence running down the left-hand side. The fence and the road appear to vanish in the distance. You could also imagine standing in the middle of a railroad track that goes straight away from you into the far distance.

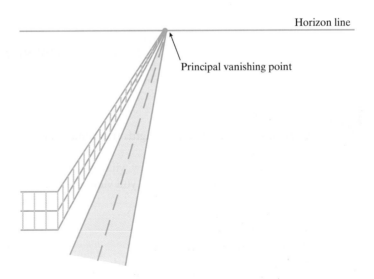

FIGURE 2-19

Another aspect of perspective is that of one's point of view of an object or set of objects. Looking at a structure such as a house or building can yield different pictures depending on which side you look at. Figure 2-20 demonstrates this concept.

74 Chapter 2 **Applications of Algebraic Modeling**

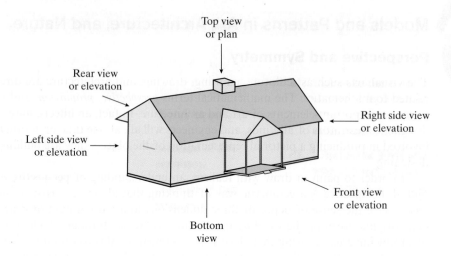

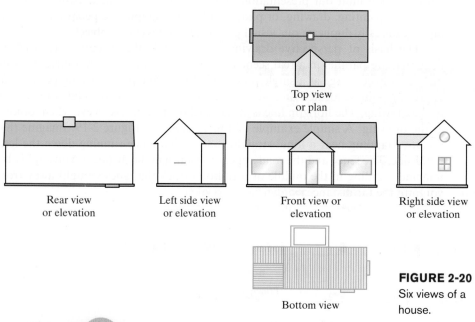

FIGURE 2-20
Six views of a house.

Example 1

Your Point of View

Look at the structure on the right. It is some kind of machine part.

Create line drawings to show what this machine part would look like if you viewed it:

(a) directly from the top
(b) directly from the front
(c) directly from the bottom
(d) directly from the right-hand end

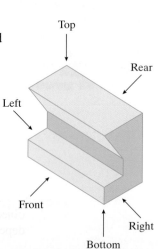

Section 2-4 **Models and Patterns in Art, Architecture, and Nature** 75

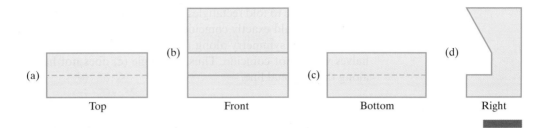

The term *symmetry* has several meanings. Symmetry implies some kind of balance. If you were to draw a vertical line from your head to your toes, a line that goes through your nose and your navel, your body is then divided into two *symmetrical* halves. They aren't identical but they are balanced. Repetitive patterns are also said to be symmetrical. For an example of this, look at the brick walkway shown in Figure 2-21.

To the mathematician, an object has symmetry if it remains unchanged after some "operation" is performed on it. Though there are many *symmetries* possible, we will only look at two that are easy to identify: reflection symmetry and rotational symmetry.

If an object has reflection symmetry, then the object and its mirror image are identical. For example, the capital letter H and its reflection in a mirror are identical. In addition to this, the letter H exhibits reflection (or mirror) symmetry when reflected across (cut by) a vertical line or a horizontal line as shown in Figure 2-22. Note that if you fold the letter H along the dotted lines, the two halves will exactly coincide with each other.

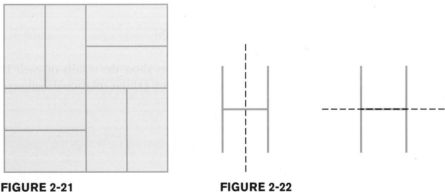

FIGURE 2-21
Brick walkway pattern.

FIGURE 2-22

Example 2 Reflection Symmetry for a Rectangle

A rectangle is pictured here with three possible lines of reflection symmetry. Does the rectangle exhibit reflection symmetry in all the cases shown? If not, which pictures do and why?

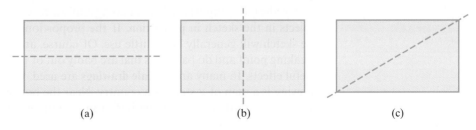

If you were to fold rectangles (a) and (b) along the dotted line of symmetry, the two halves would exactly coincide with each other. So, rectangles (a) and (b) both have reflection symmetry along the line drawn. However, in rectangle (c) the two halves would not coincide. Thus, rectangle (c) does not have reflection symmetry along the diagonal line.

If an object exhibits rotational symmetry, then when it is rotated a specified amount around some chosen point, it will coincide with the original object. For example, the letter H exhibits rotational symmetry if it is rotated 180° around its center.

Example 3 Rotational Symmetry

Which of the following objects exhibits rotational symmetry if rotated 180° around the point or line shown?

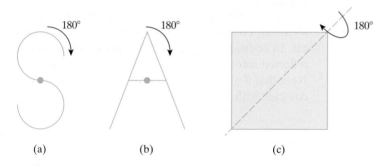

(a)   (b)   (c)

The following figures show the results of each 180° rotation. As you can see, the letter S and the square exhibit rotational symmetry under these conditions but the letter A does not.

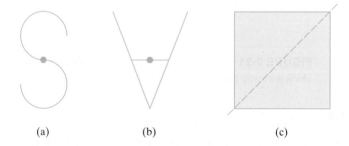

(a)   (b)   (c)

Scale and Proportion

One of the most important rules in doing realistic looking drawings or art is to keep objects in the sketch in **proportion.** If the proportions in a drawing are bad, then the sketch will generally be of little use. Of course, artists can bend the rule to the breaking point and do paintings that are badly out of proportion for comic or other useful effects. In many areas, **scale drawings** are used. One good example of a scale drawing is a map of a state or a country. Near the bottom of the map you will see the *scale factor* that was used, perhaps 1 in. = 5 mi or something similar.

Section 2-4 **Models and Patterns in Art, Architecture, and Nature**   77

Example 4

A Scale Drawing of a House

An architect is going to draw a house so that his picture is $\frac{1}{24}$ the size of the actual house. If the highest point on the house is 14 ft 6 in. and a tree in the yard is 24 ft tall, how large will each of these be in the architect's drawing? Give all your answers in inches.

Since the drawing will be drawn to scale in inches, the two measurements given need to be converted to inches.

$$14 \text{ ft } 6 \text{ in.} = (14 \text{ ft})(12 \text{ in./ft}) + 6 \text{ in.} = 174 \text{ in.}$$

$$24 \text{ ft} = (24 \text{ ft})(12 \text{ in./ft}) = 288 \text{ in.}$$

Now we will find the size to be used in the scale drawing by multiplying by the scale factor, $\frac{1}{24}$.

$$\left(\frac{1}{24}\right)(174) = 7.25 \text{ in.} \quad \text{and} \quad \left(\frac{1}{24}\right)(288) = 12 \text{ in.}$$

Thus, on the scale drawing, the tallest point on the house will be 7.25 in. and the tree will be 12 in. tall.

The next example involves *similar triangles*. If two shapes are similar, then one is an enlargement or reduction of the other. This means that the two triangles will have the same angles and their corresponding sides will be in the same proportion. For example, if triangle ABC is similar to triangle DEF (denoted $\triangle ABC \sim \triangle DEF$), then $\angle A$ is the same size as $\angle D$, $\angle B$ is the same size as $\angle E$, and $\angle C$ is the same size as $\angle F$. In addition, the ratio of side $\overline{AB}$ to side $\overline{DE}$ is the same as the ratios of side $\overline{BC}$ to side $\overline{EF}$ and side $\overline{AC}$ to side $\overline{DF}$.

Example 5

Scaling Triangles

The two triangles pictured here are similar triangles and $m\angle A = m\angle D$, $m\angle B = m\angle E$, and $m\angle C = m\angle F$.

(a) What scale factor was used to transform $\triangle ABC$ into $\triangle DEF$?
(b) How long is side $\overline{AB}$?
(c) How long is side $\overline{EF}$?

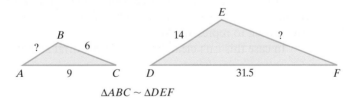

$\triangle ABC \sim \triangle DEF$

(a) Because the triangles are *similar*, the ratio of the corresponding sides of the two triangles is a constant. Thus, we will compare side $\overline{AC}$ with the corresponding side, $\overline{DF}$, in the other triangle as follows:

$$\frac{9}{31.5} = \frac{90}{315} = \frac{2}{7}$$

This means that all of the sides of the smaller triangle are $\frac{2}{7}$ the length of the sides in the larger triangle.

(b) Using our scale factor:

$$\frac{2}{7} = \frac{\overline{AB}}{14}$$

$$(2)(14) = 7(\overline{AB})$$

$$\frac{28}{7} = \overline{AB}$$

$$\overline{AB} = 4$$

Thus, side $\overline{AB}$ is 4 units long.

(c)
$$\frac{2}{7} = \frac{6}{\overline{EF}}$$

$$2(\overline{EF}) = (7)(6)$$

$$\overline{EF} = \frac{42}{2}$$

$$\overline{EF} = 21$$

Thus, side $\overline{EF}$ is 21 units long.

As we have said before, a mathematical basis underlies all of art, music, and architecture. Even in ancient times this was apparent. Ancient artists and architects used ratios or lengths to position objects and scale drawings, paintings, and buildings. One famous such ratio can be found in many ancient art works and buildings. It goes by several names: *the golden mean, the golden ratio, the golden section,* and even *the divine proportion.* This ratio dates back to about 500 BC, the time of Pythagoras, when Greek scholars were asking how to break a line segment into two pieces that would have the most visual appeal or balance to the eye. This is a question of *beauty* being asked by mathematicians, and oddly enough, there was good agreement on the answer.

Stated in mathematical terms, the Greeks said that the most visually pleasing division of the line was into two pieces of unequal length, such that the ratio of the longer piece to the shorter piece is the same as the ratio of the entire length to that of the longer piece. In the early 20th century, the Greek letter *phi*, ϕ (pronounced fi), was chosen to represent this ratio.

In case this isn't clear, here it is in pictorial and algebraic form.

$$\text{The Golden Ratio} = \phi = \frac{a}{b} = \frac{a+b}{a}$$

By cross-multiplying and generating a quadratic equation, the numerical value of the Golden Mean can be easily calculated.

Section 2-4 **Models and Patterns in Art, Architecture, and Nature**

The value of the **Golden Ratio** turns out to be an irrational number (a nonending, nonrepeating decimal because of $\sqrt{5}$) with a value of about 1.62.

The Golden Ratio has had many uses and turns up in one form or another in a lot of artwork, architecture, and in natural objects, both living and nonliving. The Greeks and others applied this idea of "pleasing lengths" to the construction of geometric figures of many kinds. A simple example is a "golden rectangle," where the ratio of the length and width is the Golden Ratio.

Example 6 — Constructing a Golden Rectangle

Suppose the rectangle shown here has a length of 5 cm, and you wish to find the width so that the ratio of the length to the width is $\frac{l}{w} = \phi$.

Here we will round off the value of the Golden Ratio to 1.62 to make the calculation more reasonable to do.

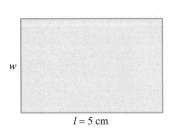

$l = 5$ cm

$$\phi = \frac{l}{w}$$

$$1.62 = \frac{5 \text{ cm}}{w}$$

$$w = \frac{5 \text{ cm}}{1.62} = 3.086419753\ldots \text{ cm}$$

So the width should be about 3.1 cm.

Many artists over the centuries have scaled sculptures using the Golden Ratio and placed objects in paintings so that they will fit within a golden rectangle. Even the front of the famous Greek Parthenon is very close to a golden rectangle (see Figure 2-23).

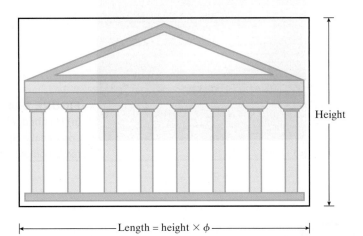

FIGURE 2-23

Many say that the "perfect" or most pleasing human faces exhibit many Golden Ratios relating to the placement of facial features and their relative sizes. Here are just a few:

$$\phi = \frac{\text{height of face}}{\text{width of face}} = \frac{\text{eye to chin}}{\text{nose to chin}} = \frac{\text{eye to mouth}}{\text{eye to nose}} = \frac{\text{nose to chin}}{\text{mouth to chin}} = \frac{\text{width of face}}{\text{distance between eyes}}$$

The Golden Mean turns up in the "art" of nature in many forms. One common example of this is based on golden rectangles. If you start with a golden rectangle, then divide the rectangle into a square and a rectangle, the new small rectangle will also be a golden rectangle. If you continue breaking each new rectangle down and then connect the corners of all the squares with a smooth curve, you will create a *logarithmic spiral*. This is shown in Figure 2-24. These spirals are called *Hambridge's Whirling Squares*. If you were to accurately measure the length of any section of this spiral, you would find that it is about 0.618 as large as the remainder of the spiral.

This logarithmic spiral can be seen in a chambered nautilus shell, the curve of ram's horns, comet tails, the curl of the surf, a parrot's beak, the roots of human teeth, a spider web, bacterial growth curves, and the list just goes on and on. The hexagonal-shaped scales on a pineapple, the "leaves" on a pinecone or an artichoke, and the seeds in a sunflower all spiral around in several directions if you look carefully. All these spirals match this same logarithmic spiral.

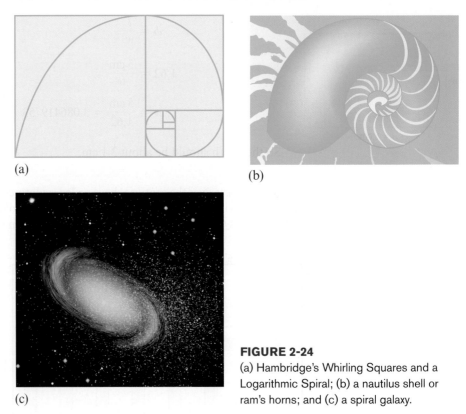

(a)

(b)

(c)

FIGURE 2-24
(a) Hambridge's Whirling Squares and a Logarithmic Spiral; (b) a nautilus shell or ram's horns; and (c) a spiral galaxy.

Practice Set 2-4

1. What is the basic idea behind the use of perspective in drawings?

2. Are all drawings perspective drawings? Why or why not?

3. What is symmetry?

4. Describe both rotational symmetry and reflection symmetry. Find four examples of symmetry in your classroom.

Section 2-4 **Models and Patterns in Art, Architecture, and Nature**   81

These letters are for use in answering problems 5 to 8:

A B C D E F G H I J K L M N O P Q R S T U V W X Y Z

5. Find letters of the alphabet that exhibit reflection symmetry about a vertical or horizontal line through their centers, and show the line of symmetry for each one. Which letters do not exhibit reflection symmetry?

6. If you rotate all the letters 90°, which ones exhibit rotational symmetry?

7. If you rotate all the letters 180°, which ones exhibit rotational symmetry?

8. Without rotating the letters 360°, which letters do not exhibit rotational symmetry for any amount of rotation?

Use these three geometric figures for problems 9 to 12.

(a)   (b)   (c)

(a) An equilateral triangle (3 equal side lengths)
(b) A square (4 equal side lengths)
(c) A regular pentagon (5 equal side lengths)

9. What is the least number of degrees that you could rotate Figure (a) around its center so that it appears to be unchanged?

10. What is the least number of degrees that you could rotate Figure (b) around its center so that it appears to be unchanged?

11. What is the least number of degrees that you could rotate Figure (c) around its center so that it appears to be unchanged?

12. All of the figures, (a), (b), and (c), are figures in which all the sides of the figure are of equal length. There is a pattern between the number of sides and the number of degrees that they may be rotated and yet remain unchanged in appearance. What is that pattern? Use the pattern to write a formula for rotating any closed figure with any number of equal sides.

13. Draw sketches of the following figure as it would look viewed from the front, one end, and looking directly down on it.

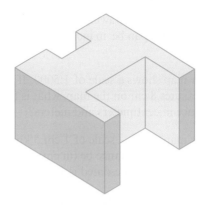

14. Draw sketches of this machine part from the front, top, and right-hand end.

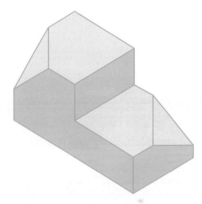

15. A $\frac{1}{16}$ scale model (1 in. = 16 ft) of an airplane has a wingspan of 21 in. What is the wingspan of the actual airplane in feet?

16. If a model of an airplane is $\frac{1}{16}$ scale (1 in. = 16 ft), how many inches long would the model be if a certain airplane were 48 ft long?

17. The larger of two similar polygons has a perimeter of 175 cm. If the scale factor between the two polygons is $\frac{3}{4}$, what is the perimeter of the smaller polygon?

18. Two similar triangles have perimeters of 45 cm and 75 cm respectively. What scale factor would relate these two triangles?

19. A $\frac{1}{800}$ scale model (1 cm = 800 cm) of the aircraft carrier USS *Eisenhower* (CVN-69) is 42.2 cm long and 8.2 cm wide at the widest point of the deck. What are the actual length and width of this ship in meters?

20. You want to do a scale drawing of a Boeing 747 passenger plane. The actual length of the plane is 225 ft, and the wingspan is 204 ft. If you did a

$\frac{1}{144}$ (1 in. = 144 in.) scale drawing, what would the length and wingspan be, in inches, on your drawing?

21. A particular map shows a scale of 1:5000. If a distance measures 8 cm on the map, what is the actual distance in centimeters and meters?

22. A particular map shows a scale of 1 cm:5 km. What would the map distance be (in cm) if the actual distance to be represented is 14 km?

23. The following is the silhouette of the battleship USS *Texas* as she appeared in 1942. The USS *Texas* is 573 ft long. Measure the length of the ship in this picture and determine the scale of the drawing (? in. = ? ft).

24. The following is the silhouette of the battleship USS *North Carolina* as she appeared in 1942. The USS *North Carolina* is actually 728 ft. Determine the scale of this drawing (? in = ? ft).

25. A road map of the state of Texas has a scale of 0.5 in. = 50 mi. If the distance between two points in the state is 600 mi, what would the equivalent distance be on the state map?

26. A building is 250 ft by 120 ft. If you were to draw this building using a scale of 1 in. = 20 ft, what would the dimensions of your drawing be?

27. A golden rectangle is to be constructed such that the shortest side is 23 ft long. How long is the other side?

28. If a golden rectangle has a width of 9 cm, what is its length?

29. If one dimension of a golden rectangle is 45.5 cm, what *two* values are possible for the length of the other side?

30. If one dimension of a golden rectangle is 36 in., what *two* values are possible for the length of the other side?

Section 2-5 Models and Patterns in Music

It is interesting to realize that mathematics and music are linked. Gottfried Wilhelm von Leibniz (1646–1716), who along with Sir Isaac Newton gets credit for developing modern calculus, said, roughly translated, "Music is the pleasure the human soul experiences from counting without being aware that it is counting." In medieval times, schools linked arithmetic, astronomy, geometry, and music together in what was called the *quadrivium*. Today, computers are often used to produce music, and thus the link between math and music is being perpetuated.

Symbols used to denote musical notes are really indicators of the duration (time) of a note in a musical composition. Table 2-2 shows some of the symbols used for notes of various times and for rests where no note is played for a time.

TABLE 2-2 Symbols and Names of Common Musical Notes

Note	Rest	American name	British name
𝅝	𝄻	whole note	semibreve
𝅗𝅥	𝄼	half note	minim
♩	𝄽	quarter note	crotchet
♪	𝄾	eighth note	quaver
𝅘𝅥𝅯	𝄿	sixteenth note	semiquaver
𝅘𝅥𝅰	𝅀	thirty-second note	demisemiquaver
𝅘𝅥𝅱	𝅁	sixty-fourth note	hemidemisemiquaver
𝅘𝅥𝅲	𝅂	hundred-twenty-eighth note	quasihemidemisemiquaver

Tempos can be 3/4 time (three quarter notes per measure), 4/4 time (four quarter notes per measure), or 3/8 time (three eighth notes per measure), for example. In a 4/4 measure, each whole note counts four beats, each half note counts two beats, each quarter note counts one beat, each eighth note counts $1/2$ beat, and each sixteenth note counts $1/4$ beat.

In addition to notes, there are also *rests* where no notes are played for a period of time. Rests can be for the same length of time as any of the notes listed. This means that a rest can last the length of a whole note, half note, quarter note, and so forth. A musician has to count beats per measure and then deal with whole notes, half notes, quarter notes, eighth notes, sixteenth notes, and rests, and so on. Music is written so that there are a fixed number of beats per measure. All those notes and rests must add up to the same number of beats for each measure of music. For example, all the notes and rests in each measure of music written in 4/4 time must add up to 4 beats. It is the same as finding common denominators for fractions so that you may add them. Look deeply enough at music and you will find such topics as ratios, periodic functions, and exponential curves.

Example 1 Adding Notes

A measure of music written in 4/4 tempo must have four beats in the measure. One way to achieve this is to use four quarter notes as shown here.

$$♩ + ♩ + ♩ + ♩ = \frac{1}{4} + \frac{1}{4} + \frac{1}{4} + \frac{1}{4} = \frac{4}{4}$$

Fill in the missing note or notes so that each of the following totals four beats.

(a) ♩ + 𝅗𝅥 + ___   (b) 𝅗𝅥 + ♩ + ♪ + ___   (c) ♩ + ♩ + ♩ + ♪ + ___ + ___

If you write each note as a fraction and then add them, the total must equal 4/4, as shown in the first set of notes.

(a) Translate into fractions. Let $x =$ missing note.

$$1/4 + 1/2 + x = 4/4$$
$$1/4 + 2/4 + x = 4/4$$
$$3/4 + x = 4/4$$
$$x = 4/4 - 3/4 = 1/4$$

Thus, one quarter note is needed.

(b) Translate into fractions. Let $x =$ missing note.

$$1/2 + 1/4 + 1/8 + x = 4/4$$
$$4/8 + 2/8 + 1/8 + x = 8/8$$
$$7/8 + x = 8/8$$
$$x = 8/8 - 7/8 = 1/8$$

Thus, one eighth note is needed.

(c) Translate into fractions. Let $x =$ missing note.

$$1/4 + 1/4 + 1/4 + 1/8 + x + x = 4/4$$
$$2/8 + 2/8 + 2/8 + 1/8 + x + x = 8/8$$
$$7/8 + 2x = 8/8$$
$$2x = 8/8 - 7/8 = 1/8$$
$$x = (1/8)(1/2) = 1/16$$

The missing notes are sixteenth notes.

The Pythagoreans in Greece are the first people known to have related music and math. They discovered that the sound (note) produced by a string when it is plucked varies with the length of the string. With a little experimentation, they found that equally tight strings would give harmonious notes if their lengths were in whole-number ratios to each other.

Vibrations produce waves that propagate through the surrounding medium. For example, a vibrating string on a violin produces sound waves that propagate (move through) the medium (air) that surrounds the violin. The **frequency** of the string's vibration is the number of times the string vibrates up and down in one second. A frequency of 200 vibrations per second is said to have a frequency of 200 Hertz, or 200 Hz.

Suppose that you have a string that will generate a sound with a frequency of 200 Hz when the whole string vibrates as one piece. This is called the *fundamental frequency* of the string. If you then hold this string down at its midpoint and pluck it, the new wave will be half as long as the original wave, but its frequency will be 400 Hz, twice that of the whole-string vibration. This is called the *first harmonic* or *first overtone*. If you hold the string down so that only one-fourth of its original length is plucked, the wave will be one-fourth the original wavelength, and the frequency will double again to 800 Hz. This is called the *second harmonic* or *second overtone*. This is shown in Figure 2-25.

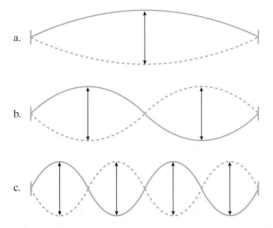

FIGURE 2-25
The fundamental frequency of a string (a) and its first (b) and second (c) harmonic.

To our ear, these different frequencies have different pitches. The higher the frequency, the higher the pitch of the sound to your ear. Also, when the frequency of a note is doubled, the pitch is said to have been raised an octave. For example, A above middle C on a piano has a frequency of 440 Hz. The next higher-pitched A on the keyboard will have a frequency of 880 Hz and is one octave higher in pitch. The next-lower pitched A below middle C will have a frequency of 220 Hz and be one octave lower in pitch. To the Greeks and to people today, the most pleasing combinations of notes are those that are the same note but an octave apart in pitch.

Figure 2-26 shows part of a piano keyboard. The white keys are the notes called C, D, E, F, G, A, B. The black keys are either called sharps, if they are to the right of the white key and thus a higher pitch; or flats, if they are to the left of a white key and thus lower in pitch. This means that all of the black keys are both sharps and flats. The symbol for D sharp is D$^\sharp$ and the symbol for D flat is D$^\flat$.

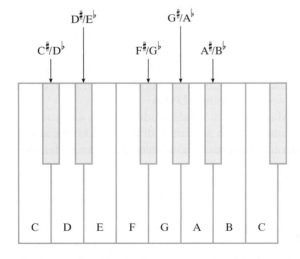

FIGURE 2-26
An octave on a piano keyboard.

If you count the keys from C to the next C on the keyboard pictured in Figure 2-26, you will count 12 keys (C$^\sharp$/D$^\flat$, D, D$^\sharp$/E$^\flat$, E, F, F$^\sharp$/G$^\flat$, G, G$^\sharp$/A$^\flat$, A, A$^\sharp$/B$^\flat$, B, C). These 12 keys produce 12 *semitones* between the two Cs that are an octave apart. To change a frequency by an octave, the frequency is multiplied by 2 as mentioned before. This leads to the most common method of tuning pianos. Each key is tuned so that it is $\sqrt[12]{2}$ (the twelfth-root of two) different from the key next to it. This means to multiply or divide by $\sqrt[12]{2} = 1.0594630943593\ldots$, or about 1.05946. Stated mathematically, the ratio between the frequencies of any two successive pitches is 1.05946.... Thus, if middle C has a frequency of 261.626 Hz, C sharp has a frequency of $C^\sharp = (261.626)(1.05946) = 277.182$ Hz. If you continue to multiply by $\sqrt[12]{2}$, the C an octave above middle C will have a frequency of 523.251, or twice that of middle C. Complete lists of frequencies of all notes of the scale may be obtained from many sources, including local music stores or the Internet at sites like answers.com. For some interesting connections between music and the Golden Ratio that you learned about in Section 2-4, refer to the suggested lab at the end of this chapter regarding sonatas written by Mozart.

If you play the same note on a violin and a flute, they sound entirely different in many respects. The reason for this has to do with *harmonics,* which were mentioned in relation to the length of plucked strings. When a note is played on a flute, only one note is heard. When a note is played on a violin or other acoustic instrument, not only the fundamental frequency is produced but many harmonics as well. For example, an A at 440 Hz played on a violin with full harmonics will produce not only the A at 440 Hz but also an A at 880 Hz with approximately half the volume of the primary A; an A at 1320 Hz and about a third of the volume of the primary A; and so forth until either the frequency gets too high for the human ear to respond or the volume gets too low to be heard.

Our discussion of music and math would not be complete without mentioning *digital* recordings. Sound is a wave and, in the past, recordings were made that actually reproduced the shape of the wave in the grooves of a vinyl record. If you are too young to remember phonograph records, ask some "old person" what a $33\frac{1}{3}$- or 45-RPM is. This method of recording music is called the *analog* method.

To produce a modern digital recording, a computer takes the analog style of a piece of music and chops it up into very small pieces, each only a fraction of a second long, measures the frequencies present in each piece; and then records them *numerically*. This method of transforming music or sound into a list of numbers (digits) is called *digitizing*. A CD player is really a small computer that reverses the digitizing process and converts the stored numbers back into sound waves that you can hear. This process leaves a very short blank space between each set of numbers on the CD, but when played back at a normal speed, these gaps are small enough that the music sounds continuous to the human ear. A similar process is used with light waves to produce DVDs.

Many modern bands produce music using *synthesizers* that can imitate many instruments, such as pianos, drums, and horns, at the push of a key. This all goes under the heading of *digital signal processing*. Here is where music and mathematics merge with each other.

Sound is a compression wave that can be mathematically analyzed as sine waves broken up into different frequencies. Frequency is the number of waves per second, which is called Hertz (Hz). If the note A has a frequency of 440 Hz, then this means that 440 waves are passing through a particular point during an interval of 1 second. The higher the frequency of a sound wave, the higher the pitch that we hear. The loudness of a sound, measured in decibels (dB), corresponds to the amplitude or height of the sine wave.

Sine waves can be used to re-create any sound. The sine wave related to a musical pitch is given by the following equation: $y = A\sin(Bt)$, where A is the *amplitude* of the sound, t is time, and B is the frequency of the note. The sound of middle C on the piano results from a wave that vibrates 256 times per second (256 Hz). Therefore, the equation of the sine wave with $A = 1$ would be $y = \sin(256t)$, where $t =$ time. If you put this equation into your graphing calculator, you can see the wave action for this tone. Look at the picture of this wave in Figure 2-27.

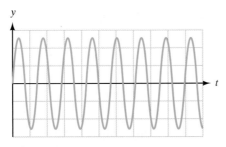

FIGURE 2-27
$y = \sin(256t)$.

As the frequencies increase, the pitch of each note is higher, and the waves will become closer together. Look at the sine wave for the note A above middle C, having a frequency of 440 Hz, in Figure 2-28.

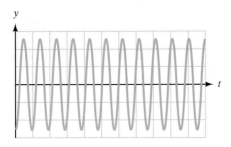

FIGURE 2-28
$y = \sin(440t)$.

When a mixture of overtones is included with the fundamental tone, derivations of the sine wave in the shape of sawtooth or square waves are formed. The wave in Figure 2-29 is a composite of several different frequencies.

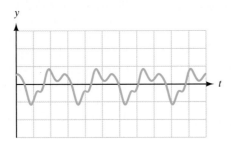

FIGURE 2-29
A composite sound with several frequencies.

As noted earlier in this section, when a note is played on a flute, only one note is heard. This is the instrument that produces the purest sound and a graph that is closest to the standard $y = \sin(x)$. The most complex sine wave is produced by a cymbal because of the many combinations of sound present. Experiment with the various notes discussed earlier in this section examining the graphs of the sine functions associated with each of them.

Practice Set 2-5

1. How are frequency and pitch related?

2. In what fundamental way does one musical note differ from another?

3. What is an analog recording of music?

4. How does an analog recording differ from a digital recording of music?

5. What does it mean to digitize sound?

6. What kinds of waves are digitized on a DVD?

7. If you start with a note with a frequency of 261.626 Hz, what frequencies are one octave higher and one octave lower than this frequency?

8. If you start with a note with a frequency of 110 Hz, what frequencies are one octave higher and one octave lower than this frequency?

9. The note that is G above middle C has a frequency of approximately 392 Hz. What is the frequency of the next highest semitone?

10. The note that is G above middle C has a frequency of approximately 392 Hz. What is the frequency of the next lowest semitone?

11. If a certain piece of music is written in 4/4 time, how many half notes are required for one measure of music?

12. If a certain piece of music is written in 3/4 time, how many eighth notes are required per measure of music?

13. If a certain piece of music is written in 3/2 time, how many half notes are required per measure of music?

14. If a certain piece of music is written in 3/8 time, how many sixteenth notes are required per measure of music?

Find the missing note(s) in each of the following so that the proper tempo is maintained.

15. Tempo = 3/4   ♩ + ♩ + ___

16. Tempo = 3/4   𝅗𝅥 + ___

17. Tempo = 3/8   ♪ + ♪ + ___

18. Tempo = 3/8   ♩ + ___

19. Tempo = 2/4   𝅗𝅥 + ___

20. Tempo = 2/4   ♪ + ♩ + ___

Use the model $y = A \sin(Bt)$, where A is the *amplitude* of the sound, t is time, and B is the frequency of the note to write equations for sounds having the following characteristics. Since A = the volume of the note, let $A = 5$ for these exercises. Then, use your graphing calculator to sketch a drawing of the resulting sound wave for problems 21 to 24.

21. Frequency of the note A = 220 Hz

22. Frequency of the note $F^{\sharp}$ = 185 Hz

23. Frequency of the note G = 196 Hz

24. Frequency of the note A at a lower pitch = 55 Hz

25. Group Project: The note A has a frequency of 110 Hz. The note $A^{\sharp}$ (A sharp) has a frequency of 116.54 Hz. Use your graphing calculator to graph both of these waves and compare them. Sketch these waves on the same graph. Pleasing chords have simple ratios such as 200 Hz:300 Hz, while dissonant chords result from more complicated ratios such as 290 Hz:300 Hz. Use your graphing calculator to graph the note A with a frequency of 110 Hz and another note having a frequency of 220 Hz. Sketch these waves on the same graph. What do you notice about the graphs of these waves? Discuss with your group members the similarities and differences between the two graphs.

26. Group Project: Describe how the sine waves of notes that are octaves apart compare. Give examples of notes and the related equations in your discussion. Sketch the waves of these notes.

Chapter 2 Summary

Key Terms, Properties, and Formulas

acute angle	hypotenuse	Pythagorean theorem
acute triangle	isosceles triangle	right angle
area	obtuse angle	right triangle
circumference	obtuse triangle	scale drawing
equilateral triangle	perimeter	scalene triangle
frequency	perspective	sine wave
Golden Ratio	proportion	symmetry

Formulas to Remember

Circumference of a Circle:

$$C = 2\pi r = \pi d$$

Area Formulas:

Square $\quad A = s^2$
Rectangle $\quad A = lw$
Triangle $\quad A = 0.5bh$
Circle $\quad A = \pi r^2$

Hero's Formula for the Area of a Triangle:

$$A = \sqrt{s(s-a)(s-b)(s-c)} \quad s = \frac{a+b+c}{2}$$

Pythagorean Theorem:

$$a^2 + b^2 = c^2$$

The Golden Ratio:

$$\phi = \frac{a}{b} = \frac{a+b}{a}$$

Sine Wave for Musical Pitch:

$$Y = A \sin(Bt) \text{ where } A = \text{amplitude}, B = \text{frequency}, \text{ and } t = \text{time}$$

Chapter 2 Review Problems

1. Find the perimeter and area of the figure shown in this diagram.

   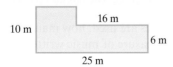

2. At $0.55/ft, how much will it cost to put molding around the window pictured?

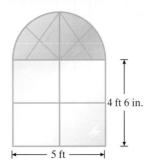

3. A triangle has two angles that measure 35° and 52°. Calculate the size of the third angle and classify the triangle as right, acute, or obtuse.

4. Find the area and circumference of a circle having a diameter of 36 cm. Round to the nearest tenth.

5. Find the area and perimeter of the triangle shown.

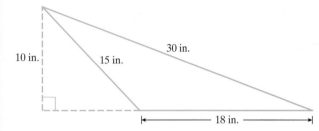

6. How many square feet of concrete surface are there in a circular walkway 8 ft wide surrounding a tree if the tree has a diameter of 3 ft? (Round to the tenths place.)

7. The lengths of the sides of a triangle are 6 in., 8 in., and 12 in. Find the area of the triangle. (Round to the nearest tenth.)

8. The siding on one side of a house was damaged in a storm and needs to be replaced. The drawing gives the dimensions of the end of the house and the windows in that end.

   (a) Find the area that needs to be covered with new siding.
   (b) If siding costs $22.25/yd², what will the cost of the siding be for this side of the house? (Note that you cannot buy a fractional part of a square yard.)

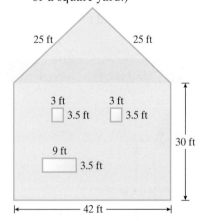

9. How long must a piece of board be to reach from the top of a building 15 ft high to a point on the ground 8 ft from the building?

10. A slow-pitch softball diamond is a square with sides that are each 60 ft long. What is the distance from first base to third base? (Round to the nearest tenth.)

11. Does a triangle with side lengths of 7.5 ft, 12.5 ft, and 10 ft form a right triangle? How do you know?

12. An isosceles triangle has a base 24 in. long and a height from the vertex angle to the base of 18 in. How long is each of the other sides of the isosceles triangle?

13. Which type of symmetry, horizontal and/or vertical reflection symmetry, does each of the following exhibit?

    (a) The letter W
    (b) The number 3
    (c) An isosceles triangle sitting on its unequal side
    (d) The number 9

14. The corresponding sides of two similar triangles have lengths of 50 cm and 75 cm. What scale factor would relate the lengths of the sides of these two similar triangles?

15. A $\frac{1}{32}$ (1 in./32 in.) scale model of an automobile has a length of 4.125 in. What is the length of the actual car?

16. A golden rectangle has longest side lengths of 32.4 cm. What is the length of the shorter sides?

17. A road map uses a scale of 1 in. = 20 mi. If the distance between two locations is actually 35 mi, what distance would this be indicated by on the map?

18. Fill in the missing two notes so there is a total of 4 beats.

    ♩ + ♪ + ____ + ____

19. What frequency is two octaves below 110 Hz?

20. If only quarter notes are used, how many are required for one measure of music written in 3/2 time?

Chapter 2 Test

1. Find the perimeter and area of the triangle shown.

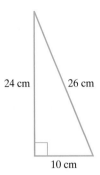

2. Find the area of the shaded portion of the diagram.

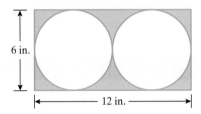

3. Maria is finalizing the set for the school play and needs to buy fringe to sew onto the edges of the three circular tablecloths needed for the final scene. Each tablecloth is 6 ft in diameter. How much fringe will she need for the three tablecloths?

4. Calculate the size of the third angle in $\triangle ABC$ and classify it as acute, obtuse, or right if $\angle A = 25°$ and $\angle B = 65°$.

5. Find the length a in the given right triangle.

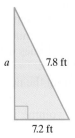

6. Find the area of the isosceles triangle shown.

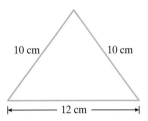

7. The lengths of the sides of a triangle are 2.6 ft, 3.4 ft, and 4.1 ft. Find the area of the triangle.

8. (a) Find the area and perimeter of the room shown in the figure.

   (b) If carpet costs $24.95 per square yard, how much will it cost to carpet the room? (Remember: 1 yd² = 9 ft²)

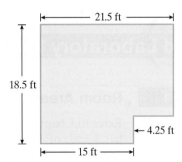

9. A 26-ft ladder is leaning against a wall, and the foot of the ladder is 4.5 ft from the foot of the wall. If the ground is level, what is the height of the point at which the top of the ladder is resting on the wall?

10. Find the width of a rectangle having a diagonal of 6.5 cm and a length of 5.8 cm.

11. A car tire stands 21 in. tall.

    (a) How far will the tire move if it completes one revolution?

    (b) There are 5280 ft in a mile. How many times will this tire turn each mile it travels?

12. What letter(s) of the alphabet exhibit(s) rotational symmetry if rotated 45°? (Consider capital letters only.)

13. What letters of the alphabet exhibit rotational symmetry if rotated 270°? (Consider capital letters only.)

14. A map has a scale that says 1 cm = 2500 m. What distance on the map would indicate an actual distance of 6.25 km?

15. An HO-scale model train is a $\frac{1}{87}$ scale. This means that each inch (or foot) of model size is equivalent to 87 in. (or 87 ft) in real life. If a model of a steam engine that is actually 75 ft long is to be made, how long will the model be in inches?

16. Does the ratio of the lengths of the sides of a rectangle that is 48 in. by 64 in. match the Golden Ratio?

17. Which is closer to being a golden rectangle, an 8.5-in. by 11-in. sheet of notebook paper or an 8.5-in. by 14-in. sheet of paper from a legal pad?

18. What frequency is three octaves above a note of 440 Hz?

19. Fill in the missing note so there is a total of 4 beats.

    ♪ + ♪ + ♩ + ♩ + ___

20. What is the difference between 3/4 time and 4/4 time in music?

Suggested Laboratory Exercises

Lab Exercise 1

Room Areas

Refer to Chapter 5, Lab Exercise 4, and assume that the floor plan pictured there has a scale of 1 in. = 8 ft. What is the size of the two bedrooms, the living/dining area, and the kitchen? What is the approximate square footage for the entire condo? How many square yards of carpet would be needed to carpet the two bedrooms, including the closets, and the living/dining area?

Lab Exercise 2

A Comparison of Calculated and Measured Volumes

In this exercise you will compare a measured quantity with a calculated value for the same quantity (i.e., comparing a ruler measure with a model or formula value).

Obtain four cans of differing sizes, a liquid measuring cup or similar item from the chemistry department with SI (metric) volume units (ml), a ruler with SI markings (mm, cm), and a source of water. You will also need your calculator.

First measure the height, h, and radius, r, of each can in centimeters (cm). Use these values to calculate the volume of each can. $V_{cylinder} = \pi r^2 h$ is the formula that "models" the volume of cylinders with circular, parallel ends. These volumes will be in cubic centimeters (cm^3). Record all of your data in a table like Table 2-3.

Now use the measuring cup and fill each can with water, recording the number of milliliters (ml) of water that it takes to fill each one.

The volume in ml can easily be converted to the equivalent volume in cm^3 because:

1 ml = 1 cm^3. Do this for each of the cans.

Now compare the calculated volumes with the measured volumes as follows:

$$\frac{\text{measured value} - \text{calculated value}}{\text{calculated value}} \times 100\%$$

This calculation will give you the percent difference between the two volume measures for each can.

TABLE 2-3

Can	Measured Height (cm)	Measured Radius (cm)	Calculated Volume (cm³)	Measured Volume (cm³)	Difference	Percentage Difference
1						
2						
3						
4						

1. Were your calculated and measured volumes exactly the same in any/all cases? If not, what might be some reasons for the difference observed?
2. If an error of difference of ±5% is acceptable, did your values fall below this level? What changes might be made in the procedure that you used so as to minimize differences?

Lab Exercise 3 — Musicians and Mathematics

If you do a little research on the Internet about the connection of music to math, you will find that many musicians, both modern and classical, used mathematics as a basis for some of their compositions. Research one of these persons and write a two-page essay on how they used mathematics.

Lab Exercise 4 — Are Golden Rectangles Really Cute?

Draw a series of four or five rectangles, only one of which is a golden rectangle. Then do a survey around your school asking which rectangle is most "pleasing" in appearance. Is the one that is the golden rectangle chosen most often? Make measurements in inches and then in centimeters. Are the resulting ratios the same?

Lab Exercise 5 — The Golden Belly Button Project

It has been theorized that, on the average, a person's total height and the height of that person's belly button from the floor approximates the Golden Ratio. Collect lots of belly button data and test this theory. What is your conclusion?

Lab Exercise 6 — Mozart Sonatas and the Golden Ratio

Table 2-4 lists the number of measures of music in the first and second parts of several of Mozart's piano sonatas. Fill in the third column by calculating the ratio of the total number of measures in Parts 1 and 2 all together to the number of measures in the longer part. Write a discussion of your results and whether or not you think that Mozart had the Golden Ratio in mind when he wrote these sonatas.

TABLE 2-4

Measures in Part 1	Measures in Part 2	Ratio
77	113	$(77 + 113)/113 = 1.681$
24	36	
39	63	
53	67	
38	62	
28	46	
15	18	
40	69	
56	102	
46	60	

Graphing

Graphs can help us visualize data and their behavior. When ordered-pair solutions of a linear equation are graphed, the result is a straight line. Its slope and direction can be meaningful in interpreting data or predicting future outcomes. In this chapter, we review the topics of graphing lines, interpreting slope, and graphing linear equations. We will also use the slope of a given line and a point on that line to write its linear equation.

> *Mathematics is a game played according to certain simple rules with meaningless marks on paper.*
> – David Hilbert

In this chapter

- **3-1** Rectangular Coordinate System
- **3-2** Graphing Linear Equations
- **3-3** Slope
- **3-4** Writing Equations of Lines
- **3-5** Applications and Uses of Graphs
- Chapter Summary
- Chapter Review Problems
- Chapter Test
- Suggested Laboratory Exercises

96 Chapter 3 **Graphing**

Section 3-1 Rectangular Coordinate System

You know how to graph a number as a point on a number line. Ordered pairs are graphed as points in a plane on what is called the **rectangular coordinate system**. The rectangular coordinate system consists of a horizontal axis called the **x-axis** and a vertical axis called the **y-axis**. The x- and y-axes intersect in a right angle. The point of intersection of the two axes is called the **origin**, and the coordinate plane is separated into four regions called **quadrants**. (See Figure 3-1.)

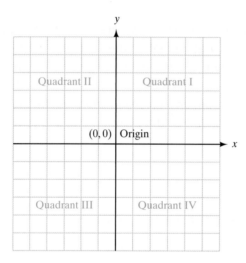

FIGURE 3-1
Rectangular coordinate system.

Each point in the plane corresponds to an ordered pair (x, y) of real numbers. The x and y values are called the **coordinates** of the point. The first number in the ordered pair is the **x-coordinate**, and it tells you how many units to the left or right the point is from the y-axis. If x is positive, you will move to the right of $(0, 0)$, and if x is negative, you will move to the left of $(0, 0)$. The second number in the ordered pair is the **y-coordinate**, and it tells you how many units up or down the point is from the x-axis. If y is positive, you will move up from the x-axis, and if y is negative, you will move down from the x-axis. The coordinates for the origin are $(0, 0)$.

Locating a point in a plane is called *plotting* the point. We will look at several examples of plotting points in the rectangular coordinate system in Example 1.

Example 1 Plotting Points on the Rectangular Coordinate System

Plot the points $(-1, 3)$, $(4, -2)$, $(0, -3)$, $(2, 5)$, $(-2, -4)$ on a rectangular coordinate system.

The point $(-1, 3)$ is one unit to the left of the vertical axis and three units up from the horizontal axis. The other four points are plotted in the same fashion. (See Figure 3-2.)

Section 3-1 **Rectangular Coordinate System** 97

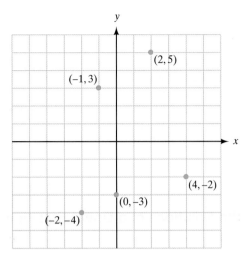

FIGURE 3-2
Example 1.

In the next example, we will determine the coordinates of the point by looking at the rectangular coordinate system.

Example 2 Identifying the Coordinates of Points in a Rectangular Coordinate System

Determine the coordinates of the points shown in Figure 3-3.

Point A lies three units to the right of the vertical axis and two units below the horizontal axis. So point A is the ordered pair $(3, -2)$. The coordinates for the other four points can be determined in the same manner. The coordinate for point B is $(1, 4)$, point C is $(-5, 0)$, point D is $(-2, -2)$, and point E is $(-1, 3)$.

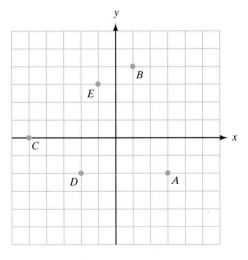

FIGURE 3-3
Example 2.

The rectangular coordinate system allows us to visualize relationships between two variables. In mathematics, relationships between variables are written as an equation. From the equation, we can construct a table of values that satisfy the given equation. In this chapter, we will be looking at a specific type of equation called a **linear equation**. The graph of a linear equation is a line. The **general form** for a linear equation is $Ax + By = C$. (Note that A and B are not both 0.) The **solutions** of a linear equation are the ordered pairs (x, y) that make the equation true. We can determine if an ordered pair is a solution of a linear equation by substituting those coordinates into the equation to verify that they give a true equality.

Example 3

Verifying the Solutions of a Linear Equation

Are the ordered pairs (3, 2) and (7, 9) solutions of the linear equation $2x - y = 5$?

(a) We will verify the solution (3, 2) by substituting it into the equation.

$$2x - y = 5$$
$$2(3) - (2) = 5$$
$$6 - 2 = 5$$
$$4 = 5 \quad \text{(false statement)}$$

Because the ordered pair (3, 2) produces a false equality, this ordered pair is not a solution of this equation.

(b) We will verify the solution (7, 9) by substituting it into the equation.

$$2x - y = 5$$
$$2(7) - (9) = 5$$
$$14 - 9 = 5$$
$$5 = 5 \quad \text{(true statement)}$$

Because the ordered pair (7, 9) produces a true equality, this ordered pair is a solution of this equation.

You can find solutions of a linear equation by giving a value to one variable and solving the resulting equation for the other variable. It is often easier to solve the equation for one of the variables, such as y, and substitute values of x to generate a set of solutions.

Example 4

Finding Solutions to a Linear Equation

Find five solutions to the linear equation $2x + y = 6$.

First, solve the equation for x or y. In this case, solving for y will be easier because y has a coefficient of 1.

$$2x + y = 6$$
$$y = -2x + 6 \quad \text{(subtract } 2x \text{ from both sides)}$$

Choose five values for x and then replace x with its value, and solve each equation for y. (See Table 3-1.)

TABLE 3-1 Example 4

x	$y = -2x + 6$	y	Solution
3	$y = -2(3) + 6$	0	$(3, 0)$
4	$y = -2(4) + 6$	-2	$(4, -2)$
0	$y = -2(0) + 6$	6	$(0, 6)$
1	$y = -2(1) + 6$	4	$(1, 4)$
2	$y = -2(2) + 6$	2	$(2, 2)$

Example 5 **Finding Solutions to a Linear Equation**

Find the indicated ordered-pair solutions to the linear equation

$$x - 6y = 18.$$

$(0, \underline{}), (-2, \underline{}), (\underline{}, 1)$

Because we are given the values of both x- and y-coordinates and asked to calculate the value of the missing coordinates, we will substitute the values into the equation as given.

(a) Find the ordered-pair solution $(0, \underline{})$.

$$x - 6y = 18$$
$$0 - 6y = 18 \quad \text{(replace } x \text{ with the value 0)}$$
$$-6y = 18$$
$$y = -3$$

This solution is $(0, -3)$.

(b) Find the ordered-pair solution $(-2, \underline{})$.

$$x - 6y = 18$$
$$-2 - 6y = 18 \quad \text{(replace x with the value } -2)$$
$$-6y = 20$$
$$y = \frac{20}{-6} = -\frac{10}{3}$$

This solution is $\left(-2, -\dfrac{10}{3}\right)$.

(c) Find the ordered-pair solution (___, 1).

$$x - 6y = 18$$
$$x - 6(1) = 18 \quad \text{(replace } y \text{ with the value 1)}$$
$$x - 6 = 18$$
$$x = 24$$

This solution is (24, 1).

Practice Set 3-1

Graph the following points on the same set of axes.

1. (5, 3)
2. (1, 2)
3. (−3, −5)
4. (−5, −2)
5. (−4, 0)
6. (0, −2)
7. (−3, 3)
8. (−4, −5)
9. (0, 0)
10. (−3, 1)
11. (4, −3)
12. (5, −4)

Problems 13 to 26 refer to the following diagram.

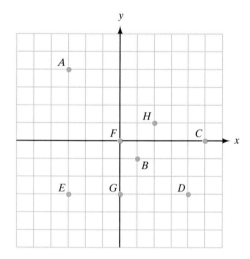

Give the coordinates and the quadrant or axis where each point is located.

13. F
14. H
15. A
16. B
17. E
18. C
19. G
20. D

Refer to the previous diagram and name the point(s) described.

21. The point is on the positive x-axis.
22. The x-coordinate is 0.
23. The y-coordinate is 0.
24. The point is on the negative y-axis.
25. The x-coordinate is the opposite of the y-coordinate.
26. The point in quadrant IV that is nearest the x-axis.
27. Is the ordered pair (3, 4) a solution of the equation $2x - 5y = 26$? Show work to support your answer.
28. Is the ordered pair (32, 0) a solution of the equation $\frac{3}{4}x + 8y = 24$? Show work to support your answer.
29. Is the ordered pair (3, −2) a solution of the equation $3x + 2y = 5$? Show work to support your answer.

Use the following table to find ordered-pair solutions for the equations in problems 30 to 40.

x	y
0	
3	
	−2
	0

30. $x - y = 6$
31. $x + y = 8$
32. $2x + y = 6$
33. $x - 3y = 12$
34. $2x + 5y = 10$
35. $x + 4y = 0$
36. $3x - 4y = 8$
37. $y = x$
38. $4x - y = 3$
39. $-x - 2y = 6$
40. $x + y = 0$

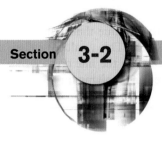

Section 3-2 Graphing Linear Equations

In the previous section, we saw that the solutions to a linear equation can be represented by points on a rectangular coordinate system. The set of all solution points of a linear equation is called its graph. In this section, we will discuss graphing linear equations by finding points for x and y and by finding two specific points called the *x-intercept* and *y-intercept*.

Example 1

Graphing a Line by Plotting Points

Sketch the graph of $2x - y = 4$.

First, let us solve the equation for y.

$$2x - y = 4$$
$$-y = -2x + 4 \quad \text{(subtract } 2x \text{ from both sides)}$$
$$y = 2x - 4 \quad \text{(divide both sides by } -1\text{)}$$

Second, choose arbitrary values for x and create a table of values. Because two distinct points determine a line, we need to find at least two solutions of the equation. Choosing more than two points allows us to see a pattern developing between the x and y coordinates. (See Table 3-2.)

Third, plot the points and draw a line connecting the five points. (See Figure 3-4.)

TABLE 3-2 Example 1

x	$y = 2x - 4$	Solution Point
3	$2(3) - 4 = 2$	$(3, 2)$
4	$2(4) - 4 = 4$	$(4, 4)$
0	$2(0) - 4 = -4$	$(0, -4)$
1	$2(1) - 4 = -2$	$(1, -2)$
2	$2(2) - 4 = 0$	$(2, 0)$

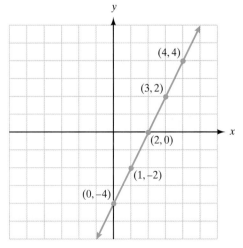

FIGURE 3-4
Example 1.

Notice that if we let x have the value 0 and find the y value that makes the equation true, we have found where the line crosses the y-axis. This y value is called the **y-intercept**. Similarly, the **x-intercept** is the x value where the line crosses the x-axis. You can find the x-intercept by giving y the value 0 and solving for x.

Rule — **Finding the Intercepts of a Line**

1. To find the x-intercept, replace y with 0 and solve the equation for x.
2. To find the y-intercept, replace x with 0 and solve the equation for y.

Example 2 — **Graphing a Line Using Intercepts**

Draw the graph of $4x + y = 3$.

To find the x-intercept, substitute 0 for y and solve for x.

$$4x + 0 = 3$$
$$4x = 3$$
$$x = \frac{3}{4} \qquad x\text{-intercept } \left(\frac{3}{4}, 0\right)$$

To find the y-intercept, substitute 0 for x and solve for y.

$$4(0) + y = 3$$
$$y = 3 \qquad y\text{-intercept } (0, 3)$$

To find another point on the line, choose any value for x and solve for y. Let $x = 1$.

$$4(1) + y = 3$$
$$4 + y = 3$$
$$y = -1 \qquad \text{point } (1, -1)$$

Graph the points $\left(\frac{3}{4}, 0\right)$, $(0, 3)$, and $(1, -1)$. Draw the graph of $4x + y = 3$. (See Figure 3-5.)

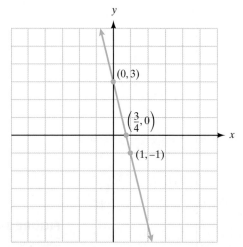

FIGURE 3-5
Example 2.

Example 3 — Graphing a Line That Passes through the Origin

Draw the graph of $x + 3y = 0$.

To find the x-intercept, substitute 0 for y and solve for x.

$$x + 3(0) = 0$$
$$x = 0 \quad \text{x-intercept } (0, 0)$$

To find the y-intercept, substitute 0 for x and solve for y.

$$0 + 3y = 0$$
$$3y = 0$$
$$y = 0 \quad \text{y-intercept } (0, 0)$$

Notice that the x- and y-intercepts are the same. This is a case when you *must* find another point to draw the graph, because two *different* points are needed to determine a line.

To find another point on the line, choose a value for y and solve for x. Let $y = 1$.

$$x + 3(1) = 0$$
$$x + 3 = 0$$
$$x = -3 \quad \text{point } (-3, 1)$$

Graph the points $(0, 0)$ and $(-3, 1)$. Draw the graph of $x + 3y = 0$. (See Figure 3-6.)

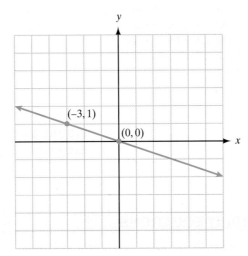

FIGURE 3-6
Example 3.

All the graphs that we have looked at up to this point have been diagonal lines. There are two special types of lines, called **horizontal** and **vertical lines**.

Example 4

The Graph of a Horizontal Line

Draw the graph of $y = -2$.

We can write this equation in general form as $0x + y = -2$. Clearly, this equation is satisfied if y has a value of -2 and x has any value. If we choose the ordered pairs $(3, -2)$ and $(-4, -2)$, we see that the equation $y = -2$ is true in both cases. By plotting these points and drawing the graph, we have a horizontal line that has a y-intercept of -2. (See Figure 3-7.)

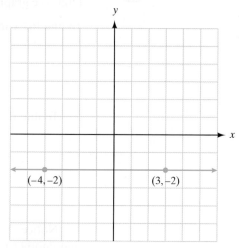

FIGURE 3-7
Example 4.

Example 5

The Graph of a Vertical Line

Draw the graph of $x = 3$.

We can write this equation in general form as $x + 0y = 3$. Again, this equation will be satisfied if x is 3 and any value for y is chosen. If we choose ordered pairs $(3, -1)$ and $(3, 4)$, we see that the equation $x = 3$ is true for both cases. By plotting these points and drawing the graph, we have a vertical line with an x-intercept of 3. (See Figure 3-8.)

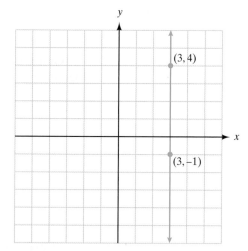

FIGURE 3-8
Example 5.

IMPORTANT EQUATIONS

Equations for Horizontal and Vertical Lines

The equation of a vertical line is $x = a$, where a is the x-intercept of the line.
The equation of a horizontal line is $y = b$, where b is the y-intercept of the line.

A good way to remember the difference in the equations of horizontal and vertical lines is to note that if an equation has only an x variable, then that equation only intersects the x-axis and does not intersect the y-axis. Similarly, if the equation only has a y variable, it intersects the y-axis and does not intersect the x-axis. The only exceptions to this rule are the two axes themselves, $x = 0$ and $y = 0$.

Practice Set 3-2

1. Explain how to find the x-intercept of a linear equation.
2. Explain how to find the y-intercept of a linear equation.
3. If you calculate the x- and y-intercepts of an equation passing through the origin, what additional information will you need before you can graph the line?
4. Which one of these equations has its graph as a horizontal line?
   (a) $2x = 6$
   (b) $2y + 6 = 0$
   (c) $x + 6y = 0$
   (d) $x - 6 = 0$

Find the missing coordinate in each of the following solutions to the given equations.

5. $x - y = 6$ (__, 0)   (0, __)   (__, −2)
6. $x + y = 8$ (__, 0)   (0, __)   (__, −2)
7. $2x + y = 6$ (__, 0)   (0, __)   (__, −2)
8. $x - 3y = 12$ (__, 0)   (0, __)   (__, −2)
9. $y = -2$ (3, __)   (0, __)   (__, −2)
10. $x = 3$ (__, 0)   (3, __)   (__, −2)
11. $2x + 5y = 10$ (__, 0)   (0, __)   (__, −2)
12. $x + 4y = 0$ (__, 0)   (0, __)   (__, −2)

Find the x- and y-intercepts of each equation and use them to sketch the graphs.

13. $-x - 2y = 4$
14. $x - 5y = 10$
15. $-x - 2y = 5$
16. $x + 3y = 6$
17. $2x + 3y = 6$
18. $-3x - 2y = 12$
19. $3x - 5y = 15$
20. $4x - 3y = 12$
21. $4x - y = 3$
22. $2x - 5y = 8$
23. $3x = 2y$
24. $4x = -2y$
25. $x + y = 0$
26. $x - y = 0$
27. $2x + 8 = -4y$
28. $2x - y - 1 = 0$
29. $x = 2y + 2$
30. $3x - 6 = -y$
31. $y = 0.2x - 4$
32. $y = -0.25x + 3$
33. $x + \frac{1}{2}y = 3$
34. $\frac{3}{4}x + y = -2$
35. $x = -3$
36. $5x - 3 = 0$
37. $3y = 2$
38. $y - 6 = 0$
39. $3y + x = 4$
40. $5y - x = 8$

| Section | **3-3** | **Slope** |

The **slope** of a line can be defined as the steepness of the line. To describe the steepness, or slope, of an airplane's flight path shortly after takeoff, you would estimate the vertical rise for every 100 meters of horizontal run and compute the ratio. Slope can also be defined as a rate of change. In this example, the slope number indicates a rate of change in the airplane's altitude. (See Figure 3-9.)

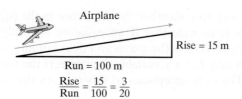

$$\frac{\text{Rise}}{\text{Run}} = \frac{15}{100} = \frac{3}{20}$$

FIGURE 3-9
Slope of an airplane's flight path.

If you want to find the slope of a straight line, choose any two points on the line and count the number of units in the rise (up or down) and the run (right or left) from one point to the other. The slope is the ratio of the rise to the run.

Example 1 Calculating the Slope of a Line from a Graph

Calculate the slope of the line shown in Figure 3-10.

The line passes through the points $A(-2, 1)$ and $B(5, 4)$. Start at point A and count up three units. Then turn right and count seven units to reach point B. The rise, or vertical change, in moving from point A to point B is $+3$. The run, or horizontal change, in moving from point A to point B is $+7$. Because slope is the ratio of the rise to the run, the slope of this line is $\frac{3}{7}$. Notice that the line rises to the *right* when the slope is *positive*.

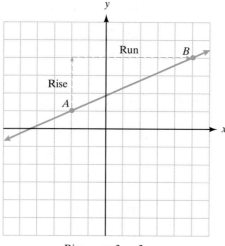

$$\frac{\text{Rise}}{\text{Run}} = \frac{\text{up 3}}{\text{right 7}} = \frac{3}{7}$$

FIGURE 3-10
Example 1.

Example 2 Calculating the Slope of a Line from a Graph

Calculate the slope of the line in Figure 3-11.

The line passes through the points $A(-3, 5)$ and $B(1, 2)$. This time we calculate the slope by beginning at point B and moving to point A. The rise is three units up $(+3)$ and the run is four units left (-4).

The slope $= \dfrac{\text{rise}}{\text{run}} = \dfrac{3}{-4} = -\dfrac{3}{4}$. Notice that the line falls as it moves from *left* to *right* when the slope is *negative*.

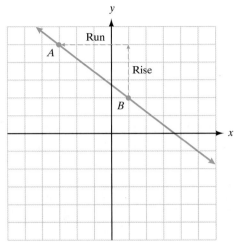

$$\frac{\text{Rise}}{\text{Run}} = \frac{\text{up } 3}{\text{left } 4} = \frac{+3}{-4} = -\frac{3}{4}$$

FIGURE 3-11
Example 2.

In general, we evaluate the behavior of a line as it moves from left to right. Therefore, lines with positive slopes are increasing as they move from left to right, and lines with negative slopes are decreasing as they move from left to right. (See Figure 3-12.)

We can now make a general definition for slope that can be used to calculate the slope of a line when given two points on the line. Given two points (x_1, y_1) and (x_2, y_2), to find the slope, subtract the y-coordinates and divide that by the difference in the x-coordinates. The variable m is generally used to represent the slope number.

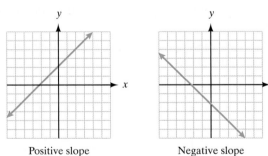

Positive slope   Negative slope   **FIGURE 3-12**

IMPORTANT EQUATIONS

Slope of a Line

$$m = \frac{\text{rise}}{\text{run}} = \frac{\text{change in } y}{\text{change in } x} = \frac{\Delta y}{\Delta x} = \frac{y_2 - y_1}{x_2 - x_1}$$

where (x_1, y_1) and (x_2, y_2) are points on the given line.

When the formula for slope is used, the *order* of subtraction is important. You must compute the numerator and denominator using the *same order* of subtraction.

$$\text{Slope} = \frac{y_2 - y_1}{x_2 - x_1} = \frac{y_1 - y_2}{x_1 - x_2} \neq \frac{y_2 - y_1}{x_1 - x_2}$$

Example 3 **Calculating the Slope of a Line Given Two Points on the Line**

Determine the slope of the line that passes through the points $(0, -5)$ and $(1, -3)$.

$$\text{Slope} = \frac{y_2 - y_1}{x_2 - x_1}$$

$$\text{Slope} = \frac{-3 - (-5)}{1 - 0}$$

$$\text{Slope} = \frac{2}{1} = 2$$

The slope of the line is 2.

Example 4 **Calculating the Slope of a Line Given Two Points on the Line**

Find the slope of the line that passes through the points $(2, 3)$ and $(2, -1)$.

$$\text{Slope} = \frac{y_2 - y_1}{x_2 - x_1}$$

$$\text{Slope} = \frac{-1 - 3}{2 - 2}$$

$$\text{Slope} = \frac{-4}{0} \qquad \text{(division by 0 is undefined)}$$

Therefore, the slope of this line is undefined and, if graphed, will be a vertical line. (See Figure 3-13.)

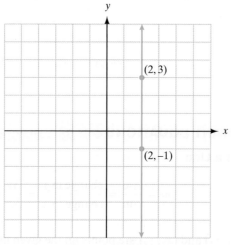

Slope = undefined

FIGURE 3-13
Example 4.

Section 3-3 **Slope** 109

Example 5

Calculating the Slope of a Line Given Two Points on the Line

Find the slope of the line that passes through the points $(-2, 4)$ and $(3, 4)$.

$$\text{Slope} = \frac{y_2 - y_1}{x_2 - x_1}$$

$$\text{Slope} = \frac{4 - 4}{3 - (-2)}$$

$$\text{Slope} = \frac{0}{5}$$

$$\text{Slope} = 0$$

If you graph the points $(-2, 4)$ and $(3, 4)$, you will see that the line is a horizontal line. (See Figure 3-14.)

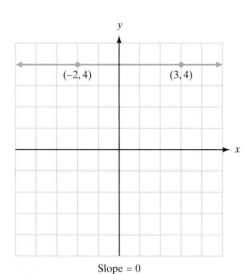

FIGURE 3-14
Example 5.

Rule

Slopes of Vertical and Horizontal Lines

The slope of a vertical line is undefined.
The slope of a horizontal line is 0.

We have discussed graphing linear equations by finding x- and y-intercepts and by finding additional points on the line. Another method that can be used to graph a linear equation is plotting a point on the line and applying the slope to find another point on the line. If we know one point on the line and the direction the line is taking (slope), then it is easy to graph it.

Example 6

Graphing Lines Using the Slope

Graph the line that passes through the point $(1, -2)$ and has a slope of -1.

(a) First, we must plot the point $(1, -2)$.

(b) Second, we will apply the slope to find another point. Recall that, for graphing purposes, slope = $\frac{\text{rise}}{\text{run}}$. Because the slope is -1, we can rewrite this as $\frac{-1}{1}$. This tells us that the rise is -1 and the run is 1. Notice that any time the slope is an integer, the run will always equal 1. To apply the slope, we will start at $(1, -2)$, and then we will move down one unit and to the right one unit to locate another point on the line. Because the slope is negative, we must go in a negative direction one time. This means that we could also move up one and to the left one to find another point on the line. (See Figure 3-15.)

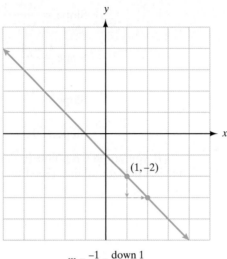

$$m = \frac{-1}{1} = \frac{\text{down 1}}{\text{right 1}}$$

FIGURE 3-15
Example 6.

Example 7 Graphing Lines Using the Slope

Graph the line that passes through the point $(-1, -1)$ and has a slope of $\frac{2}{3}$.

Plot the point $(-1, -1)$. Apply the slope by starting at $(-1, -1)$ and moving up two units and right three units. Notice that because the slope is positive, we could also move down two units and to the left three units. (See Figure 3-16.)

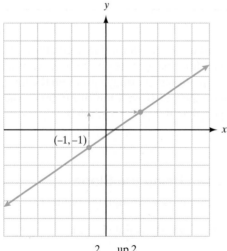

$$m = \frac{2}{3} = \frac{\text{up 2}}{\text{right 3}}$$

FIGURE 3-16
Example 7.

Recall from Section 3-1 that we discussed the general form for a linear equation. A second form for an equation of a line is called the **slope-intercept form**. This form of the equation is derived from the general form by solving it for y.

Section 3-3 **Slope** 111

> **IMPORTANT EQUATIONS**
>
> **Linear Equations**
>
> General form:
> $$Ax + By = C$$
>
> Slope-intercept form:
> $$y = mx + b$$
>
> where m = slope and b = y-intercept.

We can easily graph a line by changing the general equation of the line into its slope-intercept form. Once we have written the equation in this form, the slope number (m) is the coefficient of x, and the y-intercept is the constant (b). Always start graphing from the location of the y-intercept. Then use the slope number to help you locate another point on the line.

Example 8 **Graphing Lines Using the Slope-Intercept Form of the Equation**

Find the slope and y-intercept of the following linear equation and then graph the line.

$$3x + 2y = 3 \quad \text{(original equation)}$$

Put the equation into slope-intercept form by solving the equation for y.

$$2y = -3x + 3 \quad \text{(subtract } 3x \text{ from both sides)}$$

$$\frac{2y}{2} = \frac{-3x}{2} + \frac{3}{2} \quad \text{(divide both sides by 2)}$$

$$y = \frac{-3x}{2} + \frac{3}{2} \quad \text{(simplify)}$$

$$\text{Slope} = -\frac{3}{2} \qquad y\text{-intercept} = \frac{3}{2} = 1.5$$

Plot the y-intercept on the graph. Remember that the y-intercept is the point where the line crosses the y-axis, in this case (0, 1.5). Next, apply the slope to find another point on the line. Start at (0, 1.5) and move down three units and to the right two units. (See Figure 3-17.)

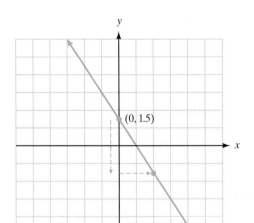

$m = \frac{-3}{2} = \frac{\text{down } 3}{\text{right } 2}$

FIGURE 3-17
Example 8.

Example 9 — Graphing Lines Using the Slope-Intercept Form of the Equation

Find the slope and y-intercept of $-\frac{1}{3}x + \frac{1}{6}y = \frac{1}{2}$ and graph the line.

$$-\frac{1}{3}x + \frac{1}{6}y = \frac{1}{2} \quad \text{(original equation)}$$

It will be easier to put this equation into slope-intercept form if the fractions are eliminated from the equation. This can be done by multiplying all terms of the equation by the lowest common denominator (LCD) of the fractions, which in this case is 6.

$$6\left(-\frac{1}{3}x\right) + 6\left(\frac{1}{6}y\right) = 6\left(\frac{1}{2}\right) \quad \text{(multiply the equation by 6)}$$

$$-2x + y = 3$$

$$y = 2x + 3 \quad \text{(add } 2x \text{ to both sides)}$$

$$\text{Slope} = \frac{2}{1} \qquad y\text{-intercept} = 3$$

Now graph the line. (See Figure 3-18.)

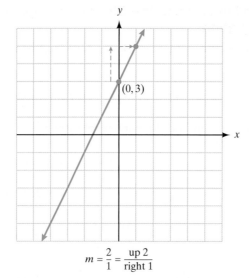

$$m = \frac{2}{1} = \frac{\text{up } 2}{\text{right } 1}$$

FIGURE 3-18
Example 9.

Parallel and Perpendicular Lines

We have been discussing the graphs of linear equations. Two lines that lie in the same plane and do not intersect are called **parallel lines**. The graphs of $y = 2x + 1$ and $y = 2x - 3$ are shown in Figure 3-19.

What do you notice about these lines? Because they do not intersect, they are parallel lines. Look carefully at the equations of these lines. The slope of $y = 2x + 1$ is 2 and the slope of $y = 2x - 3$ is also 2. This illustrates the relationship between slope and parallel lines.

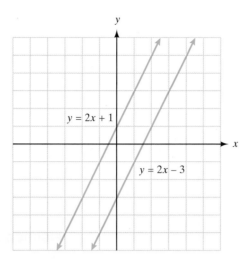

FIGURE 3-19
Two parallel lines.

Rule — The Slopes of Parallel Lines

If two lines are parallel, then they have the same slope.
If two lines have the same slope, then they are parallel.

Note that, although vertical lines have slopes that are undefined, they are parallel.

Example 10 — Finding the Slopes of Parallel Lines

Find the slope of a line parallel to the line whose equation is $-6x + 3y = 15$.

Because parallel lines have the same slope, the first step is to find the slope of $-6x + 3y = 15$. Recall that, to find the slope of a line, we put the equation into slope-intercept form ($y = mx + b$).

$$-6x + 3y = 15$$
$$3y = 6x + 15$$
$$y = 2x + 5 \qquad \text{slope is 2}$$

Therefore, the slope of any line parallel to $-6x + 3y = 15$ is 2.

The slopes of perpendicular lines also have a special relationship. **Perpendicular lines** are two lines that intersect to form right angles. The graphs of $y = -3x - 2$ and $y = \frac{1}{3}x + 1$ are shown in Figure 3-20.

Look carefully at the slopes of these two lines. The slope of $y = -3x - 2$ is -3. The slope of $y = \frac{1}{3}x + 1$ is $\frac{1}{3}$. What do you notice about the relationship of their slopes? One thing to note is that one line has a positive slope and the other has a negative slope. If they both had positive slopes or both had negative slopes, the lines would not form right angles. The directions of the lines are also reversed, and that is illustrated by the change in the slope ratio. The ratio of numbers representing the rise over the run in the given line is reversed in the perpendicular line.

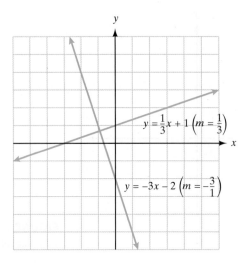

FIGURE 3-20
Two perpendicular lines.

Rule — **The Slopes of Perpendicular Lines**

The slopes of nonvertical perpendicular lines are negative reciprocals of each other. The product of their slopes is -1.
Lines are also perpendicular if one of them is vertical and the other is horizontal.

The rule about negative reciprocals does not apply to vertical or horizontal lines because the slope of a vertical line is undefined and the reciprocal of the slope of a horizontal line would also be undefined. However, it is easy for us to see that a horizontal line would be perpendicular to a vertical line and vice versa.

Example 11 — **Finding the Slopes of Perpendicular Lines**

Find the slope of a line perpendicular to the line whose equation is $x - 2y = 2$.

Because the slopes of perpendicular lines are negative reciprocals of each other, we first find the slope of $x - 2y = 2$.

$$x - 2y = 2$$
$$-2y = -x + 2$$
$$y = \frac{1}{2}x - 1 \qquad \text{slope is } \frac{1}{2}$$

Therefore, the slope of any line perpendicular to $x - 2y = 2$ is the negative reciprocal of $\frac{1}{2}$ or $-\frac{2}{1} = -2$.

Example 12 — **Slopes of Parallel and Perpendicular Lines**

Determine if the graphs of $x = 3$ and $x = 5$ are parallel, perpendicular, or neither.

Because $x = 3$ is a vertical line going through the point $(3, 0)$ and $x = 5$ is a vertical line going through the point $(5, 0)$, their slopes are both undefined. Because the slopes are the same, the lines are parallel. Notice also that if you look at the graphs of the lines $x = 3$ and $x = 5$, you will see that the lines are parallel. (See Figure 3-21.)

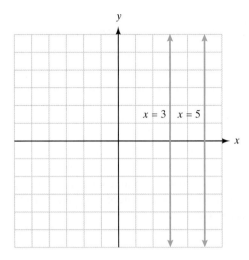

FIGURE 3-21
Example 12.

Practice Set 3-3

Determine the slopes of the lines passing through each pair of points.

1. $(6, 1)$ and $(8, -4)$
2. $(5, 7)$ and $(4, -6)$
3. $(-3, 0)$ and $(8, 2)$
4. $(-6, -3)$ and $(4, 1)$
5. $(-8, -2)$ and $(-4, 8)$
6. $(6, 1)$ and $(6, 7)$
7. $(2.5, 3)$ and $(1, -9)$
8. $\left(1\frac{3}{4}, \frac{1}{3}\right)$ and $\left(2, \frac{1}{6}\right)$

A line has a slope of 7. For each pair of points in problems 9 to 12 that are on the line, use the slope formula to find the missing coordinate.

9. $(x, 8)$ and $(4, 1)$
10. $(6, y)$ and $(2, -13)$
11. $(3, -6)$ and $(7, y)$
12. $(x, -2)$ and $(-6, -3)$

Find the slope and y-intercept and graph the following linear equations.

13. $y = -4x + 1$
14. $y = -4$
15. $3x + 2y = 6$
16. $x + 3y = 12$
17. $2x - 3y = 4$
18. $4x = 11$
19. $4y - x = 10$
20. $2x - y = 2$
21. $\frac{3}{4}x - \frac{1}{2}y = \frac{5}{8}$
22. $\frac{2}{3}x + 2y = 2$
23. $y = -x$
24. $3x - 2y = 7$
25. $4y = 12 + x$

Find the slope of a line parallel to lines with the following equations.

26. $y = -4x + 1$
27. $3x - 8y = 11$
28. $x = 3$
29. $-x + 6y = 2$
30. $y = 2$

Find the slope of a line perpendicular to lines with the following equations.

31. $x + y = 3$
32. $3x - 2y = 8$
33. $x = 2$
34. $-2x + 5y = 10$
35. $y = 2$

Determine whether the graphs of the following equations are parallel, perpendicular, or neither.

36. $x + y = 7$
    $x + y = -10$
37. $x + y = 5$
    $x - y = -5$
38. $3x + 2y = 6$
    $-2x + 3y = 4$
39. $y = 2x$
    $y = x$
40. $y = 3$
    $x = 3$

Section 3-4 Writing Equations of Lines

In previous sections, we looked at two forms of the linear equation. One we called the general form, and it was written $Ax + By = C$. The second form was called the slope-intercept form, and it was written $y = mx + b$. We used the slope-intercept form of the equation to help us graph lines. In this section, we will use this form of the equation to help us write the equation of a line if we are given specific information about the line.

In the equation $y = mx + b$, recall that m represents the slope of the line, b represents the y-intercept (where the line crosses the y-axis), and x and y represent points on the line. To write the equation of a line using $y = mx + b$, we need to find the values of m and b. Then, if necessary, we can rewrite this equation in general form $Ax + By = C$.

Example 1 — Writing the Equation of a Line Using the Slope-Intercept Form

Find an equation of a line in general form with a slope of -2 and a y-intercept of 4.

Look at the given information. Because we know the slope is -2 and the y-intercept is 4, we will substitute these values into the slope-intercept form of a linear equation.

$y = mx + b$ (slope-intercept form)

$y = -2x + 4$ (substitute -2 for the slope and 4 for the y-intercept)

Next, we rewrite the equation in general form.

$2x + y = 4$ (add $2x$ to both sides)

Example 2 — Writing the Equation of a Line Given One Point and the Slope

Find an equation of the line that passes through the point $(-2, 1)$ and has a slope of 3.

Given that the slope is 3 and the line passes through the point $(-2, 1)$, we will substitute these values into the slope-intercept form of the equation.

$y = mx + b$ (slope-intercept form)

$1 = 3(-2) + b$ (substitute $m = 3$, $x = -2$, $y = 1$)

$1 = -6 + b$ (simplify)

$7 = b$ (add 6 to both sides to solve for b)

Now write the equation of this line using the slope-intercept form by substituting the value of m and calculated value of b.

$y = mx + b$

$y = 3x + 7$ or $3x - y = -7$ (general form)

Another form of an equation of a line is the **point-slope** form. This equation can also be used to write an equation of a line. The point-slope form is $y - y_1 = m(x - x_1)$, where $m =$ the slope of the line and (x_1, y_1) is a point on the line.

Often it is convenient to use this form by filling in the missing numbers and simplifying the equation. In Example 2 we could have used the point-slope form to write the equation.

$$y - y_1 = m(x - x_1) \quad \text{(point-slope form)}$$
$$y - 1 = 3(x - (-2)) \quad \text{(substitute } x_1 = -2, y_1 = 1\text{)}$$
$$y - 1 = 3(x + 2) \quad \text{(simplify)}$$
$$y - 1 = 3x + 6 \quad \text{(add 1 to both sides)}$$
$$y = 3x + 7 \quad \text{(slope-intercept form)}$$

or

$$3x - y = 7 \quad \text{(general form)}$$

For every equation that we write, we need a minimum of two pieces of information: the slope and a point on the line. If the slope is not told to us directly, then there must be a way to calculate the slope from the information given. If we are given two points on the line, then we can use the formula $m = \dfrac{y_2 - y_1}{x_2 - x_1}$ to calculate the needed slope. If we are given the equation of another line and asked to write the equation of a line parallel to it or perpendicular to it, we can find our needed slope by using the slope-intercept form, $y = mx + b$, to find the slope of the given line for use in completing the problem.

Example 3

Writing the Equation of a Line Given Two Points on the Line

Find an equation of the line that passes through the points $(3, 2)$ and $(-1, 5)$.

First, we must find the slope of the line. Let $(x_1, y_1) = (3, 2)$ and $(x_2, y_2) = (-1, 5)$. Remember the slope formula is

$$m = \frac{y_2 - y_1}{x_2 - x_1} = \frac{5 - 2}{-1 - 3} = \frac{3}{-4} = -\frac{3}{4}$$

Second, we will substitute one point and the slope into the slope-intercept form for a linear equation.

$$y = mx + b \quad \text{(slope-intercept form)}$$
$$2 = -\frac{3}{4}(3) + b \quad \text{(substitute } m = -\frac{3}{4}, x = 3, y = 2\text{)}$$
$$2 = -\frac{9}{4} + b \quad \text{(simplify)}$$
$$2 + \frac{9}{4} = -\frac{9}{4} + \frac{9}{4} + b \quad \text{(add } \frac{9}{4} \text{ to both sides)}$$
$$\frac{8}{4} + \frac{9}{4} = b \quad \text{(common denominator)}$$
$$\frac{17}{4} = b$$

Now write the equation of the line using the slope and y-intercept that you have calculated.

$$y = mx + b$$
$$y = -\frac{3}{4}x + \frac{17}{4}$$

To check your work, substitute the x values of the original ordered pairs into this equation to determine if the y value calculated matches the original points.

Original point: (3, 2)   Let $x = 3$.   Then $y = -\dfrac{3}{4}(3) + \dfrac{17}{4}$

$$= -\dfrac{9}{4} + \dfrac{17}{4} = \dfrac{8}{4} = 2$$

Original point: (-1, 5)   Let $x = -1$.   Then $y = -\dfrac{3}{4}(-1) + \dfrac{17}{4}$

$$= \dfrac{3}{4} + \dfrac{17}{4} = \dfrac{20}{4} = 5$$

The points check, so this is the correct equation of the line that passes through these two points.

In Example 3, it does not matter which of the two points is labeled (x_1, y_1) and which is labeled (x_2, y_2). Try switching the labels to $(x_1, y_1) = (-1, 5)$ and $(x_2, y_2) = (3, 2)$ and reworking the problem to see that you obtain the same equation.

Horizontal and Vertical Lines

From the slope-intercept form of the equation of a line, you can see that a horizontal line has an equation of the form:

$y = (0)x + b$ (horizontal line has a slope of 0)

$y = b$ (equation for a horizontal line)

This is consistent with the fact that each point on a horizontal line through $(0, b)$ has a y-coordinate of b. (See Figure 3-22.)

Similarly, each point on a vertical line through $(a, 0)$ has an x-coordinate of a. (See Figure 3-23.)

Therefore, a vertical line has an equation of the form $x = a$.

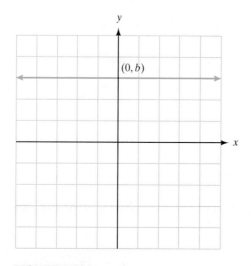

FIGURE 3-22

Graph of a horizontal line: $y = b$.

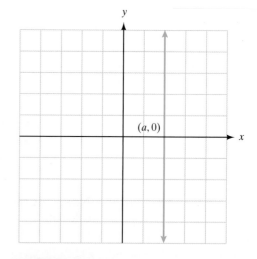

FIGURE 3-23

Graph of a vertical line: $x = a$.

Example 4 — Writing the Equations of Horizontal and Vertical Lines

Write an equation for each of the following lines.
(a) Vertical line through $(-1, 4)$
(b) Line passing through $(-4, 6)$ and $(2, 6)$
(c) Line passing through $(3, 7)$ and $(3, -9)$

(a) Because the line is vertical and passes through the point $(-1, 4)$, we know that every point on the line has an x-coordinate of -1.
 Therefore, the equation is $x = -1$.

(b) The line through $(-4, 6)$ and $(2, 6)$ is horizontal, and every point on the line has a y-coordinate of 6.
 Therefore, the equation is $y = 6$.

(c) The line through $(3, 7)$ and $(3, -9)$ is vertical, and every point on the line has an x-coordinate of 3.
 Therefore, the equation is $x = 3$.

Parallel and Perpendicular Lines

Recall from the previous section the following facts about parallel and perpendicular lines:

Rule — The Slopes of Parallel and Perpendicular Lines

Parallel lines have the same slopes.
The slopes of nonvertical perpendicular lines are negative reciprocals of each other.

Example 5 — Writing the Equations of Lines Parallel and Perpendicular to a Given Line

Find the equations of the lines that pass through the point $(2, -4)$ and are (a) parallel and (b) perpendicular to the line $x - 2y = 7$.

Write the given equation in slope-intercept form in order to determine its slope.

$$x - 2y = 7$$
$$-2y = -x + 7 \quad \text{(subtract } x \text{ from both sides of the equation)}$$
$$y = \frac{1}{2}x - \frac{7}{2} \quad \text{(divide all terms by } -2\text{)}$$

Therefore, the given line has a slope of $\frac{1}{2}$.

(a) Any line parallel to the given line must have the same slope. Therefore, the slope of our new line is $\frac{1}{2}$, and it passes through the point $(2, -4)$.

$$y - y_1 = m(x - x_1) \quad \text{(point-slope form)}$$
$$y - (-4) = \frac{1}{2}(x - 2) \quad \text{(substitute } m = \frac{1}{2}, x = 2, y = -4\text{)}$$
$$y + 4 = \frac{1}{2}(x - 2) \quad \text{(simplify)}$$

$$y + 4 = \frac{1}{2}x - 1 \quad \text{(distributive property)}$$

$$y = \frac{1}{2}x - 5 \quad \text{(slope-intercept form)}$$

or

$$x - 2y = 10 \quad \text{(general form)}$$

Notice that this equation and the original equation written in slope-intercept form, $y = \frac{1}{2}x - \frac{7}{2}$, are identical except for the values of b. All parallel lines will have identical slopes but different y-intercepts.

(b) Any line perpendicular to the given line must have a slope that is the negative reciprocal of $\frac{1}{2}$. Therefore, the slope of our second line is -2, and it passes through the point $(2, -4)$.

$$y = mx + b \quad \text{(slope-intercept form)}$$
$$-4 = (-2)2 + b \quad \text{(substitute } m = -2, x = 2, y = -4\text{)}$$
$$-4 = -4 + b \quad \text{(simplify)}$$
$$0 = b \quad \text{(solve for } b\text{)}$$

Write the equation: $y = -2x + 0$ or $2x + y = 0$.

IMPORTANT EQUATIONS

Summary of Equations of Lines

1. General form of equation of a line: $Ax + By = C$
2. Slope-intercept form of equation of a line: $y = mx + b$
3. Point-slope form of the equation: $y - y_1 = m(x - x_1)$
4. Slope of line through (x_1, y_1) and (x_2, y_2): $m = \frac{y_2 - y_1}{x_2 - x_1}$
5. Parallel lines have the same slope.
6. Perpendicular lines have slopes that are negative reciprocals of each other (with the exception of vertical and horizontal lines).
7. Equation of a vertical line: $x = a$
8. Equation of a horizontal line: $y = b$

Practice Set 3-4

Find an equation of the line given the following information. Final answers should be given slope-intercept form and general form.

1. Slope $= -2$ y-intercept $= 8$
2. Slope $= 5$ y-intercept $= -3$
3. Passes through the point $(6, -2)$ slope $= \frac{1}{2}$
4. Passes through the point $(8, 1)$ slope $= -\frac{3}{4}$
5. Passes through the point $(-8, -9)$ slope $= -2$

6. Passes through the point (2, 5)   slope = −1

7. Passes through the point (−1, 4)   slope = $\dfrac{5}{2}$

8. Passes through the point (−3, 1)   slope = $-\dfrac{1}{3}$

9. Slope = $-\dfrac{1}{3}$ x-intercept = 5

10. Slope = $-\dfrac{2}{5}$ x-intercept = −5

11. Passes through the points (4, 3) and (7, 9)

12. Passes through the points (1, 3) and (6, 10)

13. Passes through the points (5, −2) and (8, −2)

14. Passes through the points (6, 1) and (−2, 1)

15. Passes through the points (−1, 4) and (1, 2)

16. Passes through the points (6, −3) and (8, −9)

17. Passes through the points $\left(7, 1\dfrac{1}{3}\right)$ and $\left(5, \dfrac{1}{3}\right)$

18. Passes through the points $\left(5\dfrac{2}{3}, 5\right)$ and $\left(2\dfrac{2}{3}, -4\right)$

19. Passes through the points (−4, 3) and (−1, −4.5)

20. Passes through the points (5, −2) and (3, 1.5)

21. Passes through the points (3, −5) and $\left(3, 1\dfrac{1}{3}\right)$

22. Passes through the points $\left(5, 2\dfrac{3}{5}\right)$ and (5, 2)

23. Slope = undefined   passes through (1, 4)

24. Slope = undefined   passes through (−3, 5)

25. Passes through the point (1, −7) and is parallel to the line $2x + y = 6$

26. Passes through the point (−6, 5) and is parallel to the line $x − 2y = 8$

27. Passes through the point (3, 10) and is perpendicular to the line $x − 3y = −4$

28. Passes through the point (−2, 6) and is perpendicular to the line $2x − y = 4$

29. Passes through the point (−2, 9) and is parallel to the line through (2, 8) and (−1, 11)

30. Passes through the point (−1, −4) and is parallel to the line through (0, 1) and (5, −1)

31. Passes through the point (−5, 2) and is perpendicular to the line through the points (−4, 1) and (1, 3)

32. Passes through the point (3, 6) and is perpendicular to the line through the points (5, 2) and (6, −1)

33. Passes through the point (3, 2) and is parallel to the line $x = 4$

34. Passes through the point (−6, 1) and is parallel to the line $y = 8$

35. Passes through the point (−1, 3) and is perpendicular to the line $x = 4$

36. Passes through the point (−2, 0) and is perpendicular to the line $y = −9$

37. x-intercept of 3 and y-intercept of −1

38. x-intercept of −5 and y-intercept of −6

39. Passes through the origin and has a slope of −1

40. Passes through the origin and has a slope of $\dfrac{2}{3}$

Section 3-5 Applications and Uses of Graphs

In previous sections, you have graphed lines, examined the slopes of lines, and written the equation of lines given certain information about those lines. In this section, we will look at the relationships among these topics and examine some applications of linear equations in real-life situations.

122 Chapter 3 **Graphing**

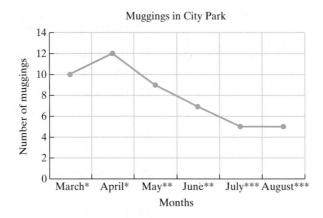

FIGURE 3-24
Number of muggings in City Park.

There are many jobs that require the collection of data for examination and analysis. Many times the graph of a set of data will show a linear trend either upward or downward. Trends in the data as seen on a graph can assist planners in predicting future events or needs in many cases. For example, assume that the police chief of a large city is trying to convince the city council that he needs more police officers on staff so he can patrol the city park more effectively to prevent muggings. He collects the following data related to the incidences of muggings in the park over a six-month period. In March and April, he assigns one officer to patrol the park, and there are 10 muggings in March and 12 in April. During the months of May and June, he pays an additional officer overtime to work in the park, and there are 9 muggings reported in May and 7 in June. During the summer months of July and August, he pays a third officer overtime to patrol the park and 5 muggings are reported in July and 5 in August. In order to present this information to the city council, he graphs the data to help illustrate the downward trend of the muggings as the number of officers was increased. Look at Figure 3-24. This is a more effective tool for his presentation than a table of numbers because it is easy for the members of the city council to visualize the trend in the data.

Not all relationships are linear—some graphs will curve or even seem unpredictable—but for graphs that have points that seem to behave in a linear fashion, we can derive a prediction equation from the data that results in a line that we call the best-fit line. The method of determining this equation from data is called linear regression and will be examined more closely in Chapter 8. The prediction equation or least-squares equation is based on the slope-intercept form of the equation that we have studied in this chapter. The slope and y-intercept are determined from the data points and used to write an equation. Let's look at a simple example of deriving a linear equation from the graph of a line.

Example 1 **Writing the Equation of a Line from a Graph**

Look at the graph of the line in Figure 3-25 and write the equation of the line shown.

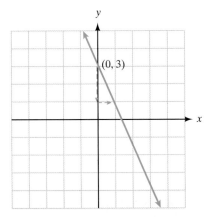

FIGURE 3-25

Remember that the slope-intercept form of the line is $y = mx + b$. In order to write the equation of the line, we need to determine the slope of the line shown and its y-intercept. The line is trending downward from left to right at a rate of two steps down for every one step right. This gives a slope of $\frac{down\ 2}{right\ 1} = \frac{-2}{1} = -2$. The line crosses the y-axis at $(0, 3)$ so $b = 3$. Using this information, then, our equation for this line is $y = -2x + 3$ or $2x + y = 3$.

Many weight-loss programs ask participants to weigh themselves only once a week, and then record the results on a graph. Participants are able to see the trend in their weight over a period of time and hopefully have an idea of the time it will take to reach their goal weight if a constant amount of weight is lost weekly. Look at the next example using results from Felecia's weight-loss program.

Example 2

Charting Weight Loss

Felecia joined the FBC Weight-Loss Program and began her diet and exercise program after weighing in on January 1st. She enrolled for a three-month period with a goal weight of 160 lbs. She recorded the following weights for January. (See Figure 3-26.) At this rate, will she reach her goal weight prior to the end of the three-month period? Answer the following questions to determine the answer to this question.

(a) What is the y-intercept of the graph and what does it tell you about Felecia's weight?

The y-intercept is $(0, 192)$. This indicates that her weight was 192 on January 1 when she began her diet and exercise program.

(b) What is the slope of this line and what does it tell you about Felecia's weight loss?

The graph indicates that she has lost 3 pounds each week so the rate of change in this graph is $\frac{loss\ of\ 3}{1\ week} = \frac{-3}{1} = -3$. Therefore, the slope is $m = -3$.

(c) Use the graph to answer the question: Will she reach her goal weight before the end of March?

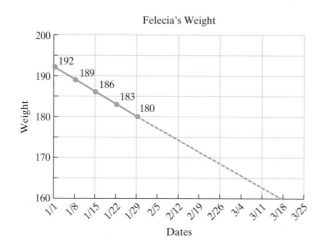

FIGURE 3-26

If you take a ruler and continue to draw the line downward at the same rate, the line will intersect the *x*-axis near the 18th of March. Since the *x*-axis is labeled 160 which is her goal weight, she should reach her goal before the three-month period is over.

If we look at Example 2, we see that Felecia's weight loss is a steady loss of 3 pounds per week, which results in a straight line graph. Because there is a steady loss, we can derive a slope of -3 and because we know her initial weight of 192, we have a *y*-intercept or value for *b*. Using these values and the slope-intercept form of the equation, we can write an equation that will help us predict her weight after a certain number of weeks.

$$y = -3x + 192.$$

For example, to predict her weight after 6 weeks on the program, we can substitute 6 for *x* in the prediction equation and calculate that she should weigh 174 pounds.

$$y = -3(6) + 192 = -18 + 192 = 174$$

Realistically, we know that a steady loss of exactly 3 pounds per week would be unusual. With real data where people lose 2 pounds one week, 1 pound the next week, and then 3 pounds, we will use a more advanced method called linear regression to derive the equation. However, the resulting equation would be used in the same way to make predictions about outcomes in the future.

Example 3 Predictions of Sales

The Acme Sales Company began its operations in 2008 selling music CDs. Look at the graph of its sales for the first five years of operation in Figure 3-27 and answer the following questions.

(a) How much have sales increased each year? Relate this to the slope of the line.

Each year, sales have increased by 0.5 million CDs. At a rate of change of $+0.5$ million every 1 year, the slope is $m = 0.5$ million.

(b) Using the graph, determine the year that sales will reach 6 million.

By continuing a straight line at the same slope, the graph indicates that the company should reach sales of 6 million in the year 2015.

Acme Sales Co.—Sales of CDs

FIGURE 3-27

(c) Using the slope-intercept form of the equation, write a prediction equation for sales at Acme assuming this steady rate of sales continues.

Sales in 2008, the first year of operation, were 2.5 million so we use that value for b in the slope-intercept form of the equation. With a slope of 0.5, we have the equation $y = 0.5x + 2.5$ where x represents the number of years after 2008 and y represents sales in millions.

(d) Predict the number of CDs that Acme will sell in the year 2016 using the prediction equation.

Since 2016 is 8 years after 2008, we will let $x = 8$ and substitute it into the equation, $y = 0.5x + 2.5$.

$$y = 0.5(8) + 2.5 = 4 + 2.5 = 6.5$$

Therefore, we predict that Acme Sales Company will sell 6.5 million CDs in the year 2016.

In the next chapter, we will look more closely at the equations of lines in the form of linear functions. You will continue to graph data as well as derive the equations of the lines formed by plotting the data using the slope-intercept form of the equation. You will also use your graphing calculator extensively to examine the behavior of linear functions. In Chapter 8, we will look at real sets of data and derive prediction equations to use in planning for the future.

Practice Set 3–5

Look at the lines drawn in each graph and write the equation of the line pictured.

1.
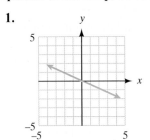

2.
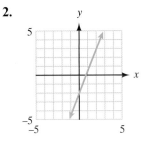

126 Chapter 3 **Graphing**

3.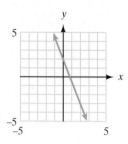

4.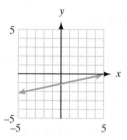

5.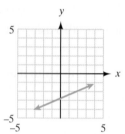

6.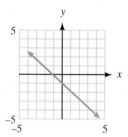

7.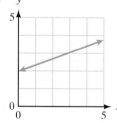

8.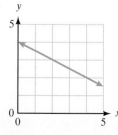

9.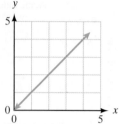

10.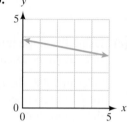

11.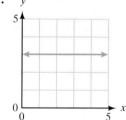

12.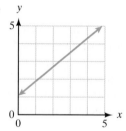

Plot the data sets in each of the following problems on a piece of graph paper, and then answer the questions. Be sure to label your graphs accurately.

13. The school board has to plan well in advance for increased enrollment in a school district. The maximum capacity of one high school is 800 students. Enrollment figures for the past 4 years are given below, and the population is increasing at a steady rate.

Year	2009	2010	2011	2012
Enrollment	650	662	674	686

(a) Plot this data on a graph and label the x-axis through the year 2020. Connect the given data points and extend the line using a dotted line

to completely cross the graph. Based on this steady growth rate, will the population exceed the capacity of the school by the year 2020?
(b) What is the rate of change or slope for this data?
(c) Using the enrollment figure from the year 2009 as your initial value (y-intercept on your graph), write the equation of a prediction line for this data set letting x = the number of years after 2009.
(d) Use your equation to predict the enrollment at this school in the year 2017. Verify your answer by looking at the graph.

14. The number of aldermen on the town council is related to the population of the town. The town charter allows for 1 alderman for every 250 citizens in the town. In 2009, the town's population was 1440 with 5 aldermen on the town council. When the population of the town dips below 1250, one seat will be lost and only 4 aldermen will be required for the town council. The table below gives the population figures for the number of citizens within the town limits for the past 5 years.

Year	2009	2010	2011	2012	2013
Population	1440	1426	1412	1398	1384

(a) Plot this data on a graph and label the x-axis through the year 2019. Connect the given data points and extend the line using a dotted line to completely cross the graph. Based on this growth rate, will number of aldermen required for the town council be reduced by the year 2019?
(b) What is the rate of change or slope for this data?
(c) Using the population figure from the year 2009 as your initial value (y-intercept on your graph), write the equation of a prediction line for this data set letting x = the number of years after 2009.
(d) Use your equation to predict the population of this town in the year 2017. Verify your answer by looking at the graph.

15. Shelly took a new job at a large corporation in 2009 making $29,500 per year. She received yearly performance raises and decided to chart her annual incomes to try and project her salary in five years. The table lists her salaries.

Year	2009	2010	2011	2012	2013
Income	$29,500	$30,237.50	$30,993.44	$31,768.27	$32,562.48

(a) Plot this data on a graph and label the x-axis through the year 2019. Connect the given data points and extend the line using a dotted line to completely cross the graph. Based on this growth rate, will her salary be more than $35,000 in 2019?
(b) What is the rate of change or slope for this data? Is it constant?
(c) Using her initial salary from the year 2009 as your initial value (y-intercept on your graph), write the equation of a prediction line for this data set letting x = the number of years after 2009.
(d) Use your graph to predict her salary in 2018.

16. An electrical circuit has a variable voltage source and as the voltage (volts, V) in the circuit increases, so does the current (milliamps, mA) flowing in the circuit. The table below gives voltages and the corresponding currents, based on direct measurements of the circuit.

volts	5	6	7	8	9
milliamps	7.4	9.7	12.0	14.3	16.6

(a) Plot a graph of this data with voltage as the independent variable.
(b) Write the equation of the resultant line.
(c) What current would flow if a voltage of 3.0 V was supplied to this circuit?
(d) Approximately what voltage would be necessary to produce a current of 25 mA?
(e) If the voltage doubles in this circuit, what happens to the current?

17. Since 2008, a California city's total waste collection in tons has decreased due to efforts by the city encouraging residents to recycle glass and plastics. The table below lists the tons of refuse taken to the city dump for the last three years.

Year	Total Refuse (in tons)
2008	119,145
2009	111,892
2010	104,639

Plot these values on a graph and extend the x-axis through the year 2017. What is the rate of change for this set of data? Do you believe the rate of change for this problem will remain constant through the year 2017? Why or why not?

18. The University of North Carolina at Chapel Hill nurses have received a 3% raise over the last three

years. Amy works at the UNC hospital and her salaries for the last three years are listed in the table. (*Source*: www.allnurses.com)

Year	Annual Salary
2009	$49,500.00
2010	$50,985.00
2011	$52,514.55
2012	$54,089.99

Plot these values on a graph. Do they make a straight line when graphed? Calculate the difference between the salary in 2009 and the salary in 2010. Do the same for 2010 and 2011. What can you say about the rate of change for this problem?

19. Use the Internet to research postal rates in the United States from 1885 until today, and then graph them. One site that provides this information is http://www.prc.gov/rates/stamphistory.htm. Do these rates have a linear trend? That is, do they have a constant rate of change that you can calculate? Why do you think this is true?

20. Use the Internet to research the stopping distances of cars on wet pavement at different speeds, and then graph them. One site that provides this information is http://www.driveandstayalive.com/info%20section/stopping-distances.htm. Does your graph show a linear relationship? Is the stopping distance constant as you increase your speed in 10-mile per hour increments?

Chapter 3 Summary

Key Terms, Properties, and Formulas

coordinate
general form
horizontal line
linear equation
origin
parallel lines
perpendicular lines

point-slope form
quadrant
rectangular coordinate system
slope
slope-intercept form
solutions
vertical line

x-axis
y-axis
x-coordinate
y-coordinate
x-intercept
y-intercept

Points to Remember

The slope of a line can be found by using two points on the line or by putting the equation of a line into slope-intercept form.

The x-intercept of a line is found by substituting a 0 for y and solving for x. The y-intercept of a line is found by substituting a 0 for x and solving for y.

Lines that have the same slope are parallel.

Lines that are perpendicular (except for vertical and horizontal lines) have slopes that are negative reciprocals of each other.

The slope of a horizontal line is 0. The slope of a vertical line is undefined.

An equation of a line can be found from the slope and the y-intercept, the slope and any point on the line, or from two points on the line.

- Slope of a line: $m = \dfrac{\text{rise}}{\text{run}} = \dfrac{\Delta y}{\Delta x} = \dfrac{y_2 - y_1}{x_2 - x_1}$

- General form (linear equation):

$$Ax + By = C$$

where A, B, and C are integers, and A and B are not both 0.

- Slope-intercept form (linear equation):

$$y = mx + b$$

where m is the slope and b is the y-intercept.

- Point-slope form (linear equation):

$$y - y_1 = m(x - x_1)$$

where m is the slope and (x_1, y_1) is a point on the line.

Chapter 3 Review Problems

Graph each ordered pair in problems 1 to 4 and name the quadrant or axis on which it lies.

1. $(2, 4)$
2. $(-3, 4)$
3. $(-2, -1)$
4. $(0, 6)$

5. Is the point $(1, -4)$ a solution to the equation $2x - y = 6$? Show work to support your answer.

Determine the missing coordinate that satisfies the linear equation.

6. $2x - y = 8$ $(-2, \underline{})$
7. $y = -7x + 2$ $(6, \underline{})$
8. $3x + 2y = 8$ $(\underline{}, 4)$
9. $y = 7$ $(\underline{}, 7)$
10. $5x - 3y = 11$ $(0, \underline{})$

Find the x- and y-intercepts and the slope of the following linear equations. Sketch the graph for each equation.

11. $y = \frac{1}{3}x - 1$
12. $x = -4$
13. $2x + y = 5$
14. $-3x + 2y = 6$
15. $x - 3y = -3$
16. $y = 3$

Determine the slope of the line passing through each pair of points.

17. $(1, 4)$ $(2, 7)$
18. $(-2, 4)$ $(5, -10)$
19. $(-3, 6)$ $(-3, 8)$
20. $(-4, 6)$ $(-6, -1)$

Write an equation of a line given the following information.

21. Slope = 2 y-intercept = -3
22. Passes through the points $(-2, 6)$ and $(-3, -4)$
23. Slope = -2 passes through the point $(3, 4)$
24. Passes through the point $(-2, 5)$ and is perpendicular to the line $y = 2x - 5$
25. y-intercept = 4   parallel to the line $3x - 4y = 9$

26.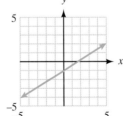

27.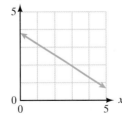

28. Production of blueberries is steadily increasing in the United States. In 2007, 139,000 tons of blueberries were produced. In 2009, 177,000 tons of blueberries were produced. Find the rate of change of the production of blueberries. Write the equation of a line that fits this data. Use the

data from 2007 as your initial value. (*Source*: www.usda.com)

29. The following chart shows data for a car driven in the city.

Gallons of Gasoline Consumed	Number of Miles Driven
2	40
4	80
6	120
8	160

(a) Find the rate of change or slope for this data. What does the rate of change represent for this data set?
(b) Write the equation of a line for this data set.
(c) Use the equation to determine the number of miles driven on 18 gallons of gas.

30. The following chart shows the amount of money in billions of dollars spent on holiday online shopping in the United States. Use the year 2000 as your initial value.

Year	Online Holiday Spending, in Billions of Dollars
2000	$8
2003	$20
2006	$32
2009	$44

(a) Find the rate of change or slope for this data. What does the rate of change represent for this data set?
(b) Write the equation of a line for this data set.
(c) If this trend continues, what is the predicted amount of online spending in 2016?

Chapter 3 Test

Graph each ordered pair in problems 1 to 3 and name the quadrant in which it lies.

1. $(-2, 3)$ 2. $(3, -1)$ 3. $(-3, -4)$

4. Is the ordered pair $(-3, 5)$ a solution of the equation $3x + 2y = -1$? Show work to support your answer.

Graph each linear equation. State the *x*- and *y*-intercepts and the slope.

5. $-4x + y = -4$ 6. $2x + y = -4$
7. $x = 5$ 8. $x + y = 3$
9. $y = -2$

Find the slope of the following lines.

10. A line through the points $(2, -1)$ and $(-3, -2)$
11. A line perpendicular to a line whose slope is -3
12. A vertical line
13. The line with the equation $y = x$

Write the equation of a line given the following information.

14. Slope $= -3$ passes through $(-1, 6)$

15. Slope $= \dfrac{1}{2}$ passes through $(-6, 1)$

16. Passes through $(0, -4)$ and $(1, 2)$

17. *y*-intercept $= 4$ and is perpendicular to the line $2x - y = 3$

18. Passes through the point $(2, -3)$ and is parallel to the line $x = 5$

19.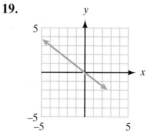

20.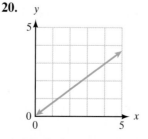

21. The world population, in billions, is estimated and projected by the linear equation $y = 0.072x + 4.593$, where x represents the number of years since 1980. ($x = 0$ corresponds to 1980.) (*Source*: www.census.gov)

    (a) What is the rate of change for this equation?
    (b) Using this equation, estimate the world population in 2015 and 2030.
    (c) In what year would we estimate the world population to be 8.049 billion?

22. The median annual income for realtors has declined in recent years. Their income can be approximated by the equation $y = -1698x + 52{,}620$ where x is the number of years since 2002. (*Source*: www.realtor.org)

    (a) What is the rate of change for this equation?
    (b) Using this equation, estimate the median income in 2006 and 2011.
    (c) If this rate of decline continues, in what year will the median income be \$32,244?

23. A bookstore recorded approximately \$7.9 billion in retail sales during the year 2000. In the year 2007, the same store recorded \$12.2 billion. Write a linear model (equation of a line) for the sales (in billions of dollars) at the bookstore during 2000–2007. Using this equation, estimate the retail sales for the year 2016.

24. The value of office equipment depreciates in value from \$35,000 to \$4,500 in 7 years.

    (a) Using this information, write a linear model (equation of a line) of the depreciation in value of this equipment. Use $x = 0$ for the time when the equipment was bought.
    (b) What is rate of change in this equation? What does it mean?
    (c) What was the value of the equipment after 5 years?
    (d) Using this model, after how many years will the value of this piece of office equipment be \$0?

Suggested Laboratory Exercises

Lab Exercise 1   Slope Activity

Purpose: Slope refers to the steepness of a line or an incline. In this exercise, you will measure the rise and run of several staircases and calculate their slopes. The values that you obtain will then be compared to building code guidelines to determine if the stairs in question meet the building code safety standards.

Procedure: Choose several different stairways in your classroom building or nearby buildings or at home. If all of the individual steps are of equal height and width, carefully measure the height (rise) of one step and the width (run) of one step. The choice of units should not affect the value of the slope, because slope is a simple ratio of the amount of rise to the amount of run. To confirm this, measure the values of rise and run in both English (USCS) and metric (SI) units. Record these data in a table similar to the one that follows. Calculate the slope using both sets of units.

Note: If individual steps are of different heights or widths, you will have to measure all the heights and add them to get the total rise for the stairs, and then measure all the widths and add them to get the total run for the stairs.

Data:

Staircase #1: Location: _____.

Units	Rise	Run	Slope Fraction	Decimal
Metric (SI)				
USCS				

Staircase #2: Location: _____.

Units	Rise	Run	Slope Fraction	Decimal
Metric (SI)				
USCS				

Analysis: Now you must find out the local building code for the type of building in which the stairs are located. Call the office of the local building inspector for your city or county to get this value. Some information may be available on local municipal websites. Or use this value as a general guide: Staircases for general use by the public should have a slope of 0.545 or less. Do the staircases you measured meet the building code?

Lab Exercise 2 — Linear Equations Practice Activity

Do each of the following on graph paper. Answer all questions posed on this lab sheet.

1. Using graph paper and a ruler, plot the points $(-6, 2)$ and $(6, 10)$.
2. Sketch the line that goes through these two points. Be sure that the line extends past both points and intersects both the x- and y-axes.
3. Based on the graph alone, should the slope of this line be positive or negative?
4. Using the coordinates of the two points given in question 1, use the slope formula to calculate the slope of this line.
5. By counting blocks vertically and horizontally on your graph, verify the slope you calculated in question 4.
6. What are the x- and y-intercepts of this line? Look at your graph.
7. Find the slope and y-intercept of the line $2x - 3y = -18$. Is this consistent with the line you have drawn on your graph?
8. Using the equation in question 7, complete the following ordered pairs:
   $(12, \_\_)$ $(-3, \_\_)$ $(3, \_\_)$ $(9, \_\_)$.
9. Plot the ordered pair solutions you calculated in question 8 on the graph. They should lie on your line.

Lab Exercise 3 — Linear Equations on the TI-84 Plus® Graphing Calculator

The purpose of this lab activity is to help you to become familiar with a graphing calculator. There are several different brands of graphing calculators in use today, and they all have similar functions. This exercise is written for the TI-84 Plus, but is adaptable to other brands. The main point is to learn to graph using the calculator. Other functions of the calculator will be covered if necessary or useful.

All equations *must* be written in slope-intercept form ($y = mx + b$) before entering them into the calculator. Be sure that you have reviewed Section 3-3 of this chapter before you begin this exercise.

Now get your calculator and begin:

1. Turn the calculator on and check to see if the screen is dark enough to easily see the blinking cursor in the upper left-hand corner of the screen. If not, press the 2nd key and then hold down the up ▲ or down ▼ arrow to darken or lighten the screen. Now press the WINDOW key. If the list shown does not look like the following list, then use the down arrow ▼ arrow key followed by the CLEAR key to remove the numbers, and then enter the ones shown below. This sets the size of the graph to run from 10 to -10 on both axes.

$X_{min} = -10 \quad X_{max} = 10 \quad X_{scl} = 1 \quad Y_{min} = -10 \quad Y_{max} = 10 \quad Y_{scl} = 1$

2. Press the Y= button. You should see a list that goes from Y_1 to Y_8. This screen will allow you to enter up to eight equations or functions to graph on one screen.
3. Type in "x" using the X,T,O key so that $Y_1 = x$ is shown. If you type this incorrectly, press DEL to erase it and start over. Once you have the equation typed in, press GRAPH. Record the graph as seen on your calculator in the space given here. Call this line number 1 on your drawing. Now hit the Y= key again, followed by the down arrow key and type in $5x$ so that $Y_2 = 5x$. Press the GRAPH key and you should see both lines graphed. Record the second line on this paper as line number 2. Now go back to the Y= screen as before and let $Y_3 = 0.2x$. Remember that $0.2 = \frac{1}{5}$. GRAPH and record this line as number 3. *If you use a fractional coefficient, you must use parentheses around the fraction for the calculator to interpret it correctly: $Y = (1/5)x$.*

Based on the lines shown on your calculator screen, you should have three lines drawn in Figure 3-28 that represent these equations.

1. $Y_1 = x$
2. $Y_2 = 5x$
3. $Y_3 = 0.2x$

   or

   $Y_3 = \frac{1}{5}x$

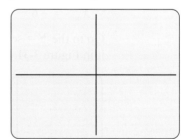

FIGURE 3-28

Use the graphs of these three lines to answer each of the following questions.

(a) Which line is the steepest?
(b) How does the coefficient of x in the equation $Y = mx + b$ affect the graphs?
(c) What do these three lines have in common?

4. Go back to the Y= screen and press CLEAR to erase each equation. Now enter each of the following equations. GRAPH them and record them in Figure 3-29.

1. $Y_1 = 2x + 5$
2. $Y_2 = 2x - 5$
3. $Y_3 = 2x$

FIGURE 3-29

Answer the following questions about these graphs.
(a) What do these three graphs have in common?
(b) How do the graphs differ from each other?
(c) Explain how the value of b in $Y = mx + b$ affects each graph.
(d) What do you know about the slopes of these lines?

5. Go back to the Y= screen, clear all the equations, and enter equations $Y_1 = 2x - 5$ and $Y_2 = -2x - 5$. Then GRAPH and draw them in Figure 3-30.

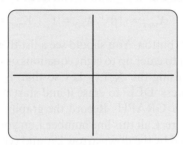

FIGURE 3-30

Answer each of the following questions about these graphs.
(a) What do these two graphs have in common?
(b) Which graph is steeper, or are they both of the same steepness?
(c) How does the sign of the coefficient of x in the equation $Y = mx + b$ affect the graph of a line?

6. Go back to the Y= screen and clear all equations. Press the WINDOW key and set the values as follows:

$X_{min} = -6 \quad X_{max} = 6 \quad X_{scl} = 1 \quad Y_{min} = -4 \quad Y_{max} = 4 \quad Y_{scl} = 1$

Go to the Y= screen and set $Y_1 = \frac{2}{3}x + 1$. GRAPH the equation and record it in Figure 3-31.

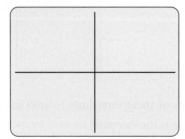

FIGURE 3-31

Using the rules about the slopes of perpendicular lines, find the equation of a line that has the same y-intercept as the given line ($y = \frac{2}{3}x + 1$) but is *perpendicular* to it. This is line Y_2. Write it here: $Y_2 = $ _____. Now enter the equation into the calculator and GRAPH it. Does Y_2 look perpendicular to Y_1?

7. Clear the Y= screen and reset the WINDOW as below:

$X_{min} = -10 \quad X_{max} = 10 \quad X_{scl} = 1 \quad Y_{min} = -10 \quad Y_{max} = 10 \quad Y_{scl} = 1$

Now set $Y_1 = 3x - 6$ and GRAPH it. Now press the TRACE key. A small flashing box will appear on the line and values of X and Y will be shown at the bottom of the screen. If you now press the left ◀ and right ▶ arrow keys, you will see the little box move along the line and the values of X and Y at the bottom of the screen will change. By moving the box, locate the x- and y-intercepts of this line. What are they shown to be on the calculator screen? According to the TRACE function:

(a) the x-intercept is _____.
(b) the y-intercept is _____.

Are these exact values or are they just approximations? _____
Find the x- and y-intercepts algebraically:

(a) x-intercept = _____.
(b) y-intercept = _____.

8. Return to the Y= screen and enter each of the following equations. Then GRAPH these equations and record them in Figure 3-32.

1. $Y_1 = 3$
2. $Y_2 = 7$
3. $Y_3 = -3$
4. $Y_4 = -7$

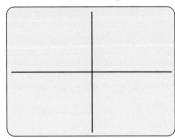

FIGURE 3-32

Now find a single equation that is perpendicular to all of these lines (do not use the y-axis). Can you graph your equation using the calculator?

Lab Exercise 4

Recycling

Over time, according to records kept by the Environmental Protection Agency (found at www.epa.gov), recycling rates have increased from 10% of municipal solid waste generated in 1980 to 16% in 1990, to 29% in 2000, and to 34% in 2009. Meanwhile, disposal of waste to a landfill has decreased from 89% of the amount generated in 1980 to 54.3% of solid waste in 2009. The prediction equation (or least-squares prediction equation) for the percent of solid waste recycled by Americans is

$$y = 0.932x - 1837$$

where y = percent of solid waste recycled and x is the year.

(a) Graph these data (percent recycled vs. year), leaving room at both ends for past and future years. Draw the line given by the prediction equation on your graph also.
(b) Is the slope positive or negative? What does the slope tell us about the rate of recycling over the past 30 years?
(c) Use the prediction equation to determine the x-intercept. What does this value tell you? Use your graph to help you verify your answer by plotting this point.
(d) During what year will recycling theoretically reach 50% according to this model? Plot this point on your graph.
(e) Theoretically, will recycling ever reach 100%? Practically speaking, is that really possible?

4 Functions

> *Our federal income tax law defines the tax y to be paid in terms of the income x; it does so in a clumsy enough way by pasting several linear functions together, each valid in another interval or bracket of income. An archeologist who, five thousand years from now, shall unearth some of our income tax returns together with relics of engineering works and mathematical books, will probably date them a couple of centuries earlier, certainly before Galileo and Vieta.*
>
> — Hermann Weyl

There are many quantities in everyday life that are related to one another. In this chapter, we will define and discuss a special kind of relationship between two quantities called a function. We will interpret graphs of functions and examine both linear and nonlinear functions. We will also look at application problems involving direct, inverse, and joint variation.

In this chapter

- **4-1** Functions
- **4-2** Using Function Notation
- **4-3** Linear Functions as Models
- **4-4** Direct and Inverse Variation
- **4-5** Quadratic Functions and Power Functions as Models
- **4-6** Exponential Functions as Models

  Chapter Summary

  Chapter Review Problems

  Chapter Test

  Suggested Laboratory Exercises

Section 4-1 Functions

Many quantities in everyday life are related to one another or depend on one another. A fundamental idea in mathematics and its applications is that of a **function**, which tells how one thing *depends* on another. For example, the amount of bacteria in a growing culture is dependent on the amount of time that passes. The amount of sales tax owed on a purchase is dependent on the total cost of the items purchased. As the value of the first quantity changes, a corresponding change will occur in the related quantity.

When the value of one quantity uniquely determines the value of the second quantity, we say the second quantity is a function of the first. One way to express these relationships is using ordered pairs (x, y). The x variable is called the **independent variable** and the y variable is called the **dependent variable**, because its value is dependent on the value of x. For example, if the sales tax rate in an area is 6% on purchases made, a $5 purchase will be taxed $0.30. This can be written as the ordered pair ($5, $0.30). The sales tax owed (y) is uniquely determined by the amount of the purchase (x). Sales tax is a function of the amount of the purchase. The tax table shown in Table 4-1 illustrates this concept.

Generally, a function can be represented in four ways:

- Description or rule
- Table of values
- Graph
- Equation or formula

In the case of sales tax, state and local governments set the rule for calculating sales tax. In this example, they have determined that the sales tax will be calculated at 6% of the purchase amount. A table can be created listing all possible purchase amounts and the associated taxes to be charged. Many small businesses or individual entrepreneurs use tax tables provided by the state to determine the amount of sales tax to charge for a specific purchase amount. A graph of this function can be drawn and its properties will show that this is a linear function because its graph is a straight line (Figure 4-1).

Finally, the calculation of sales tax can be expressed as an equation or formula. The formula $y = 0.06(x)$ gives a way of calculating the tax (y in dollars) based on the purchase price (x in dollars). We will look at these four ways of expressing functions throughout this book.

In mathematics, a **relation** associates members of one set with members of another set. A function is a special kind of relationship between two quantities.

TABLE 4-1

Purchase Amount	Sales Tax at 6%
$0.50	$0.03 = 6%(0.50)
$1.00	$0.06 = 6%(1.00)
$1.50	$0.09 = 6%(1.50)
$2.00	$0.12 = 6%(2.00)
$2.50	$0.15 = 6%(2.50)
$3.00	$0.18 = 6%(3.00)

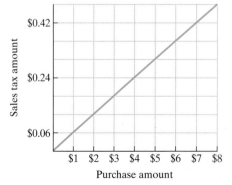

FIGURE 4-1

Specifically, in a function it must be true that for each value of the independent variable, there is *exactly* one value for the dependent variable. The calculation of sales tax fulfills this requirement because, for each purchase amount possible, there will be only one resulting sales tax.

Definition Function

A function is a relation in which, for each value of the first component of the ordered pairs, there is exactly one value of the second component.

Example 1 The Cricket Function

The snowy tree cricket is called the thermometer cricket because an accurate estimate of the current temperature can be made using its chirp rate. One can estimate the temperature in degrees Fahrenheit by counting the number of times a snowy tree cricket chirps in 15 seconds and adding 40. Use R as the chirp rate per minute and describe this function with words, a table, a graph, and a formula.

Words: Because R represents chirps per minute, we will count the value for R and divide it by 4 to get the number of chirps in 15 seconds. Then add 40 to that number to estimate the temperature.

Table: See Table 4-2. Notice that each 20 chirps per minute increase corresponds to a 5° increase in temperature.

Graph: The values in the table are plotted on the graph in Figure 4-2. The graph illustrates that if more chirps per minute are counted, the corresponding temperature is warmer.

TABLE 4-2

R (chirps/min)	T (temperature, °F)
20	45
40	50
60	55
80	60
100	65

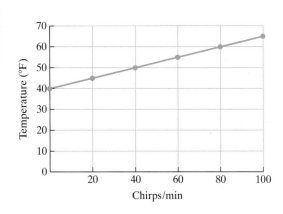

FIGURE 4-2 Example 1.

Formula: An algebraic rule can be developed using the description as follows:

$$\text{Estimated temperature} = \frac{1}{4}\left(\frac{\text{chirps}}{\text{min}}\right) + 40$$

$$T = \frac{1}{4}R + 40$$

We can use this formula to calculate the estimated temperature for any number of chirps per minute.

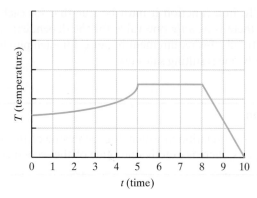

FIGURE 4-3
Graph of temperature versus time.

All these descriptions of the cricket function give the same information in different forms. Tables are easy to read, but not all values can be included in a table. The graph gives a picture of the function and may reveal certain patterns, but obtaining exact readings from the graph may be difficult. The formula lets you calculate the exact value of the dependent variable given any value of the independent variable, but patterns in the general function may be difficult to discover if only the formula is used. The best method for presenting a function will vary according to the requirements of the problem.

Graphs are often used to picture numerical data so that relationships can be analyzed and, if possible, trends can be forecast. For example, a scientist may use the graph in Figure 4-3 to indicate the temperature, T, of a certain solution at various times, t, during an experiment. The graph shows that the temperature increased gradually from time $t = 0$ to time $t = 5$. The temperature remained steady from time $t = 5$ to $t = 8$, and then decreased rapidly from time $t = 8$ to $t = 10$. Graphs of this type help us determine the trend of the temperature more clearly than a long table of numerical values.

Functions can often be found in magazines, newspapers, books, or other printed material in the form of tables or graphs. Graphs are usually an easier way to analyze and present lists of data. Closing prices on the stock market, daily high temperatures, and monthly unemployment rates are just a few of the data sets that you may find represented by graphs in newspapers.

Example 2

Interpreting Graphs

The graph in Figure 4-4 shows the sunrise time for Asheville, North Carolina, for the year.

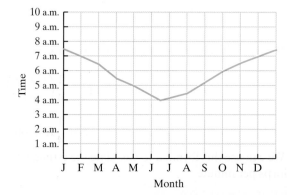

FIGURE 4-4
Example 2.

Use this graph to answer the following questions.

(a) What is the approximate time of sunrise on February 1?
(b) What are the approximate dates when the Sun rises at 5 a.m.?
(c) During which month does the Sun rise the earliest?

(a) To approximate the time of sunrise on February 1, find the mark on the horizontal axis that corresponds to February 1. From this mark, move vertically upward until the graph is reached. From the point on the graph, move horizontally to the left until the vertical axis is reached. The vertical axis reads 7 a.m.
(b) To approximate the dates when the Sun rises at 5 a.m., find 5 a.m. on the time axis and move horizontally to the right. Notice that you will intersect the graph twice, corresponding to two dates for which the Sun rises at 5 a.m. Follow both points on the graph vertically downward until the horizontal axis is reached. The Sun rises at 5 a.m. on approximately May 1 and August 20.
(c) June contains the lowest point on the graph, indicating the earliest sunrise.

The key concept in a functional relationship is that *there is only one value of the dependent variable (y) for each value of the independent variable (x)*. Think of the independent variable as the *input* of the function. This is the variable that you control—the one that you choose or determine. Once you have chosen the value of the independent variable, there will be only one dependent variable or *output* that corresponds to it. (See Figure 4-5.) When trying to determine which variable is a function of which, ask yourself which variable "depends" on the other. This will help you determine the nature of the functional relationship.

FIGURE 4-5

Example 3 — Determining Functional Relationships

Determine the independent and dependent variables in the following functions. Give a statement that illustrates the functional relationship described.

(a) The altitude and temperature when climbing a mountain

The altitude is the independent variable (input) and the temperature at that altitude is the dependent variable (output). As the altitude of the climb increases, the surrounding temperature will decrease. This is illustrated in Figure 4-6.

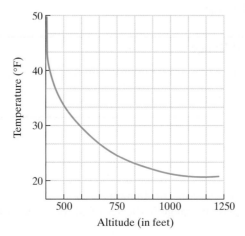

FIGURE 4-6

TABLE 4-3	
Speed of Car (mph)	Stopping Distance (ft)
45	146
50	175
55	206
60	240
65	276

TABLE 4-4	
Time (sec)	Velocity (ft/sec)
0	0
1	16
2	19.2
3	19.84
4	19.97

(b) The speed of a car and its stopping distance

The stopping distance of a car is dependent on the speed of the car when the brakes are applied. (This is the reason that policemen measure the length of the skid marks at the scene of an accident.) Therefore, the speed of the car is the independent variable (the one that you choose) and the braking distance is the dependent variable (a result of the speed that the car was traveling). The faster the car goes, the longer the stopping distance will be. Table 4-3 gives some stopping distances for cars traveling at different speeds on dry pavement.

(c) The speed of a baseball dropped from the top of a tall building as shown in Table 4-4.

If a baseball is dropped from the top of a tall building, the speed of the baseball is uniquely determined by the amount of time that it has been falling. Therefore, its velocity is a function of the time that has passed since it was dropped. As the length of time since it was dropped increases, the baseball's velocity continues to increase.

(d) The formula for calculating the area of a circle: $A = \pi r^2$

The area of a circle (A) is dependent on the radius of the circle (r) (see Figure 4-7). As the radius of the circle changes, a corresponding change in the area will occur. So the radius is the independent variable (input), and the area is the dependent variable (output) in this function.

FIGURE 4-7

Often a function is defined only for certain values of the independent and dependent variables. For instance, in Example 3, several of the independent variables such as the speed of a car, time, and length of the radius of a circle can never be negative numbers. Similarly, the stopping distance of a car, velocity of a falling object, and the area of a circle can only have positive values. This idea leads to the following definition.

Definition **Domain and Range**

In a function, the set of all possible values of the independent variable (x) is the domain, and the set of all possible values of the dependent variable (y) is the range.

Consider the set $S = \{(1, 5), (2, 6), (3, 7)\}$. The independent variables or x values in this set form the domain of S, which is $\{1, 2, 3\}$. The y values of the ordered pairs, $\{5, 6, 7\}$, make up the range of the function. If a function is expressed as an equation, the **domain** of the function is the set of all possible input values, whereas the **range** is the set of all possible output values. If the domain of a function is not specified, we usually assume it to be all possible real numbers that make sense for a specific function. If we are using a function to model a real-world situation, there are usually some limitations on the values of the domain and range that will occur.

Example 4 Identifying the Domain and Range of a Function

Consider each of the following functions and identify the values of the domain and range of the function.

(a) The sales tax function

The independent variable in this function is the amount of the purchase. Therefore, the domain of the function is all real numbers greater than or equal to 0 (rounded to the nearest hundredth or penny). The dependent variable is the amount of sales tax, and therefore, the range is also all real numbers greater than or equal to 0 (rounded to the nearest penny).

(b) $y = 2x + 3$

There are no apparent restrictions on the values of x and y because this is a general linear equation. Therefore, both the domain and the range are all real numbers.

(c) Figure 4-8

The independent variable in this function is time in hours and the dependent variable is the temperature in degrees Celsius. The values of the domain (time in hours, h) as shown in the graph are $0 \leq h \leq 17$ and the values for the range (temperature, t) are $8° \leq t \leq 28°$.

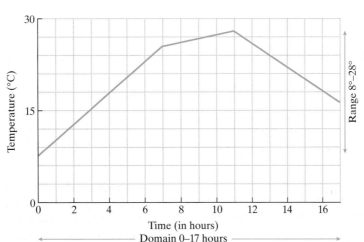

FIGURE 4-8

In mathematical applications, functions, in any form, are often representations of real events. Thus, we say they are **models**. Using a function to act as a model for a problem is one of the keys to understanding the world around us. We use models in many areas of our lives, including physical science, business, and social science. Models are used to illustrate gains or losses in investments, stopping times for cars traveling at certain speeds, and data gathered in a national census. In this chapter, we will examine the concept of a function, and in later chapters, we will concentrate on applications of functions or models.

Practice Set 4-1

1. What is a function? How do we decide which variable is the *independent* variable and which is the *dependent* variable in a function?

2. In your own words, define the domain and range of a function.

3. The following table shows the daily high temperature in degrees Fahrenheit for a week in April.

Date in April	20	21	22	23	24	25	26
High Temperature	73°	77°	74°	68°	70°	68°	72°

   (a) What was the high temperature on April 23?
   (b) Is the daily high temperature a function of the date or is the date a function of the daily high temperature?
   (c) Which variable is dependent?
   (d) What is the domain?

4. The following table lists the fine a driver must pay when she is caught for speeding on a street where the speed limit is 45 mph.

v (mph)	55	60	65	70	75
Fine	$75	$90	$105	$125	$150

   (a) What is the fine for someone speeding in a 45-mph zone if she is going 60 mph?
   (b) Is the fine a function of the speed or is the speed a function of the fine?
   (c) Which variable is dependent?
   (d) What is the range?

5. The following line graph illustrates the growth in the world's population from 1980 to 2010.

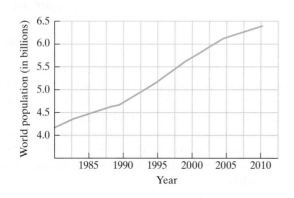

   (a) Identify the independent and dependent variables of the function illustrated in the graph.
   (b) Give the approximate domain and range of the function.

6. The following graph illustrates the growth in value of a $1000 deposit in a bank account paying 6% interest.

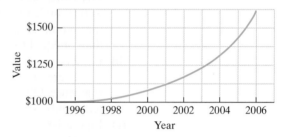

   (a) Identify the independent and dependent variables of the function illustrated in the graph.
   (b) Give the domain and range of the function.

7. At the end of the semester, students' grades are posted in a table that lists each student's ID number in the left column and the student's grade in the right column. Which quantity must necessarily be a function of the other?

8. Advertisements in newspapers for term life insurance list the applicant's age in the left column of a table and the corresponding monthly insurance premium in the right column. Which quantity is a function of the other?

For problems 9 to 14, write a short statement that expresses a possible function between the quantities. Example: (age, shoe size); solution: As a person's age increases, shoe size also increases up to a point. After that point, shoe size remains relatively constant.

9. (weight of a bag of potatoes, price of the bag)

10. (demand for a product, price of a product)

11. (slope of a hill, speed of a scooter going uphill)

12. (distance from Earth, strength of gravity)

13. (age, target heart rate when exercising)

14. (time, distance an object falls from the top of a building)

The following graph shows the sunset times for Seward, Alaska. Use the graph to solve problems 15 to 20.

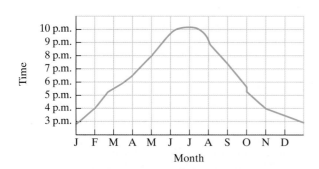

15. Approximate the time of sunset on June 1.

16. Approximate the time of sunset on November 1.

17. Approximate the date(s) when the sunset is 3 pm.

18. Approximate the date(s) when the sunset is 9 pm.

19. Approximate the times when the latest and earliest sunsets occur.

20. What are the values of the range of this function?

For each function in problems 21 to 24, use your intuition or additional research, if necessary, to make a rough sketch of the function. Label the axes. Remember that in an ordered pair (x, y) the x variable is the independent variable and should be plotted on the horizontal axis.

21. (amount of food purchases, 2% sales tax)

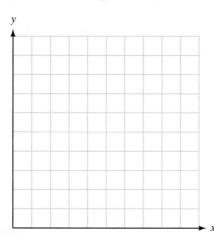

22. (interest rate, number of cars sold per year at a dealership)

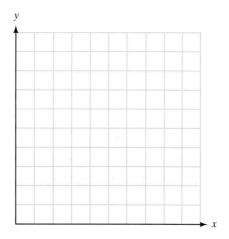

23. (years from 1985 to 2005, percent of homes with a personal computer)

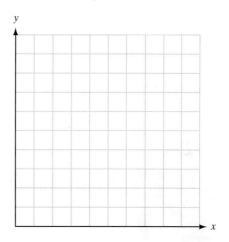

24. (minutes talking on the phone, monthly telephone bill)

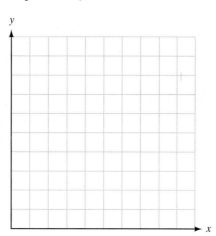

25. Human beings grow rapidly during childhood, and then their growth rate slows during the late teens until it eventually levels off. Sketch the graph of a function that represents height as a function of time.

26. A light is turned off for several hours. It is then turned on. After a few hours, it is turned off again. Sketch a possible graph of the lightbulb's temperature as a function of time.

27. Table 4-3 lists some braking distances based on the speed of the car when the brakes are applied. The formula for calculating the braking distance needed when traveling at a certain speed is $d = \frac{x^2}{20} + x$, where x is the speed of the car and d is the stopping distance. Calculate the stopping distance required when a car is traveling 70 mph. Use this result and the results in Table 4-3 to graph this function. Discuss the shape of the graph and its meaning.

28. The ideal weight of a man is a function of his height. One formula that gives this relationship is $w = 110 + 5(h - 60)$, where h is the man's height in inches and w is the weight in pounds. Calculate the ideal weights for men who are 68 in., 69 in., 70 in., 71 in., 72 in., and 73 in. List these results in a table, and then use them to graph this function. Discuss the shape of the graph and its meaning.

29. The amount of your monthly house payment is a function of several things: the interest rate, the number of years of financing required, and the amount of your loan. Assume that you are planning to borrow $125,000 at a fixed interest rate for 30 years. Find a mortgage calculator on the Internet such as the one located at http://www.bankrate.com. Use the online calculator to determine your monthly payment based on interest rates of 5.5%, 5.75%, 6%, 6.25%, 6.5%, 6.75%, and 7%. Create a table of these values and then graph this function. Discuss the shape of the graph and its meaning.

30. Use a newspaper to find an example of a functional relationship. Remember that functions may be given in the form of tables or graphs. Submit your example and answer the following questions.

    (a) What is the independent variable in this function?
    (b) What is the domain of the function?
    (c) What is the dependent variable in this function?
    (d) What is the range of the function?

Section 4-2 Using Function Notation

To indicate that P is a function of x, we use the notation

$$P = f(x)$$

We read this as "P equals f of x." The $f(x)$ symbol is another symbol for the variable P. We use the value of x as the input value for the function, and the value of P is the output value.

As defined in the previous section, the x value is the independent variable because its value can vary. The dependent variable is P because its value depends on what value of x we choose. The f is used to represent the dependence of P on x.

Example 1 — Sales Tax Function

Let $T = f(p)$ be the amount of sales tax owed on purchases p. Explain in words what the following statement tells you about sales tax:

$$f(\$25) = \$1.25$$

Given $T = f(p)$, the input of the function is p, the amount of the purchase, and T is the output of the function or the amount of sales tax owed. Therefore, $f(\$25) = \$1.25$ says that a purchase amount of $25 would require sales tax of $1.25.

Section 4-2 **Using Function Notation** 147

Example 2

Velocity and Distance Functions

If an object is dropped from the top of a building and the acceleration of gravity is 32 ft/sec^2, then its downward velocity v after t seconds would be described by the function $v(t) = 32t$. The distance d that the object has fallen would be described by the function $d(t) = 16t^2$. Name the independent variable and the domain of the function.

In both of these functions, time (t) is the independent variable. The velocity (v) of the object and the distance (d) it has fallen both depend on the time that has elapsed since the object was dropped. The domain and range for these functions would be all real numbers greater than or equal to 0.

Many algebraic equations involve the use of two variables, x and y, in the equation. Many of these equations are functions and can be written using **function notation**. This is done by solving the equation for the y variable and replacing it with the notation $f(x)$. Look at the following example.

Example 3

Function Notation

Write the equation $2x - 3y = 9$ in function notation.

$$2x - 3y = 9$$
$$-3y = -2x + 9 \quad \text{(subtract } 2x \text{ from both sides)}$$
$$y = \frac{2}{3}x - 3 \quad \text{(divide both sides by } -3\text{)}$$
$$f(x) = \frac{2}{3}x - 3 \quad [\text{replace } y \text{ with } f(x)]$$

By writing an equation in function notation, we have, in essence, created a formula that allows us to input a value for the independent variable and calculate the value of the dependent variable. For example, in the function $f(x) = \frac{2}{3}x - 3$, the name of the function is "f," the independent variable is "x," and the value of the dependent variable will be "$f(x)$." So, if I want to calculate the value of the function at -3, I write

$$f(x) = \frac{2}{3}x - 3$$
$$f(-3) = \frac{2}{3}(-3) - 3$$
$$f(-3) = -5$$

This is called *evaluating the function*, and we say that the function f has a value of -5 when $x = -3$. If we want to calculate the value of the function at 5, then we write $f(5)$, read as "f of 5," and substitute 5 into the formula and calculate the answer.

Remember:

Name of the function Independent variable

$$y = f(x) = \frac{2}{3}x - 3$$

Value of the function   Function formula

Although $f(x)$ is the standard notation for a function, other variables are also used to denote functions. For example, $v(x)$ could be used to denote a volume function or $t(x)$ could denote a function used to calculate tax.

Example 4 — Evaluating Functions

Let $g(x) = 3x^2 - 8$.

(a) Find the value of the function if $x = -2$. We write

$$g(-2) = 3(-2)^2 - 8$$
$$g(-2) = 3(4) - 8 = 12 - 8 = 4$$

Therefore, the value of the function when $x = -2$ is 4, or $g(-2) = 4$.

(b) Evaluate $g\left(-\frac{1}{2}\right)$.

$$g\left(-\frac{1}{2}\right) = 3\left(-\frac{1}{2}\right)^2 - 8$$

$$g\left(-\frac{1}{2}\right) = 3\left(\frac{1}{4}\right) - 8 = \frac{3}{4} - 8 = -7\frac{1}{4}$$

Therefore, $g\left(-\frac{1}{2}\right) = -7\frac{1}{4}$.

Many formulas can be written in function notation and easily evaluated. The equation $y = f(x) = 0.06x$ represents the function that calculates the sales tax owed for the amount of the purchases made by the buyer using a 6% tax rate. Using this equation, the sales tax can be computed for purchases in any amount. Sales tax tables used in stores are generated using this equation. Of course, the exact amounts are rounded to the nearest cent to be practical.

Example 5 — Calculating Sales Tax

Calculate the sales tax owed for the purchases shown in Table 4-5 if the sales tax rate is 6%.

Function equation:

$$f(x) = 0.06x$$

where x is the purchase amount and $f(x)$ is the amount of tax rounded to the nearest cent.

(a) $f(\$12.50) = 0.06(\$12.50) = \$0.75$
(b) $f(\$29.95) = 0.06(\$29.95) = \$1.797 \approx \$1.80$
(c) $f(\$135.47) = 0.06(\$135.47) = \$8.1282 \approx \$8.13$

See Table 4-5.

TABLE 4-5

Purchases (x)	Exact Tax (y)	Rounded Tax
$12.50	$0.75	$0.75
$29.95	$1.797	$1.80
$135.47	$8.1282	$8.13

Note that the set {$12.50, $29.95, $135.47} contains elements of the domain of this function. The set {$0.75, $1.80, $8.13} contains elements of the range of the function.

Example 6 Using a Function in a Word Problem

Forensic anthropologists (specialists who work for the police and use scientific techniques to solve crimes) can determine a woman's height in life by using the skeletal remains of her humerus bone. (This is the long bone that forms the upper arm.) The formula is $3.08x + 64.47$, where x represents the bone length in centimeters. Write this formula in function notation and find the estimated height in centimeters of a woman whose humerus measures 31.0 cm. Then convert this measurement to inches.

Because the formula is designed to measure a woman's height, we will use the variable h to represent the name of the function and x, the length of the humerus, will be the independent variable.

(a) Function notation:
$$h(x) = 3.08x + 64.47$$

(b) Evaluate the function if the length of the humerus is 31.0 cm.
$$h(31.0) = 3.08(31.0) + 64.47 = 159.95 \text{ cm}$$

Therefore, the predicted height of this woman, based on the length of her humerus, is 159.95 cm.

(c) There are approximately 2.54 cm in 1 in., so we can convert this answer to inches by dividing.
$$159.95 \text{ cm} \div 2.54 \text{ cm/in.} \approx 62.97 \text{ in. or 63 in. (5 ft 3 in.)}$$

Because graphs and tables are also used to represent functions, we can evaluate them using function notation. Look at the following examples.

Example 7 Evaluating Functions Using a Table

The cost of first-class delivery of a letter by the U.S. Postal Service is a function of the weight of the letter. Table 4-6 gives a list of the prices of first-class letters effective

TABLE 4-6 First-Class Mail

LETTERS

Weight Not Over (ounces)	Single-Piece
1	0.45
2	0.65
3	0.85
4	1.05
5	1.25
6	1.45
7	1.65
8	1.85
9	2.05
10	2.25
11	2.45
12	2.65
13	2.85

January 12, 2012. Use the table to evaluate the function for the following weights: 3 oz and 4.2 oz. Use the single-piece rates.

(a) Evaluate $f(3\text{ oz})$. Using the postal rate table, we locate the 3-oz row and look at the next column to determine a cost of $0.85. Therefore, $f(3\text{ oz}) = \$0.85$.

(b) Evaluate $f(4.2\text{ oz})$. The heading on the first column indicates that a first-class letter with a weight of 4.2 oz would require the customer to pay the rate that applies to the row labeled 5 oz because the letter weighs more than 4 oz. Therefore, $f(5\text{ oz}) = \$1.25$.

Example 8 — Evaluating Functions Using Graphs

Use the graph in Figure 4-9 to evaluate the following values of the domain.

(a) Find $g(1)$ on the graph.
(b) Find $g(2)$ on the graph.

(a) Locate 1 on the horizontal axis. At $x = 1$, the value of y on the graph is 3. Therefore, $g(1) = 3$.

(b) Locate 2 on the horizontal axis. At $x = 2$, the value of y on the graph is 2. Therefore, $g(2) = 2$.

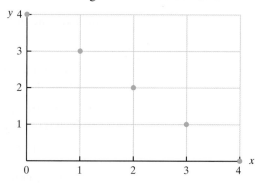

FIGURE 4-9
Example 8.

Practice Set 4-2

Let the function $g(x) = 3x - 5$. Evaluate $g(x)$ by finding the following values:

1. $g(0)$
2. $g(-4)$
3. $g(10)$
4. $g\left(-\dfrac{1}{2}\right)$

Write the following equations in function notation. Then find $f(-2)$ for each function.

5. $y = 2x + 1$
6. $2y = 4x - 8$
7. $3x^2 - y = 1$
8. $1 = 2x^2 - y$
9. $5x - 2y = 8$
10. $2x - 3y = 9$

11. A ball is thrown downward from the top of a 90-story building. The distance, d, in feet, the ball is above the ground after t seconds is given by the function equation $d = f(t) = -16t^2 - 80t + 800$. Determine $f(2)$ and explain what it means.

12. If a ball is dropped from a building and hits the ground in 3 s, how many meters tall is the building? Neglecting air resistance, the ball will reach a constant acceleration of 9.8 m/s. Therefore, the function for this problem will be $d(t) = \dfrac{1}{2}(9.8)t^2$ or $d(t) = 4.9t^2$. Determine $d(3)$.

13. The femur is the long bone in your leg that connects your hip to your knee. A man's height in centimeters can be predicted from the length of his femur using the function $h(x) = 2.32x + 65.53$, where x is the length of the femur in centimeters. Calculate the approximate height of a man whose femur is 46.5 cm.

14. A woman's height in centimeters can be predicted from the length of her femur using the function $h(x) = 2.47x + 54.10$, where x is the length of the femur in centimeters. Calculate the approximate height of a woman whose femur is 45.5 cm.

15. Alec's height is approximately 1.83 m. Calculate the approximate length of his femur. (Refer to problem 13.)

16. Jennifer's height is approximately 1.55 m. Calculate the approximate length of her femur. (Refer to problem 14.)

17. A checkerboard contains 64 squares. If two pennies are placed on the first square, four pennies on the second square, eight pennies on the third square, 16 pennies on the fourth square, and so on, the number of pennies on the nth square can be found by the function $A(n) = 2^n$.

    (a) Find the number of pennies placed on the 30th square.
    (b) What is this amount in dollars?

18. A house initially costs $85,000. The value, V, of the house after n years if it appreciates at a constant rate of 4% per year can be determined by the function $V = f(n) = \$85,000(1.04)^n$. Determine $f(8)$ and explain its meaning.

19. A movie theater seats 200 people. For any particular show, the amount of money the theater makes is a function of the number of people, n, who buy tickets for a show. Each ticket costs $7.50.

    (a) If 128 tickets are sold for the late show, how much money does the theater make?
    (b) If a ticket costs $7.50, find a formula that expresses the amount of money the theater makes, $M(n)$, as a function of the number of tickets sold, n.
    (c) What is the domain of this function? What is the range of the function?

20. A cab company advertises $1.95 per mile (or part of a mile) with no fixed charge. Therefore, the amount of money a cabbie will make in a day is a result of the number of miles that he drives his customers.

    (a) If a cabbie drives his customers a total of 115 miles on Friday, how much money did he earn?
    (b) Write a function that will represent a method of calculating his total earnings based on miles driven.
    (c) What is the independent variable in this function? The dependent variable?

21. While exercising, a person's recommended target heart rate is a function of age. The recommended number of beats per minute, y, is given by the function $y = f(x) = -0.85x + 187$, where x represents a person's age in years. Determine the number of recommended heartbeats per minute for a 20-year-old. Then calculate $f(60)$ and explain any relationship that your two answers suggest concerning age and target heart rates for people who are exercising.

22. Suppose that the future population of a town t years after January 1, 1995, is described in thousands by the function $P(t) = 120 + 4t + 0.05t^2$. Calculate the value of $P(10)$ and explain its meaning.

23. The Texas Education Agency lists its minimum salary schedule for public school teachers on its website. The following table represents annual salaries for teachers with 0–5 years of teaching experience. Find $f(3)$ and explain what it means. What are the independent and dependent variables in this function?

Teacher Salary Schedule (10-month contract) 2010–2011

Years of Experience	Annual Salary
0	$27,320
1	$27,910
2	$28,490
3	$29,080
4	$30,320
5	$31,560

Source: www.tea.state.tx.us

24. The monthly salaries for 10-month teachers holding master's degrees in the North Carolina Public School System are listed in the following table. Find $f(6)$ and explain its meaning. What are the independent and dependent variables in this function?

Master's Degree Certified Teacher Salary Schedule 2010–2011

Years of Experience	Monthly Salary
5	$4021
6	$4193
7	$4359
8	$4518
9	$4646
10	$4705

Source: www.dpi.state.nc.us/fbs/finance/salary

25. The Astronomical Applications Department of the U.S. Naval Observatory in Washington, DC, has a website that lists the sunrise and sunset times for locations throughout the United States. The following table gives these times for Denver, Colorado, for the first part of January 2012. (Time is given in military time.) Consider sunrise times to be the function $f(x)$ and sunset times to be the function $g(x)$. Find $f(4)$ in this table and explain its meaning. Then find $g(10)$ and explain its meaning. What are the independent and dependent variables in these functions?

	January	
	Rise	Set
Day	h m	h m
01	0721	1646
02	0721	1647
03	0721	1648
04	0721	1649
05	0721	1649
06	0721	1650
07	0721	1651
08	0721	1652
09	0721	1653
10	0721	1654

26. The National Climatic Data Center compiles weather data for locations throughout the United States. The average total monthly snowfall in inches (including sleet and ice pellets) for the city of Asheville, North Carolina, is given in the following table. The "T" in the table indicates a month during which there was a trace amount of snowfall. Find $f(02)$ and explain its meaning. What are the independent and dependent variables of this function?

Month	01	02	03	04	05	06
Inches	4.8	4.1	2.9	0.6	T	T

Month	07	08	09	10	11	12
Inches	T	T	0	T	0.7	2.1

27. A cyclist can average 15 mph on a long trip. The following graph shows the distance he can cover as a function of time.

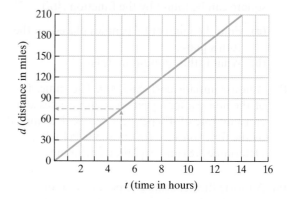

(a) Find the distance he can cover in 5 hr, $f(5)$.
(b) Find $f(8.5)$.

28. An absolute value function is shown in the following graph. Use the graph to evaluate the following.

(a) $f(0)$
(b) $f(-2)$
(c) $f(4)$
(d) $f(-4)$

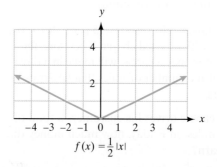

$f(x) = \frac{1}{2}|x|$

29. The cost of a rental car is $25 per day plus $0.05 per mile. The following graph illustrates this cost function. Use the graph to determine $f(100)$ and explain its meaning. What is the domain of this function? The range?

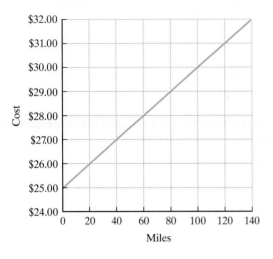

30. A car's value is a function of many things, including the total mileage on the car. The following graph illustrates the depreciation of a car as more miles are driven. Find the value of $f(50,000)$ and explain its meaning. What is the range of this function?

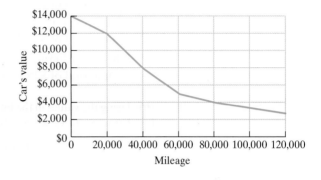

Section 4-3 Linear Functions as Models

Many functions can be grouped together based on similar characteristics. Functions whose formulas are similar, whose graphs have the same shape, and whose tables display similar patterns form **families of functions**. In this section, we will study linear functions or functions whose graphs are lines.

A function whose *rate of change* or *slope* is constant is called a **linear function**. Earlier in this chapter, we looked at the sales tax function $T(p) = 0.075p$. This is an example of a linear function. The tax, T, is a function of the amount purchased, p. For every additional dollar a customer spends, her tax will be increased by a constant rate of $0.075. This amount is referred to as the rate of change or slope of the function.

Problems that have constant rates of change can be modeled with linear functions. The cost of renting a car based on mileage is one example of a linear model. The following example illustrates this idea.

Example 1 Renting a Van

Tran wishes to rent a van for one day to take his baseball team to a tournament in Abilene. The rental cost of the van is $15 per day plus $0.05 for every mile that is driven.

(a) Make a table that gives the rental cost of the van using mileage figures from 0 mi to 200 mi.
(b) Graph the values in the table.
(c) Find a formula for cost (C) in terms of miles (m).

(a) Let us calculate the costs for every 50 mi of the trip and display these in the table.

At the beginning of the trip with $m = 0$, the cost of the van is $15.

If Tran drives 50 mi, the cost will be $15 plus an additional $(50)(\$0.05) = \$2.50$, or a total of $17.50.

If he drives 100 mi, the cost will be $15 plus an additional $(100)(\$0.05) = \$5.00$, or a total of $20.00.

If he drives 150 mi, the cost will be $15 plus an additional $(150)(\$0.05) = \$7.50$, or a total of $22.50.

If he drives 200 mi, the cost will be $15 plus an additional $(200)(\$0.05) = \$10.00$, or a total of $25.00. (See Table 4-7.)

(b) Using the values in the table, our function will form a straight line. (See Figure 4-10.)
(c) From part (a), we see that the cost is calculated using the formula

$$\text{Cost} = \text{rental fee} + \text{mileage charge}$$
$$\text{Cost} = \$15.00 + (\$0.05 \text{ per mile}) \times \text{miles}$$

Therefore, in function format using m for miles, the equation is

$$C(m) = \$15 + \$0.05m$$

TABLE 4-7
Cost of Rental Van

Miles	Cost
0	$15.00
50	$17.50
100	$20.00
150	$22.50
200	$25.00

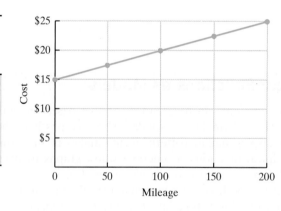

FIGURE 4-10
Example 1(b).

As you can see from Example 1, the graph of the function models a straight line. This is a fundamental relationship between a linear function and a straight line. In Example 1, the value of the function at $m = 0$ was $15. This is the y-intercept of the graph and is often referred to as the *initial value of the function*. The slope, m, of the line is the *rate of change of the function*, $0.05 per mile. In general, a linear function can be modeled using the slope-intercept form of the equation ($y = mx + b$), where m *is the rate of change* of the function and b *is the initial value* of the function.

Definition | **Linear Functions**

If $y = f(x)$ is a linear function, then the formula for the function is

$$y = mx + b \quad \text{or} \quad y = b + mx$$

Written in function notation, we have

$$f(x) = \text{initial value} + (\text{rate of change})(x)$$

Example 2

Long-Distance Rates as a Linear Function

A special discount telephone company offers customers the opportunity to make long-distance calls by paying a $0.39 connection fee and $0.03 per minute used.

(a) Use the definition of a linear function to write a function model for this telephone company's long-distance charges.
(b) Calculate the cost of a 25-minute phone call.
(c) Draw a graph of this function.

(a) The linear function model is $f(x) =$ initial value $+$ (rate of change)(x). In this problem, the connection fee of $0.39 is the initial value because you will pay that charge just for dialing the long-distance call. The rate of change is the $0.03 per minute. This is a constant rate for all minutes regardless of the length of the phone call. Therefore, using this model, the function is

$$f(x) = \$0.39 + \$0.03x$$

where $x =$ number of minutes.

(b) To calculate the cost of a 25-minute phone call, we will evaluate the function at $x = 25$.

$$f(25) = \$0.39 + \$0.03(25) = \$1.14$$

(c) The graph of the function can be drawn using the model by remembering how to graph linear equations of the form $y = mx + b$. In this case, the y-intercept, b, is the initial value $0.39 and the slope, m, is the rate of change of the function, $0.03 per minute. (See Figure 4-11.)

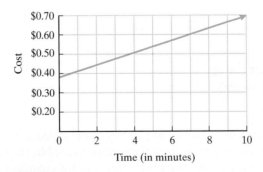

FIGURE 4-11
Example 2.

In Example 2, the linear function has a constant rate of change, which is $0.03 per minute. The longer the phone call is, the higher the charges will be. This is a direct relationship and is called an increasing function. Note that the graph shows the increase as we look from left to right. Not all linear functions are increasing functions, however. If the rate of change represents a constant loss, it will be negative, and therefore the slope of the line will be negative. When graphed, the line will fall as it moves from left to right, indicating a decreasing function. This concept is illustrated in Example 3.

Example 3: Depreciation as a Linear Function

Suppose you buy a new computer for your business. The cost of the computer is $2500. The value of the computer drops, or depreciates, by $400 each year following your purchase.

(a) Write a formula to estimate the value of the computer based on its age.
(b) Use functional notation to show the value of the computer after 3.5 years.
(c) Make a graph of the function.
(d) What is the meaning of the y-intercept? The x-intercept?

(a) The rate of change for this problem is $-\$400$, a loss in value of $400 per year. The initial value of the function is $2500, the purchase price of the computer. Therefore, the current value $V = V(t) = \$2500 - \$400t$, where t is time in years.
(b) $V(3.5) = \$2500 - \$400(3.5) = \$1100$. This is the value of the computer after 3.5 years.
(c) See Figure 4-12.
(d) The y-intercept is $2500, which is the initial value of the computer at purchase. The x-intercept is 6.25. This tells us that in 6.25 years the computer will have a value of $0.

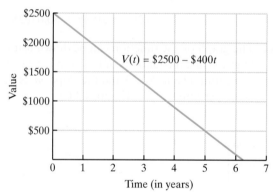

FIGURE 4-12
Example 3(c).

Linear functions may not always be presented as graphs or equations. Tables of data can also represent linear functions. In a table, the function will be linear if the changes in the y value are proportional to the changes in the value of x. Therefore, if the value of x goes up by equal amounts in the table of a linear function, then the value of y should also go up (or down) by equal amounts.

For example, in Table 4-8, the x values increase by 2 and the y values increase by 4. The slope of any line is the rate of change defined by the ratio

$$\text{Rate of change} = \frac{\text{change in } y}{\text{change in } x}$$

TABLE 4-8

x	y
-1	-3
1	1
3	5
5	9

This ratio corresponds to the slope formula presented in Chapter 3: $m = \dfrac{y_2 - y_1}{x_2 - x_1}$. For the values in Table 4-8, the slope $= \dfrac{4}{2} = 2$.

To write the equation of a linear function from a table of values, we use the slope-intercept form of the equation, $y = mx + b$. Use the calculated slope and one pair of points from the table to calculate the value of b.

$$\text{Slope} = 2 \quad \text{Point} = (1, 1)$$

$$y = mx + b$$

$$1 = (2)(1) + b$$

$$1 = 2 + b$$

$$-1 = b$$

Therefore, the function $f(x) = mx + b$ is $f(x) = 2x - 1$.

Example 4

Deriving a Linear Function from a Table

Many fitness centers display tables that list your desirable target heart rate while exercising based on your age. It is recommended that your heart rate not exceed 190 beats per minute at any age. For a 70-year-old, the recommended upper limit is 120 beats per minute. Use the information in Table 4-9 to derive a function or formula for calculating your desirable target heart rate based on age.

TABLE 4-9

Age	Target Heart Rate
20	170
30	160
40	150
50	140
60	130
70	120

To write the function, we need to find the rate of change first. We can use any two sets of ordered pairs in our table for the calculations.

$$\text{Rate of change} = m = \frac{y_2 - y_1}{x_2 - x_1} = \frac{170 - 160}{20 - 30} = \frac{10}{-10} = -1$$

Our next step is to calculate the value of b using the rate of change, m, and one set of ordered pairs from the table.

$$\text{Slope} = -1 \quad \text{Ordered pair} = (20, 170)$$

$$y = mx + b$$

$$170 = -1(20) + b$$

$$170 = -20 + b$$

$$190 = b$$

Using this information, we can write the following function formula to calculate your target heart rate.

$$f(x) = -1x + 190 \quad \text{or} \quad f(x) = 190 - x$$

where x represents a person's age.

Practice Set 4-3

1. (a) Suppose a taxi driver charges $1.75 per mile. Fill in the following chart with the correct values for $f(x)$, the cost of a trip of x miles.

x	$f(x)$
1	
3	
5	
7	
9	

   (b) What is the initial value of this function? What is the rate of change of this function?
   (c) Write a linear function that gives a rule for the amount charged.
   (d) Evaluate $f(8)$ and explain what it means.

2. (a) A long-distance company charges a monthly fee of $4.95 and $0.02 per minute. Fill in the following chart with the correct values for $f(x)$, the monthly long-distance bill, where x represents total minutes of long-distance time during the month.

x	$f(x)$
0	
15	
30	
45	
60	

   (b) What is the initial value of this function? What is the rate of change of this function?
   (c) Write a linear function that gives a rule for the amount charged.
   (d) Evaluate $f(110)$ and explain what it means.

3. Suppose $P(t)$, the population (in millions) of a country in year t, is defined by the formula $P(t) = 5.5 + 0.02t$.

   (a) What is the country's initial population?
   (b) What is the population growth rate of the country?
   (c) Predict the population of the country 5 years from now. (Let $t = 5$.)

4. A plumber's fee schedule can be modeled with a linear function using his service charge to come to your house and his hourly rate for fixing the problem. The fee schedule for J&B Plumbers is $C(h) = \$50 + \$20h$, where h represents the number of hours that the plumber works at your house.

   (a) What is the service charge (initial value) at J&B Plumbers?
   (b) What is the hourly rate that J&B charges?
   (c) Calculate the value of $C(0)$ and explain its meaning.

5. Josh's daughter is getting married in June and he is trying to calculate the cost of the wedding reception. The cost will be a function of the number of guests who attend the reception. The rent for the banquet hall is $525 for the evening. The caterer is charging $17.50 per person for the food.

   (a) Write a linear function to model this problem.
   (b) How much will the reception cost if 200 people attend?
   (c) If Josh has only budgeted $3150 for the reception, approximately how many guests can be invited?

6. Business A and Business B are printing companies. The cost equations for printing wedding invitations at each business are given by

   $$\text{Cost}_A = \$200 + \$1.75x$$
   $$\text{Cost}_B = \$150 + \$2.00x$$

   where x = number of invitations.

   (a) Describe in words what these equations tell us about the costs of printing invitations at each business.
   (b) If we need 500 wedding invitations, which business will be less expensive?

7. Brandi bought a computer in 2012 for $1800. She estimated the useful life of the computer to be 5 years with a remaining value of $100 at the end of this period. She used the straight-line method of depreciation to estimate her yearly loss in value during this time period.

   (a) What is the rate of change for this function?
   (b) Write a function that can be used to estimate the value of the computer, V, at any time during the 5-year period.
   (c) Calculate $V(4)$ and explain its meaning.

8. Green Giant Daycare Center purchased new furniture for its toddler room. The estimated useful life of the furniture is 10 years. The total cost was $3650 and the value at the end of 10 years is expected to be $250.

   (a) What is the rate of change for this function?
   (b) Write a function that can be used to estimate the value of the furniture, V, at any time during the 10-year period.
   (c) Calculate $V(8)$ and explain its meaning.

Use the data in the tables to write a linear function equation.

9. See Table 4-10.

TABLE 4-10

x	5	10	15	20	25
y	-1	-2	-3	-4	-5

10. See Table 4-11.

TABLE 4-11

x	-2	0	4	6
y	$-\frac{5}{2}$	-1	2	$\frac{7}{2}$

11. See Table 4-12.

TABLE 4-12

x	2	-1	-4	-7	-10
y	5	5	5	5	5

12. See Table 4-13.

TABLE 4-13

x	0	10	20	30
y	-5	-10	-15	-20

13. The population of a small town grew from 450 in 2001 to 650 in 2011.

    (a) Assume that the growth is linear and calculate the rate of change.
    (b) Let $x = 0$ correspond to the year 2001. Write a linear function that can be used to predict future population after 2001.
    (c) Calculate $f(15)$ and explain its meaning.

14. The deposits at Money Masters Bank grew from $150 million in 2007 to $205 million in 2012.

    (a) Assume that deposits grew at a rate that is linear. Calculate the rate of change.
    (b) Let $x = 0$ correspond to the year 2007. Write a linear function that can be used to predict future deposits of the years after 2007.
    (c) Calculate $f(10)$ and explain its meaning.

15. A businessman decides to purchase a second facility after his annual sales reach $1,500,000. In 2006, annual sales were $600,000, and in 2011, they were $1,000,000.

    (a) Assume that the growth is linear and calculate the rate of change.
    (b) Let $x = 0$ correspond to the year 2006. Write a linear function that can be used to predict future annual sales after 2006.
    (c) Use this function equation to determine the year that his annual sales will reach $1,500,000.

16. The cost of manufacturing goods is a linear function that includes fixed costs, such as utilities and rent, and per unit costs of manufacturing each individual item. In a month, Marvin Stephens Manufacturing Company produces 100 units for $8400 and 250 units for $9000.

    (a) Assume that the cost of production is a linear function and find the rate of change.
    (b) If the fixed costs of production are $8000 per month, write a linear function to model the monthly production of x units.
    (c) How many units can be produced for a cost of $10,000?

17. The average yearly in-state tuition and fees charged by U.S. public universities for the years 2008–2010 are shown in Table 4-14.

TABLE 4-14

Year	2008	2009	2010
Average Tuition	$3536	$4115	$4694

   (a) What is the rate of change or slope of this function?
   (b) Write the linear function model for these data. Let $3536 be the initial value of the function with 2008 as year 0.

(c) Plot the data points and use them to draw the line representing the linear function.
(d) Use the linear function and your graph to predict the average tuition cost in 2014.

18. Table 4-15 records the growth in population of Maryville.

TABLE 4-15

Year	2010	2011	2012	2013	2014
Population (in thousands)	124	127	130	133	136

(a) What is the rate of change or slope of this function?
(b) Write a linear function for these data. Let 124 be the initial value of the function and let $t = 0$ correspond to the year 2010.
(c) Plot the data points and use them to draw the line representing the linear function.
(d) Use the linear function to predict the population in 2018.

19. There is a functional relationship between the height and weight of a person. For women who are 5 ft or taller, a function equation can be written to give the ideal weight of a woman based on her height. The rule used for many years by insurance companies stated that a woman 60 in. tall should weigh 100 lb and for those taller, an additional 5 lb should be added for every inch over 60 in.

(a) Write a linear function to predict the ideal weight of a woman 60 in. or taller based on this insurance rule.
(b) According to this rule, what is the ideal weight for a woman who is 5 ft 7 in.?

20. The 2011 Tax Rate Schedule X for single taxpayers states that a taxpayer who is single with taxable income greater than $31,850 but less than $77,100 should pay $4386.25 plus 25% of the amount over $31,850.

(a) Use this information to write a function equation for this tax rule.
(b) If a single taxpayer's taxable income was $35,450, how much tax did he owe the federal government in 2011?

Section 4-4 Direct and Inverse Variation

Many functions involve the concepts of direct or inverse proportionality. For example, if a car is traveling at a constant rate of 50 miles per hour, then the distance d traveled in t hours is given by the equation $d = 50t$. The change in distance is directly affected by the change in the time. You can see that as the time increases, the distance traveled also increases. This is an example of a **direct variation**. This equation could also be written $d(t) = 50t$, because distance is a function of time.

In the equation $d = rt$, the relationship between the variables d and t is such that if t gets *larger* then d also gets *larger*, and if t gets *smaller* then d also gets *smaller*. Because of this relationship, we say that d is **directly proportional** to t. Another way to read this equation is that d varies directly as t. The number 50 in the equation $d = 50t$ is called the **constant of variation**, or **constant of proportionality**.

IMPORTANT EQUATIONS

Direct Variation

y varies directly as x means that
$$y = kx$$
where k is a nonzero positive constant.

Section 4-4 **Direct and Inverse Variation** 161

Example 1

Direct Variation

Suppose y varies directly as the square of x. Find the constant of variation if y is 16 when $x = 2$ and use it to write an equation of variation.

Because this is a direct variation, the general equation will be
$$y = kx^2$$
Replacing y with 16 and x with 2 gives
$$16 = k(2)^2$$
Solve this equation for k.
$$16 = 4k$$
$$4 = k$$
Therefore, the constant of variation is 4. Substituting 4 for k, we get the variation formula for y:
$$y = 4x^2$$

Boyle's law for the expansion of gas is $V = \frac{K}{P}$, where V is the volume of the gas, P is the pressure, and K is a constant. This formula relates a pressure and a volume. As the pressure, P, increases, the volume of gas, V, decreases, and as the pressure, P, decreases, the volume of the gas, V, increases.

This is an example of **inverse variation**. We say that "V *varies inversely* as P" or "V is **inversely proportional** to P."

Definition

Inverse Variation

y varies inversely as x means that
$$y = \frac{k}{x}$$
where k is a nonzero positive constant.

Example 2

Inverse Variation

Suppose y varies inversely as the square root of x. Find the constant of proportionality if y is 15 when x is 9 and write the equation of variation.

Because this is an inverse variation, the general equation will be
$$y = \frac{k}{\sqrt{x}}$$
Replacing y with 15 and x with 9 gives
$$15 = \frac{k}{\sqrt{9}}$$
Solve this equation for k.
$$15 = \frac{k}{3}$$
$$15(3) = k$$
$$45 = k$$
Therefore, the constant of proportionality is 45. The variation equation is
$$y = \frac{45}{\sqrt{x}}$$

When working with variation problems, it is important that you pay attention to the words of the problem to determine if the relationship is a direct variation ($y = kx$) or an inverse variation ($y = \frac{k}{x}$). Remember, in direct variation, an increase in one variable causes a corresponding increase in the other. In inverse variation, an increase in one variable causes a corresponding decrease in the other variable. Look at Example 3 for some practice in determining whether a situation involves direct or inverse variation.

Example 3 — Direct or Inverse Variation?

Use your intuition and the appropriate definitions to determine whether the variation between these variables is direct or inverse.

(a) The time traveled at a constant speed and the distance traveled
(b) The weight of a car and its gas mileage
(c) The interest rate and the amount of interest earned on a savings account

(a) The time traveled at a constant speed and the distance traveled is an example of a direct variation. The longer you travel at a constant speed, the more distance you will cover.
(b) The weight of a car and its gas mileage is an example of inverse variation. Although many things affect a car's gas mileage, in general, the heavier the car, the lower the miles per gallon for that car.
(c) The interest rate and the amount of interest earned on a savings account is an example of a direct variation. The higher the interest rate, the more interest you will earn on your account.

Two additional types of variation equations are **joint variation** and **combined variation**.

Definition — Joint Variation

y varies jointly as x and z means that

$$y = kxz$$

where k is a nonzero positive constant.

Definition — Combined Variation

y varies jointly as u and v and inversely as w and x means that

$$y = \frac{kuv}{wx}$$

where k is a nonzero positive constant.

Example 4 — Variation Equations

Write the following statements as variation equations.

(a) w varies jointly as y and the cube of x.
(b) x is directly proportional to y and inversely proportional to z.

(a) Using the definition for joint variation,
$$w = kyx^3$$
(b) Using the definition for combined variation,
$$x = \frac{ky}{z}$$

Example 5 **Variation Problem**

Suppose y is directly proportional to x and inversely proportional to the square root of z. Find the constant of variation if y is 4 when x is 8 and z is 36 and write an equation of variation. Determine y when x is 5 and z is 16.

General variation equation:
$$y = \frac{kx}{\sqrt{z}}$$

Substitute 4 for y, 8 for x, 36 for z, and solve for k.
$$4 = \frac{k \cdot 8}{\sqrt{36}}$$

Solve for k.
$$4 = \frac{k \cdot 8}{6}$$
$$4(6) = 8k$$
$$24 = 8k$$
$$3 = k$$

Variation equation:
$$y = \frac{3x}{\sqrt{z}}$$

Substituting 5 for x and 16 for z (from the second set of data), we can find the value of y.
$$y = \frac{3 \cdot 5}{\sqrt{16}} = \frac{15}{4} = 3.75$$

We can use variation equations to solve applied problems. The following guidelines will help you in solving application problems involving variation.

Rule **Steps for Solving Applied Variation Problems**

1. Write a general variation equation expressing the stated relationship. Make sure to include a constant of variation, k.
2. Substitute values from the given data into the variation equation and solve for k.
3. Obtain a specific variation formula by replacing k in the variation equation with the value obtained for k in step 2.
4. Solve for the missing value of any variable in the formula you have written using additional data from the problem.

Example 6 — Applied Variation Problem

The distance a car travels at a constant speed varies directly as the time it travels. Find the variation formula for the distance traveled by a certain car if it is known that the car traveled 220 mi in 4 hr at a constant speed. How many miles will the car travel in 7 hr at that speed?

Let d = distance traveled and t = time.
General variation equation:
$$d = kt$$

Substitute 220 for d and 4 for t.
$$220 = 4k$$

Solve for k.
$$55 = k$$

In this problem, k represents the speed of the car. Therefore, we now know that the car is traveling at a constant speed of 55 mph.
Write the variation equation:
$$d = 55t$$

Now substitute 7 for t and solve for the distance, d.
$$d = 55(7)$$
$$d = 385 \text{ mi}$$

Example 7 — Ohm's Law

Ohm's law says that the current, I, in a wire varies directly as the electromotive force, E, and inversely as the resistance, R. If I is 11 A when E is 110 V and R is 10 Ω, find I if E is 220 V and R is 11 Ω.

General variation equation:
$$I = \frac{kE}{R}$$

Substitute 11 for I, 110 for E, and 10 for R.
$$11 = \frac{k \cdot 110}{10}$$

Solve for k.
$$11(10) = 110\,k$$
$$110 = 110\,k$$
$$1 = k$$

Variation equation:
$$I = \frac{1 \cdot E}{R}$$

Now substitute 220 for E, 11 for R, and solve for I.
$$I = \frac{1 \cdot 220}{11} = \frac{220}{11} = 20 \text{ A}$$

Practice Set 4-4

Use your intuition and the appropriate definitions to determine whether the variation between these variables is direct or inverse.

1. The diameter of a bicycle wheel and its circumference
2. The number of people in line at the grocery store and the time required to check out
3. The time required to fill a swimming pool and the amount of water coming in from the hose
4. The number of people mowing a lawn and the time it takes to finish the job
5. The amount of water used during a month and the monthly water bill
6. The outside temperature and the time it takes for the snow to melt

Write a variation equation for the following statements.

7. y varies directly as z.
8. p varies inversely as q.
9. a varies jointly with b and c.
10. m varies directly as the square of n.
11. d varies jointly with e and the cube of f.
12. f varies jointly with g and h and inversely with the square of j.
13. m varies directly as the square root of n and inversely as the cube of p.
14. v varies inversely as the square root of g.

Express the following statements as variation equations and find the values of the constants of proportionality and the indicated unknowns.

15. y varies directly as x; $y = 8$ when $x = 24$. Find y when $x = 36$.
16. p varies directly as q; $p = 6$ when $q = 120$. Find p when q is 100.
17. y varies inversely as x; $y = 9$ when $x = 6$. Find y when $x = 36$.
18. r varies inversely as t; $r = 6$ when $t = 2$. Find r when $t = 8$.
19. y varies directly as the square root of x; $y = 24$ and $x = 16$. Find y when $x = 36$.
20. y varies inversely as the square root of x; $y = 6$ when $x = 100$. Find y when $x = 144$.
21. y varies jointly as x and the square of z; $y = 150$ when $x = 3$ and $z = 5$. Find y when $x = 12$ and $z = 8$.
22. y varies jointly as x and the square root of z; $y = 48$ when $x = 2$ and $z = 36$. Find y when $x = 3$ and $z = 25$.
23. p varies directly as q and inversely as the square of r; $p = 40$ when $q = 20$ and $r = 4$. Find p when $q = 24$ and $r = 8$.
24. y varies directly as r and inversely as the product of x and t; $y = 4$ when $r = 6$, $x = 3$, and $t = 1$. Find y when $r = 14$, $x = 1$, and $t = 6$.

Solve the following applied variation problems.

25. The area of a circle varies directly as the square of its radius. If the area of a circle is 78.5 cm^2 when its radius is 5 cm, find the area of the circle when the radius is 7 cm.
26. Hooke's law says that the force required to stretch a spring is directly proportional to the distance stretched. If a force of 10 lb is required to stretch a spring 5 in., how much force will be required to stretch the spring 10 in.?
27. The distance traveled by a ball rolling down an inclined plane varies directly as the square of the time. If the inclination of the plane is such that the ball traveled 27 ft at the end of 3 sec, what is the distance traveled at the end of 2 sec?
28. The distance that a car travels after the brakes have been applied varies directly as the square of its speed. If the stopping distance for a car going 30 mph is 54 ft, what is the stopping distance for a car going 70 mph?
29. The period of a pendulum varies directly as the square root of its length. If a pendulum of length 16 in. has a period of 0.5 sec, find the length of a pendulum with a period of 0.75 sec.
30. The interest on an investment varies directly as the rate of interest. If the interest is $48 when the interest rate is 5%, find the interest when the rate is 4.75%.
31. The time required to fill a tank varies inversely as the square of the diameter of the pipe used. If it

takes 20 hr to fill a given tank with a 2-in. pipe, find the diameter of the pipe needed to fill the same tank in 5 hr.

32. Boyle's law states that the volume of a fixed mass of gas under constant temperature is inversely proportional to the pressure. If the volume is 10 ft^3 when the pressure is 15 lb/in.^{2}, find the volume when the pressure is 60 lb/in.^{2}

33. The weight of a body above the surface of the Earth varies inversely as the square of the distance of the body from the center of gravity of the Earth. If an astronaut weighs 180 lb on the surface of the Earth, how much would she weigh 500 mi above the surface of the Earth? Assume the radius of the Earth to be 4000 mi.

34. The time it takes to drive a certain distance varies inversely as the speed traveled. If it takes 3 hr at 60 mph to drive a certain distance, how long would it take to drive the same distance at 65 mph?

35. The electric power, P, used in a circuit varies directly with the square of the voltage drop, V, and inversely with the resistance, R. If 180 W of power in a circuit are used by a 90-V drop with a resistance of 45 Ω, find the power used when the voltage drop is 120 V and the resistance is 30 Ω.

36. The weight of a cylindrical container filled with a given substance varies jointly as the height of the container and the square of its radius. If a container with a radius of 10 cm and a height of 30 cm weighs 60 kg, what is the weight of a container filled with the same substance when its radius is 12 cm and its height is 60 cm?

37. The expansion of metal varies jointly as the temperature and the length of the metal. A brass rod expands 2.7 in. when its length is 100 ft and the temperature is 225°F. Find the expansion when the length of the brass rod is 40 ft and the temperature is 150°F.

38. The maximum load that a cylindrical column with a circular cross section can hold varies directly as the fourth power of the diameter of the cross section and inversely as the square of the height. If a 10-m column that is 1.5 m in diameter can support 32.805 t, how many metric tons can be supported by a column 9 m high and 1 m in diameter?

39. The volume, V, of a sphere is given as a function of its radius, r, by

$$V(r) = \frac{4\pi r^3}{3}$$

(a) Is this a direct or inverse variation?
(b) What is the constant of proportionality?
(c) Evaluate $V(9 \text{ in.})$.

40. The area of a circle, A, is given as a function of its radius: $A(r) = \pi r^2$.

(a) Is this a direct or inverse variation?
(b) What is the constant of proportionality, k?
(c) Evaluate $A(3 \text{ in.})$.

Section 4-5 Quadratic Functions and Power Functions as Models

We examined functions that are linear in the previous section. Now let us look at two types of nonlinear functions: quadratic functions and power functions.

Quadratic Functions

The first family of nonlinear functions we will discuss is the **quadratic function**.

Definition **Quadratic Function**

A quadratic function is a function f defined for all real numbers x by an equation of the form:

$$f(x) = ax^2 + bx + c$$

in which a, b, and c are constants and the leading coefficient $a \neq 0$.

Section 4-5 **Quadratic Functions and Power Functions as Models** 167

The graph of a quadratic function is a parabola. One of the simplest quadratic functions is defined by $f(x) = x^2$. To graph this equation, choose several values for x and solve for $f(x)$. (See Table 4-16.)

Now plot these points on a coordinate plane. (See Figure 4-13.) This is the graph of the quadratic function $f(x) = x^2$.

TABLE 4-16

x	$f(x) = x^2$	$f(x)$	Solution
-2	$f(-2) = (-2)^2$	4	$(-2, 4)$
-1	$f(-1) = (-1)^2$	1	$(-1, 1)$
0	$f(0) = (0)^2$	0	$(0, 0)$
1	$f(1) = (1)^2$	1	$(1, 1)$
2	$f(2) = (2)^2$	4	$(2, 4)$

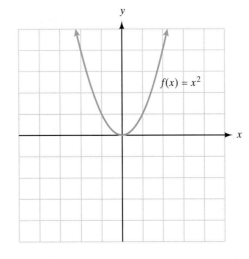

FIGURE 4-13
Graph of quadratic function $f(x) = x^2$.

You will notice that the graph in Figure 4-13 is symmetric to the y-axis. This means that the graph is the same on both sides of the y-axis. The y-axis acts as a mirror.

Example 1 Quadratic Functions

Sketch the graph of $f(x) = -x^2$.

First, choose several values for x and solve for $f(x)$. (See Table 4-17.) Second, plot these points on a coordinate plane. (See Figure 4-14.) This figure shows the graph of $f(x) = -x^2$.

TABLE 4-17

x	$f(x) = -x^2$	$f(x)$	Solution
-2	$f(-2) = -(-2)^2$	-4	$(-2, -4)$
-1	$f(-1) = -(-1)^2$	-1	$(-1, -1)$
0	$f(0) = -(0)^2$	0	$(0, 0)$
1	$f(1) = -(1)^2$	-1	$(1, -1)$
2	$f(2) = -(2)^2$	-4	$(2, -4)$

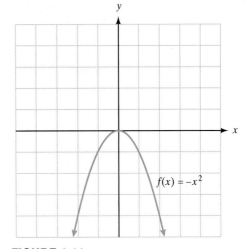

FIGURE 4-14
Graph of quadratic function $f(x) = -x^2$.

You will notice that the graph of $f(x) = -x^2$ opens down. If the leading coefficient is positive, then the parabola will open upward. If the leading coefficient is negative, the parabola will open downward.

IMPORTANT EQUATIONS

Quadratic Function

$$f(x) = ax^2 + bx + c$$ (if a > 0, then the graph opens up)

(if $a < 0$, then the graph opens down)

Each parabola has a minimum point or a maximum point. The vertex of a parabola is the high or low point of the graph—the maximum (max) or minimum (min) of the function. If the parabola opens downward, the function has a maximum value that is the y-coordinate of the vertex. If the parabola opens upward, the function has a minimum point that is also the y-coordinate of the vertex. The maximum and minimum points are important in many application problems. Another important aspect of a quadratic function are the **roots** or zeros of the function. The roots of a quadratic function are the same as the x-intercepts. Remember, the x-intercept of any equation is where the graph crosses the x-axis.

CALCULATOR MINI-LESSON

Using a Graphing Calculator to Find Max/Min and Roots

A TI-84 Plus graphing calculator can be used to easily find the maximum or minimum value of a parabolic function and the roots of the function. Follow these steps.

Zeros—these are the x-intercepts or roots of the equation.

1. Select **2:zero** from the CALCULATE menu (located above the TRACE button). The graph will be displayed with the words **Left Bound?** at the bottom of the screen.
2. Move the cursor to the left of the location of one of the x-intercepts using the arrow keys. Press ENTER. A small arrow will appear at the top of the screen marking the left boundary. At the bottom of the screen, you should see **Right Bound?** Move the cursor right, past the x-intercept, and press ENTER again. Another arrow should appear at the top of the screen marking the right boundary.
3. The word **Guess?** is now displayed at the bottom of the screen. Press ENTER one more time and the coordinates of the x-intercept or root should be displayed at the bottom of the screen.

Max/Min—depending on the equation, the graph will have either a maximum point *or* a minimum point.

1. Select **3:minimum** or **4:maximum** from the CALCULATE menu. The graph of the function will be displayed.
2. Set the left and right boundaries for the maximum or minimum point following the same steps that were used to find the zeros.

To find the x-intercepts, we substitute a 0 for y and solve the equation for x. The x-intercepts or roots of a quadratic function can also be determined using a graphing calculator.

Example 2

Finding Roots of Quadratic Functions

Graph the quadratic function $f(x) = -2x^2 + 3x + 2$ using a graphing calculator. Find the value of the maximum or minimum and determine the roots of the function.

Enter the equation using the $\boxed{Y=}$ function as

$$Y_1 = -2x^2 + 3x + 2$$

Use the $\boxed{\text{GRAPH}}$ button to display the parabola. Use the $\boxed{\text{TRACE}}$ function to move the cursor to the highest point on the graph. This is the maximum point of the function. Record the coordinates shown at the bottom of the screen. Now use the $\boxed{\text{TRACE}}$ function to move the cursor to each of the x-intercepts. These are the roots of the equation. Record the coordinates displayed on your screen. If you have trouble getting the exact answer using the $\boxed{\text{TRACE}}$ function, follow the instructions in the Calculator Mini-Lesson on page 166 to use the $\boxed{\text{CALCULATE}}$ function. (See Figure 4-15.)

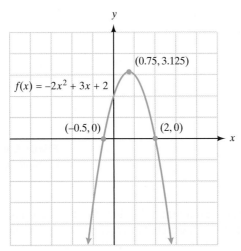

FIGURE 4-15
Example 2.

Note that the roots are $(-0.5, 0)$ and $(2, 0)$, and the maximum value is 3.125.

Quadratic functions are used to describe many scientific applications, such as throwing objects or shooting projectiles. The force of gravity and the initial velocity of a moving object are used to write a quadratic function that describes a particular situation. Look at the following situation and the function used to describe it.

Example 3

Finding the Maximum Height of a Quadratic Function

A rock is thrown upward from the ground. Its height in feet above ground after t seconds is given by the equation

$$f(t) = -16t^2 + 15t$$

Find the maximum height of the rock and the number of seconds it took for the rock to reach its maximum height.

Enter the equation into a graphing calculator. Use the $\boxed{Y=}$ function:

$$Y_1 = -16t^2 + 15t$$

Use the $\boxed{\text{GRAPH}}$ button to display the parabola. Move the cursor to the highest point on the graph using the $\boxed{\text{TRACE}}$ function. The coordinates of the cursor will be displayed at the bottom of your screen. In this application, the t variable is plotted on the x-axis, so the x-coordinate of the maximum represents the time. The y value represents the height. Use the $\boxed{\text{CALCULATE}}$ feature to verify these points. (See Figure 4-16.)

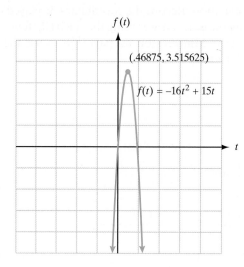

FIGURE 4-16
Example 3.

Note that the maximum is (.46875, 3.515625). This means that the rock's maximum height is 3.5 ft, which was reached in approximately 0.5 sec.

Example 4

Profit Function

A company's profit (in thousands of dollars) is given by the equation $P(s) = 230 + 20s - 0.5s^2$, where s is the amount (in hundreds of dollars) spent on advertising.

Complete the table with the designated values for s and use the graphing calculator to help you draw a sketch of the function. (You will need to adjust the window settings to view the entire graph.) Use the $\boxed{\text{TABLE}}$ or $\boxed{\text{TRACE}}$ function to verify the values that have been calculated in Table 4-18. The graph of this function is shown in Figure 4-17.

TABLE 4-18

$s =$ Amount Spent on Advertising (in $100)	Profit (in $1000)
0	$P(0) = 230 + 20(0) - 0.5(0)^2 = 230$
8	$P(8) = 230 + 20(8) - 0.5(8)^2 = 358$
15	$P(15) = 230 + 20(15) - 0.5(15)^2 = 417.5$
25	$P(25) = 230 + 20(25) - 0.5(25)^2 = 417.5$
40	$P(40) = 230 + 20(40) - 0.5(40)^2 = 230$
50	$P(50) = 230 + 20(50) - 0.5(50)^2 = -20$

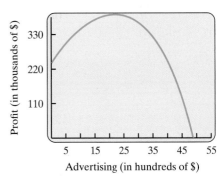

FIGURE 4-17
Example 4.

Using the graph, and table, how much would you spend on advertising to maximize profit?

The following application problems are examples of problems that may be solved by factoring.

Example 5 Electrical Currents

A variable voltage in a given electrical current is given by the formula $V = t^2 - 12t + 40$. At what values of t (in seconds) is the voltage V equal to 8.00V?

$$8 = t^2 - 12t + 40$$
$$0 = t^2 - 12t + 32$$
$$0 = (t - 4)(t - 8)$$
$$t - 4 = 0 \quad \text{or} \quad t - 8 = 0$$
$$t = 4 \text{ sec} \quad \text{or} \quad t = 8 \text{ sec}$$

Example 6 Shooting Projectiles

A projectile is shot vertically upward at an initial velocity of 96 ft/sec. Its height h in feet after t seconds may be expressed by the formula $h = 96t - 16t^2$.

(a) Find the height of the projectile at 7 sec.

$$h = 96t - 16t^2$$
$$h = 96(7) - 16(7)^2$$
$$h = 672 - 784 = -112 \text{ ft}$$

Because we have a negative height, the projectile must have landed on the ground before 7 seconds had passed. (See Figure 4-18.)

(b) Find t when $h = 0$.

$$0 = 96t - 16t^2$$
$$0 = 16t(6 - t)$$
$$16t = 0 \quad \text{or} \quad 6 - t = 0$$
$$t = 0 \text{ sec} \quad \text{or} \quad t = 6 \text{ sec}$$

The projectile was on the ground at the initial take-off ($t = 0$) and back on the ground 6 seconds later.

(c) Figure 4-18 shows the height of the projectile from $t = 0$ to $t = 6$ sec. Note that it is back to an altitude of zero before 7 seconds have passed [see Part (a) of this example].

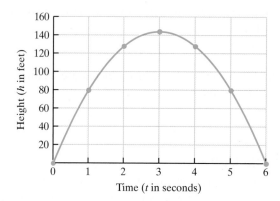

FIGURE 4-18

Not all quadratic equations can be factored. If an equation is not factorable (or even if it is!) you can find the solutions of the equation using the quadratic formula.

IMPORTANT EQUATIONS

Quadratic Formula

If $ax^2 + bx + c = 0$, $a \neq 0$, then

$$x = \frac{-b \pm \sqrt{b^2 - 4ac}}{2a}$$

When using the quadratic formula to solve a problem, remember that the equation must be set equal to zero (0) before designating the values of a, b, and c. The signs of each term must also be used in order for the formula to give correct solutions. If the value of the expression under the radical is negative, this indicates a nonreal (imaginary) solution. (These nonreal or imaginary number solutions are beyond the level of this text.) The basic use of the quadratic formula is reviewed in Example 7, and it is applied in Example 8.

Example 7: Solving a Quadratic Equation Using the Quadratic Formula

Solve the following equation using the quadratic formula:

$$3x^2 - 5x - 4 = 2$$

The first step in using the quadratic formula is to set the equation equal to 0 on one side. This is the same as if you were going to try to *factor* the equation in order to solve it.

$$3x^2 - 5x - 6 = 0$$

For use in the quadratic formula, once the equation is set equal to 0, the coefficient of the square term is a, the coefficient of the first power term is b, and any number that has no variable is c. (Please refer to the Important Equations box containing the quadratic formula.) Therefore, we will match up the numbers in our equation to the letters used:

$$a = 3, \quad b = -5, \quad \text{and} \quad c = -6$$

Now we will fill in the quadratic formula:

$$x = \frac{-b \pm \sqrt{b^2 - 4ac}}{2a}$$

$$x = \frac{-(-5) \pm \sqrt{(-5)^2 - 4(3)(-6)}}{2(3)}$$

$$x = \frac{5 \pm \sqrt{25 + 72}}{6}$$

We can find the two possible solutions as follows:

$$x = \frac{5 + \sqrt{97}}{6} \quad \text{or} \quad x = \frac{5 - \sqrt{97}}{6}$$

$$x = 2.4748 \ldots \quad \text{or} \quad x = -0.8081 \ldots$$

Answers in real applied problems will often be long decimal numbers. You will need to round the answers appropriately to match the scale of the problem and the type of answer required.

Example 8: Electrical Current (An Application of the Quadratic Formula)

The following equation models the relationship among the electrical current i measured in amperes, voltage E measured in volts, and resistance R measured in ohms for a particular circuit.

$$i^2R + iE = 4000$$

If the resistance in the circuit is 20 Ω and the voltage is 50 V, what is the current flowing in the circuit?

First, substitute the known values:

$$i^2(20) + i(50) = 4000$$

This is a quadratic equation, so we will rearrange it and set it equal to 0.

$$20i^2 + 50i - 4000 = 0$$

Because of the large value of 4000 compared with the other coefficients, this equation is not factorable. Thus, we will use the *quadratic formula* to solve it. In using the quadratic formula, first set the equation equal to zero as if you were going to factor it. We have already done this.

Next assign values to *a*, *b*, and *c* in the quadratic formula. In this application, $a = 20$, $b = 50$, and $c = -4000$. Now, plug these values into the quadratic formula.

$$x = \frac{-b \pm \sqrt{b^2 - 4ac}}{2a}$$

Here we are solving for the value of the current, *i*.

$$i = \frac{-50 \pm \sqrt{50^2 - 4(20)(-4000)}}{2(20)}$$

$$i = \frac{-50 \pm \sqrt{2500 - (-320{,}000)}}{40}$$

There are two solutions as follows:

$$i = \frac{-50 + \sqrt{322{,}500}}{40} \quad \text{or} \quad \frac{-50 - \sqrt{322{,}500}}{40}$$

$$i = 12.947 \ldots \text{ amperes} \quad \text{or} \quad i = -15.447 \ldots \text{ amperes}$$

Because there is no such thing as a negative current value, we choose the positive answer of about 12.9 amperes for the current.

Power Functions

A **power function** is an equation that takes the form $f(x) = cx^k$. The number *k* is called the *power* and is the most significant part of a power function. For most applications, we are interested only in positive values of the variable *x* and of *c*, but we allow *k* to be any number.

In a power function, larger values of the exponent, *k*, cause the function to increase faster when *k* is positive. (See Figure 4-19.) If *k* is negative, the functions

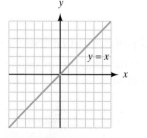
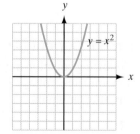
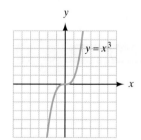

FIGURE 4-19
Power functions.

decrease toward 0. These functions are easily graphed using a graphing calculator, and the differences in their behaviors can be examined by graphing several functions with increasing powers of the exponent. Using only the first quadrant for graphing, this property is illustrated in Figure 4-20.

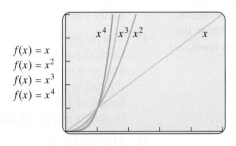

FIGURE 4-20
Comparing graphs of power functions.

Example 9

Power Function

When a rock is dropped from a tall structure, it will fall $D = 16t^2$ ft in t sec. Make a graph that shows the distance the rock falls versus time if the building is 70 ft tall. How long does it take the rock to strike the ground?

(a) Using your graphing calculator, enter the equation $D = 16t^2$, letting $D = y_1$, the distance on the vertical axis; x will equal time, which is represented on the horizontal axis. Be sure that the window of the graph has a maximum y (y max) greater than 70, because the building is 70 ft tall. (See Figure 4-21.)

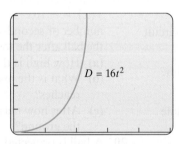

FIGURE 4-21
Graph of distance fallen versus time.

(b) We want to know how long it takes the rock to strike the ground. Therefore, we want to solve the equation letting $D = 70$, because it will strike the ground after traveling 70 ft.

$$16t^2 = 70$$
$$t^2 = \frac{70}{16}$$
$$t^2 = 4.375$$
$$t = 2.09 \text{ sec}$$

We can verify this result using the $\boxed{\text{TRACE}}$ function on the calculator.

Practice Set 4-5

Sketch the graphs of the following quadratic functions using a graphing calculator. State if the function has a maximum or a minimum. Determine the maximum or minimum value to the nearest tenth. Determine the roots of the quadratic function to the nearest tenth.

1. $f(x) = x^2 + 6x + 5$
2. $f(x) = x^2 - 6x + 5$
3. $f(x) = x^2 + 5x + 4$
4. $f(x) = -x^2 - 4$
5. $f(x) = -3x^2 + 2x + 5$
6. $f(x) = -2x^2 + 5x - 6$

Use your graphing calculator to sketch a graph of the following power functions. Calculate the value of $f(-3)$ for each function in problems 7 to 10.

7. $f(x) = 3x^2$
8. $f(x) = -4x^2$
9. $f(x) = \frac{1}{3}x^3$
10. $f(x) = 2x^4$

For each problem, write a mathematical model and factor it to find the correct answer.

11. A variable voltage is given by the formula $V = t^2 - 14t + 48$. At what time after the circuit begins to function will the voltage be 3 V?

12. Using the formula in problem 11, at what time will the voltage be 35 V?

13. Using the formula in problem 11, at what time will the voltage be 0 V?

14. Using the formula in problem 11, at what time will the voltage be 8 V?

15. If you stand on top of a building 240 ft high and throw a rock upward with a speed of 32 feet per second, by the laws of physics, the equation $h = 240 - 32t - 16t^2$ gives the height of the rock above the ground t seconds after it is thrown. How many seconds does it take the rock to reach the ground? Use your graphing calculator to sketch this graph and verify your solution.

16. If you stand on top of a building 240 ft high and drop a rock, the equation $h = 240 - 16t^2$ gives the height of the rock above the ground t seconds after it is dropped. How many seconds does it take the rock to reach the ground? Use your graphing calculator to sketch this graph and verify your solution.

17. The height h of a ball after it is thrown vertically upward is given in feet by the function $h(t) = 4 + 6t - 16t^2$. Determine the time when the ball is at its peak. Determine the maximum height of the ball. Use your graphing calculator to sketch this graph and verify your solution.

18. Marcus shoots an arrow straight upward. The velocity of the arrow is 160 ft/sec. The function that models the height of the arrow is $f(x) = -5 + 160x - 16x^2$, where x is the number of seconds that have passed since the arrow was shot. Use your graphing calculator to answer the following questions.
    (a) How high is the arrow after 5 sec?
    (b) What is the maximum height that the arrow reaches before it falls back to Earth?
    (c) After how many seconds did the arrow reach the ground?

19. A ball is projected off of a 400-ft-tall building at a velocity of 45 ft/sec. The function $f(x) = -16x^2 + 45x + 400$ (where x is the number of seconds elapsed) models the height of the ball after the toss.
    (a) How high is the ball 2 sec after it is thrown?
    (b) What is the maximum height that the ball reaches?
    (c) After how many seconds did the ball reach the ground?

20. A ball is projected off of a 320-ft-tall building at a velocity of 16 ft/sec. The function $f(x) = -16x^2 + 16x + 320$ (where x is the number of seconds elapsed) models the height of the ball after the toss.
    (a) How high is the ball 2 sec after it is thrown?
    (b) What is the maximum height that the ball reaches?
    (c) After how many seconds did the ball reach the ground?

21. The function that relates the price (in dollars) of a certain quality diamond to its weight is $p(w) = 920w^2$, where w is the weight of the diamond in carats. Sketch this function and use it to find the price of a 1.5-carat diamond.

22. The distance required for a certain automobile to come to a stop after the brakes have been applied can be modeled with the function $d(s) = 0.0273s^2$, where s is the speed the car was traveling when the brakes were applied. Sketch this function and use it to find the estimated stopping distance for a car traveling 45 mph.

Use the quadratic formula to solve the following equations. (Round to two decimal places.)

23. $x^2 - 2x - 15 = 0$
24. $2x^2 + 3x - 4 = 0$
25. $3x^2 + 1 = 7x$
26. $x^2 - 18 = 0$
27. $2x^2 - 3x = 9$
28. $7x = 2x^2$

The formula $h = rt - 16t^2$ gives a good approximation of the height h in feet that an object will reach in t seconds, when it is projected upward with an initial speed of r feet per second. Write a model for each of the following situations in problems 29 to 32. Solve each equation, and then use your graphing calculator to sketch the function and verify your answer.

29. If a ball is thrown vertically upward with an initial velocity of 112 ft/sec, at what times will it be 160 feet above the ground?

30. A rifle bullet is fired upward with an initial velocity of 3200 ft/sec. After how many minutes does it hit the ground?

31. A toy rocket is fired upward with an initial velocity of 560 ft/sec. A person in a hot air balloon 1056 ft high sees it pass him on the way up. How many seconds after launch does it pass him on the way down?

32. An object is thrown upward from the ground with an initial velocity of 32 ft/sec. Find the number of seconds it takes the object to reach the ground. What is the maximum height obtained by this object?

Solve the following problems by factoring or using the quadratic formula.

33. In business, the cost of producing a product may be represented by an equation that is quadratic in nature. This is called a *cost equation*. A cost for manufacturing a certain item is defined by the equation $C = x^2 + 16x + 114$, where x represents the units manufactured per week. Determine the number of units that can be manufactured with a weekly cost of $6450.

34. Rock & Roll Chair Company has determined that the cost, $C(x)$, of producing a given number, x, of one model of its chairs can be accurately estimated by the following equation: $C(x) = 2400 - 40x + 2x^2$. If the company has $53,400 to invest in manufacturing these chairs, how many will it be able to manufacture with the money that is available?

35. An appliance manufacturer has determined that the profit, $P(x)$, that it makes in manufacturing and selling x washing machines can be determined from the following equation: $P(x) = 3x^2 + 240x - 1800$. How many machines must the company manufacture and sell in order to make a profit of $70,200?

36. Merlin owns an ice cream stand and has determined that the cost in dollars of operating his business is given by the function $C(x) = 2x^2 - 176x + 160$ where x is the number of ice cream cones sold daily. If the cost of operations for Saturday was $520, how many cones did he sell that day?

37. Any good physics text will explain that as you dive deeper and deeper under water the pressure increases and, as you climb higher and higher above the Earth, the air pressure decreases. The accompanying table gives some data about depth, altitude, and pressure. These numbers have been rounded for convenience but are realistic.

Depth Below Surface of Water		Altitude Above the Earth's Surface	
Depth Below Surface (in ft)	Pressure (in atms)	Altitude (in thousands of ft)	Pressure (in atms)
0	1	0	1.00
66	3	20	0.70
167	5	30	0.30
300	10	50	0.11
500	16	60	0.07

(a) Are either (or both) sets of data approximately linear? How did you decide this?

(b) If you decided that there is a linear relationship for one (or both) sets of data, determine what the slope would be for the line. Do this several times using different pairs of points each time. Are all of the slopes the same? If they aren't exactly equal then explain why.

Section 4-6 Exponential Functions as Models

Mathematical models can be used to predict numbers in growth situations or decay situations using nonlinear functions called **exponential functions**. The growth of bacteria in a culture, compound interest, the growth in demand for electricity, and the decay of radioactive elements are all examples of situations that can be modeled with exponential functions. Exponential functions such as $f(x) = 2^x$ have a fixed positive number other than one as the base, with the variable in the exponent. Exponential functions increase at such a fast rate that it is usually most efficient to graph these functions using a graphing calculator.

Definition Exponential Function

$$f(x) = b^x \quad \text{(if } b \neq 0 \text{ or } 1, \text{ and } x \text{ is a real number)}$$

Example 1 Evaluating an Exponential Function

Evaluate the exponential function $f(x) = 2^x$ using a table and using a graphing calculator.

To evaluate the function using a table, we choose values for x and substitute them into the function.

x	$y = f(x) = 2^x$
-1	$y = f(-1) = 2^{-1} = \dfrac{1}{2}$
0	$y = f(0) = 2^0 = 1$
1	$y = f(1) = 2^1 = 2$
2	$y = f(2) = 2^2 = 4$
3	$y = f(3) = 2^3 = 8$
4	$y = f(4) = 2^4 = 16$

Note that the values of the function double each time the x value increases by 1. Next, let's look at how this affects the graph of the function.

Enter the equation into your graphing calculator using the $\boxed{Y=}$ function as

$$Y_1 = 2 \boxed{\wedge} x$$

Use the $\boxed{\text{GRAPH}}$ button to display the exponential function. (See Figure 4-22.)

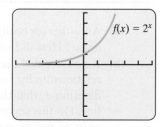

FIGURE 4-22
Example 1.

The graph illustrates how quickly the values of the function increase as the values of the domain x increase.

The function $f(x) = 2^x$ doubles each time the value of x increases by 1. This is the nature of exponential growth—starting slowly but growing faster over time. Predicting the growth of a population or the decay of a radioactive element requires the use of an exponential function called the **natural exponential function** having a base e. The number represented by the symbol e is discussed following the next equation box.

IMPORTANT EQUATIONS

Natural Exponential Function

$$y = f(x) = e^x, \text{ where } x \text{ is a real number}$$

Exponential Growth Equation

$$y = Ae^{rn}$$

Where r = rate of change, A = initial amount, n = time interval, and y = new amount. If $r > 0$, the value of y is increasing and if $r < 0$, the value of y is decreasing.

CALCULATOR MINI-LESSON

e^x

Any *standard scientific calculator* has a second function that can be used for problems that contain a factor with a base e. The $\boxed{e^x}$ function is located over the $\boxed{\ln}$ button on your calculator. In order to calculate the value of e^2, you should enter 2 followed by the $\boxed{\text{2nd}}$ button and then the $\boxed{\ln}$ key (allowing you access to the $\boxed{e^x}$ function). The value 7.389056099 will be displayed without using the $\boxed{=}$ key.

A *graphing calculator (TI-84 Plus)* also has the $\boxed{e^x}$ function above the $\boxed{\ln}$ key. However, you must access the $\boxed{e^x}$ function before you enter the exponent in order to calculate the correct value. After entering the exponent, you must press $\boxed{\text{ENTER}}$ in order to see the correct answer.

The symbol e is an irrational number and is approximately equal to 2.71828. This number has its origin and definition in calculus, so don't worry too much about where it comes from. Suffice it to say that just as the number π occurs as part of the very nature of circles, e occurs as part of the very nature of things that grow or decay exponentially. The formula for exponential growth shown previously can be used to predict bacterial growth rates or the rate of population growth of cities and states. This formula can also calculate rates of decay or decline in growth. If the formula is used for calculating a "growth," then the exponent used will be a positive number ($r > 0$), as in the example that follows. Whenever the formula is used for calculating "declining growth," or "decay," the exponent will be a negative number ($r < 0$). This is shown in an example as well.

Example 2: Bacteria Growth

A bacteria culture contains 120,000 bacteria. It grows at the rate of 15% per hour. How many bacteria will be present in 12 hours?

Use $y = Ae^{rn}$, where $A = 120{,}000$, $r = 0.15$, and $n = 12$.

$$y = 120{,}000e^{(0.15)(12)}$$
$$y = 120{,}000e^{1.8}$$
$$y = 725{,}957.6957$$
$$y \approx 726{,}000 \text{ bacteria}$$

The graph of this equation shown in Figure 4-23 makes it evident that the growth rate of bacteria is very rapid after a time. This is called **exponential growth** and is true for the rate of increase of many living things, even the population of humans on the Earth.

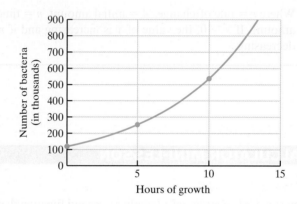

FIGURE 4-23 Example 2.

Example 3: Population Decline

According to government estimates (from www.census.gov), the population of persons aged 5–17 years in the state of West Virginia is declining at an average rate of 0.62% per year. If there were 300,000 persons in this age range in West Virginia in the year 2005, estimate the number of persons in this age range in the year 2015.

Use $y = Ae^{rn}$, where $A = 300{,}000$, $r = -0.0062$, and $n = 10$.

$$y = 300{,}000e^{(-0.0062)(10)}$$
$$y = 300{,}000e^{-0.062}$$
$$y = 281{,}965$$

This shows a decline in this age group of about 20,000 over 10 years.

Example 4: Radioactive Decay

As a result of the Chernobyl nuclear accident on April 26, 1986, radioactive debris was carried throughout the atmosphere. One immediate concern was the impact the debris had on the water supply. The percent y of radioactive material in water after t days is estimated by $y = f(t) = 100e^{-0.1t}$. Estimate the expected percent of radioactive material in the water after 30 days.

To evaluate, substitute 30 in place of t.

$$f(30) = 100e^{(-0.1)(30)} = 4.98$$

This would indicate that about 5% of the radioactive material still contaminated the water supply after 30 days.
Use your graphing calculator to verify this result as follows:
Enter the equation using the Y= function as

$$Y1 = 100 \boxed{e^\wedge} - 0.1x$$

Use the GRAPH button to display the function. You will need to adjust the WINDOW setting to view the graph. (See the settings that follow and graph shown in Figure 4-24 below.) Use the TRACE function to move the cursor to where the x value is 30 on the graph. (In this application, the t variable is plotted on the x-axis.) The coordinates of the cursor will be displayed at the bottom of your screen. Approximate the value of y to the nearest tenth. You can also use the CALCULATE feature or the TABLE feature to verify these points.

WINDOW
$X_{min} = 0$ $Y_{min} = 0$
$X_{max} = 50$ $Y_{max} = 100$
$X_{scl} = 10$ $Y_{scl} = 20$

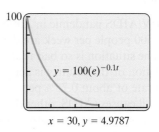

FIGURE 4-24
Example 4.

The values at the bottom of the screen are $x = 30$ and $y = 4.9787...$, so this graph verifies that approximately 5% of the radioactive material still contaminated the water supply after 30 days.

Practice Set 4-6

Draw a sketch of the following exponential functions using a graphing calculator. Find the value of $f(2)$ for each function in problems 1 to 10.

1. $f(x) = 5^x$
2. $f(x) = -3^x$
3. $f(x) = 3^x - 2$
4. $f(x) = 2^x - 1$
5. $f(x) = \left(\dfrac{1}{2}\right)^x$
6. $f(x) = \left(\dfrac{1}{4}\right)^x$
7. $f(x) = e^x$
8. $f(x) = 10e^x$
9. $f(x) = e^{-0.5x}$
10. $f(x) = e^{-0.25x}$

11. Suppose that the future population of Pikeville t years after the year 2010 is modeled by $f(t) = 15,000(1.005)^t$, where $t = 0$ corresponds to the year 2005. Sketch a graph of this model and use it to predict the population of Pikeville in the year 2020.

12. At 6% interest compounded monthly, the future value of a $5000 investment can be calculated using the formula $A = 5000(1.005)^x$, where x represents the number of months that the money is invested. Sketch the graph of this equation and find the value of the investment after 5 years (60 months).

For each problem, write a mathematical model and use it to find the correct answer.

13. The population in Loving County is 15,000 people. The growth rate of the area is estimated to be 6.5% per year. Predict the population in

10 years. Use a graphing calculator to help you sketch the graph of this function.

14. The population of Jefferson City is 95,000. Assuming a growth rate of 2.5% per year, what will its population be in 15 years? Use a graphing calculator to help you sketch the graph of this function.

15. The population of deer in rural Culberson County, Texas, is declining at an estimated rate of 0.8% per year. If this year's deer population is an estimated 12,500 deer, what will the deer population be in 7 years?

16. The growth rate of a colony of rabbits is 40% per month. If the initial population of the colony is 100 rabbits, how many will be in the colony at the end of 6 months?

17. Due to the HIV/AIDS pandemic in Zimbabwe, Africa, over 3500 people per week are dying from AIDS. The situation is so bad that the overall population of the country is actually declining at a rate of about 0.75% per year when deaths due to all causes (AIDS, other diseases, internal warfare, starvation, etc.) are taken into account. If the population of Zimbabwe was about 12,576,800 in January 2004, what will be the population of the country in 10 years? In 50 years? Use a graphing calculator to draw the graph of this function and sketch it.

18. The public school enrollment of a city is 13,500. Enrollment is declining at a rate of 4.5% per year. Predict the enrollment in 5 years. Predict the enrollment in 10 years. Use a graphing calculator to draw the graph of this function and sketch it.

19. According to the Population Reference Bureau (www.prb.org), Russia's annual growth rate during the next 15 years is projected to be -0.05%. If the population of Russia in mid-2010 is approximately 141,900,000 people, use the projected growth rate to estimate the population in mid-2025.

20. According to the Population Reference Bureau (www.prb.org), the Ukraine's annual growth rate during the next 40 years is projected to be -0.66%. If the population of the Ukraine in mid-2010 is approximately 45,900,000, use the projected growth rate to estimate the population in mid-2050.

21. The amount of decay of a certain radioactive element is given by $y = y_0 e^{-0.04t}$, where t equals time in seconds, -0.04 is the decay rate per second, y_0 is the initial amount, and y is the amount remaining. If a given sample has a mass of 155 g, how many grams remain undecayed after 30 seconds? After 2 minutes?

22. Nuclear energy derived from radioactive isotopes can be used to supply power to some spacecraft. If the output of the radioactive power supply for a spacecraft is given by $y = 40e^{-0.004t}$, where y is power in watts and t is time in days, how many watts are generated in the initial output (where $t = 0$)? How many watts are being generated after 2 months (60 days)?

23. A sample of 300 g of lead-210 decays to polonium-210 and is described by the equation $y = 300e^{-0.032t}$, where t is time in years. Find the amount in this 300 g sample after 10 years.

24. The amount (in grams) of a radioactive substance present at time t is $y = 500e^{-0.045t}$, where t is measured in days. Find the amount present after 30 days.

25. Use the exponential growth formula to calculate the amount of money you will have in the bank (y) after 5 years (n) if you deposit \$25,000 (A) into an account that pays 4.5% (r) interest compounded continuously.

26. Mary invests \$3500 at $8\frac{1}{4}\%$ compounded continuously. How much money will be in the account after 6 years?

27. Exponential functions can be used to model the increase in the cost of items given a particular inflation rate. If a loaf of bread currently costs \$2 (initial amount) and the annual inflation rate remains constant at approximately 3.5%, predict the cost of a loaf of bread in 5 years. Construct a table (or use the table function on your calculator) to determine approximately how many years it will be before bread costs more than \$3 a loaf.

28. Exponential functions can be used to model the increase in the cost of items given a particular inflation rate. If a gallon of milk currently costs \$4.95 and the annual inflation rate remains constant at approximately 3.2%, predict the cost of a gallon of milk in 3 years. Construct a table (or use the table function on your calculator) to determine approximately how many years it will be before milk costs more than \$5.50 a gallon.

Chapter 4 Summary

Key Terms, Properties, and Formulas

combined variation
constant of proportionality
constant of variation
dependent variable
direct proportion
direct variation
domain
exponential function

exponential growth
  function
function notation
independent variable
inversely proportional
inverse variation
joint variation
linear function

models
natural exponential
  function
power function
quadratic function
range
relation
roots

Points to Remember

A function is a relation in which, for each value of the independent variable (domain) of the ordered pairs, there is exactly one value of the dependent variable (range).

The domain of a function is the set of all possible values of the independent variable.

The range of a function is the set of all possible values of the dependent variable.

A linear function is written in the form $f(x) = mx + b$, and its graph is a straight line.

A quadratic function $f(x) = ax^2 + bx + c$ will have a maximum value if the graph opens down and a minimum value if the graph opens up.

$$\text{Quadratic formula: } x = \frac{-b \pm \sqrt{b^2 - 4ac}}{2a}$$

Direct variation $y = kx$ Joint variation $y = kxz$

Inverse variation $y = \dfrac{k}{x}$ Combined variation $y = \dfrac{kuv}{wx}$

Power function: $f(x) = cx^k$ where c and k are constants and x is a real number

Exponential function: $f(x) = b^x$ if $b > 0$, $b \neq 1$, and x is a real number

Exponential growth $y = Ae^{rn}$ where $r =$ rate of change, $A =$ Initial amount, $n =$ time interval and $y =$ new amount

Natural exponential function: $f(x) = e^x$ where x is a real number

Chapter 4 Review Problems

1. Let $T(p) = 0.06p$ be a function that calculates sales tax. Explain the meaning of $T(\$12) = \$0.72$.

2. Explain in your own words the meaning of the domain and range of a function.

3. Are there any restrictions on the domain and range of the tax function in problem 1?

For each problem, write a short statement that expresses a possible functional relationship between the quantities. Example: age; shoe size. Solution: As a person's age increases, shoe size also increases up to a point. After that point, shoe size remains relatively constant.

4. (age of a person, height of a person)

5. (time, number of bacteria in a culture)

6. (speed of a car, time it takes to drive from home to school)

Use the following graph to solve problems 7 and 8.

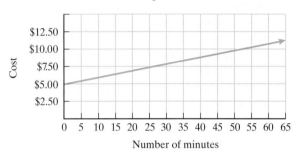

7. Identify the independent and dependent variables of the function illustrated in the graph.

8. Give the domain and range of the function.

9. The formula for calculating the braking distance needed when traveling at a certain speed is $d = \dfrac{x^2}{20} + x$, where x is the speed of the car and d is the stopping distance. Calculate the stopping distance required when a car is traveling 20 mph, 40 mph, 60 mph, and 80 mph. Graph these values and discuss the shape of the graph and its meaning.

10. Let the function $f(x) = 2x - 6$. Find the following values of $f(x)$.

    (a) $f(-2)$ (b) $f(0)$ (c) $f(4)$ (d) $f\left(\dfrac{1}{2}\right)$

Write the following equations in function notation.

11. $y = -3x + 2$
12. $-3y = 2x + 6$
13. $x - y = 8$
14. $3x + 2y = 6$
15. $4x = -y + 1$
16. Let $f(x) = 2x^2 - 1$. Complete Table 4-19 for this function.

TABLE 4-19

x	-2	-1	0	1	2
$f(x)$					

17. The graph that follows defines $g(x)$. Find the following values.

    (a) $g(0)$
    (b) $g(3)$
    (c) $g(-2)$

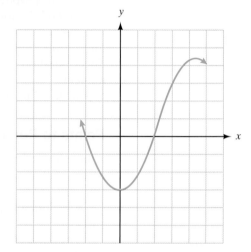

18. The cost of renting a car at Rental City is given by the function $C(m) = \$29 + \$0.06m$, where $m =$ the number of miles driven by the customer. Find $C(120)$ and explain what your answer means.

19. The local community college charges students a $75 activity fee plus $30 for every credit hour a student takes during the semester.

    (a) Write a formula in function notation that can be used to calculate a student's charges.
    (b) Find the cost for a student who enrolls in 12 credit hours for a semester.

20. MidTexas Printing Company prints paperback novels for aspiring young authors. The company charges a $1000 setup fee and $5.50 per book printed. The minimum number of books it will print is 250.

    (a) Write a linear function that models these charges.
    (b) What is the domain for this function?
    (c) Calculate $f(375)$ and explain what it means.

21. Married taxpayers who filed their 2011 taxes jointly used Schedule Y of the taxpayer guide. Those whose taxable incomes were greater than $14,000 but less than $56,800 were required to pay $1400 plus 15% of the amount over $14,000.

    (a) Write a function equation for this tax rule.
    (b) Find $f(\$32,355)$ and explain what your answer means.

22. Kim sells vacuum cleaners. Her salary plan provides her with $1000 per month plus 25% of all sales that she makes during the month.

    (a) Write a linear function to model this problem.
    (b) How much will she make this month if she sells $2500 worth of vacuum cleaners?

23. Use the data in Table 4-20 to write a linear function equation.

    TABLE 4-20

x	−1	0	1	2	3
y	−5	−3	−1	1	3

24. Use the data in Table 4-21 to write a linear function equation.

    TABLE 4-21

x	−4	0	2	6	8
y	2	4	5	7	8

25. The length of your fingernails increases 5 mm per week. Name the independent and dependent variables in this function. Is this a direct or inverse variation problem?

26. The amount of time it takes for a snowman to melt depends on the temperature. Name the independent and dependent variables in this function. Is this a direct or inverse variation problem?

Write a variation equation for each of the following statements.

27. s varies directly as t.

28. z varies inversely as p.

29. m varies jointly as r and s.

30. x varies directly as the cube root of z and inversely as y squared.

Express the following statements as variation equations and find the values of the constants of proportionality and the indicated unknowns.

31. Suppose u varies directly as the cube of v; $u = 8$ when $v = 1$. Find u when $v = 2$.

32. Suppose x varies inversely as the square of y; $x = 2$ when $y = 3$. Find x when $y = 5$.

33. Suppose s varies jointly as t and g; $s = 10$ when $t = 3$ and $g = 2.5$. Find s when $t = 7$ and $g = 6$.

34. N varies directly as the square of L and inversely as the cube of M; $N = 9$ when $L = 2$ and $M = 1$. Find N when $L = 4$ and $M = 2$.

35. The distance a stone falls when dropped off a cliff is directly proportional to the square of the time it falls. If the stone falls 64.4 ft in 2 sec, how far will the stone have fallen in 3 sec?

36. The gravitational force between two objects varies inversely as the square of the distance between the objects. If a force of 50 units results from two objects that are 12 units apart, how much force results from two objects that are 20 units apart?

37. Sketch the graphs of the following quadratic functions, giving the maximum or minimum of the function and the roots of each function.

    (a) $f(x) = 2x^2 - 4x - 6$
    (b) $g(x) = -x^2 - 2x + 8$

38. Sketch the following power functions and evaluate each function at −3.

    (a) $f(x) = \frac{1}{2}x^4$
    (b) $g(x) = -2x^3$

39. Solve this equation using the quadratic formula: $x^2 - 10x - 4 = 0$

40. An object is thrown downward with an initial velocity of 5 ft/sec. The relation between the distance s it travels and time t is given by $s = 5t + 16t^2$. How long does it take the object to fall 74 feet? Use a graphing calculator to draw a sketch of this problem and verify your answer.

41. The weekly profit p (in thousands of dollars) obtained from manufacturing telephones is given by the equation $p = -x^2 + 16x - 24$, where x is the number of telephones in thousands produced weekly. How many telephones should be produced to generate a weekly profit of $40,000?

42. Sketch the following exponential functions and evaluate each function at −2.

    (a) $f(x) = 3^x$
    (b) $g(x) = \left(\frac{1}{2}\right)^x$

43. In 2010, the population of Mexico was 110.7 million, and it was growing at a projected rate of 0.7% annually. (*Source:* Population Reference Bureau, www.prb.org.) The function that expresses this growth is $P(t) = 110.7(1.007)^t$

where t is the number of years after the year 2010. Calculate the predicted population of Mexico in 2025 ($t = 15$) using this function.

44. A bacteria culture contains 25,000 bacteria and grows at the rate of 7.5% per hour. How many bacteria will be present in 24 hours?

45. A bacteria culture contains 500,000 bacteria and is dying off at the rate of 5% per hour. How many bacteria will be left after 48 hours?

46. A city has a population of 850,000, and the city loses population at a rate of 2% per year. Using the exponential population growth model, find its population in 5 years.

Chapter 4 Test

1. The value of an automobile after t years is given by the function $V(t) = \$25{,}000 - \$1500t$. What is the significance of the $25,000 and of the $1500?

2. Write a short statement that expresses a possible functional relationship between these quantities: the square footage of a house and the cost of the house. Identify the independent and dependent variables.

3. The following graph illustrates the number of widgets produced during the first shift on Monday at Daniel's Widget Shop.

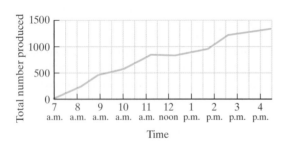

(a) Identify the independent and dependent variables in this graph.
(b) What is the initial value of the function? Is this a linear function?

Write the following equations in function notation. Find $f(-4)$ for each problem.

4. $2x - y = 4$
5. $y = 5$
6. $3x^2 + y = -2$

Use the linear function model to answer problems 7 to 9.

7. A new Ford costs $22,500. The car's value depreciates $2200 per year. Write a formula that expresses its value, V, in terms of its age, t, in years.

8. The cost of a cab ride is modeled by the function $C(d) = 1.50 + 0.75d$, where d is the number of miles traveled and C is the cost in dollars. Evaluate $C(12)$ and explain what the answer tells us.

9. For students at Alvin Community College taking 12–15 credit hours, the charges include a $21 registration fee and $28 per credit hour.
   (a) Write a linear function that models this fee scale.
   (b) What are the domain and range for this function?
   (c) Calculate $f(14)$ and explain what it means.

10. Write a linear function using the data in Table 4-22.

    TABLE 4-22

x	−1	0	1	2
$f(x)$	5	2	−1	−4

Use the definitions of direct and inverse variation to complete problems 11 and 12.

11. The time it takes a bicycler to complete a race depends on his speed. Name the independent and dependent variables in this function. Is this a direct or inverse variation problem?

12. The diameter of a tree is growing by 0.2 in./yr. Name the independent and dependent variables in this function. Is this a direct or inverse variation problem?

Express the following statements as variation equations and find the values of the constants of proportionality and the indicated unknowns.

13. Suppose t varies directly as B and inversely as the square of P; $t = 7.5$ when $B = 6$ and $P = 2$. Find t when $B = 63$ and $P = 3$.

14. Charlotte's earnings as an ICU nurse are directly proportional to the number of hours worked. If she earns $260.50 for 8 hr of work, how much money does she earn for 35 hr of work?

15. The mass of a circular coin varies jointly as its thickness and the square of the radius of the coin. A silver coin 2 cm in radius and 0.2 cm thick has a mass of 26.4 g. What is the mass of a silver coin 4 cm in radius and 0.3 cm thick?

Solve the following problems involving nonlinear functions.

16. Sketch the graph of the quadratic function $f(x) = -x^2 + 9$. Find its maximum or minimum and roots.

17. Sketch the graph of the function $f(x) = -2^x$. Calculate $f(4)$.

18. Suppose that the annual inflation rate is 4.5%. College tuition at State University is currently $5200 per year. Using this inflation rate, we can write a function to predict future tuition costs: $C(t) = 5200(1.045)^t$, where t = number of years from now. Predict the tuition cost at State University for a newborn 18 years from now.

19. In 2011, population of the United States was approximately 312 million. If the population is growing at an annual rate of 1%, predict the population of the United States in 50 years.

20. Use the quadratic formula to calculate the solutions for the equation: $3x^2 - 11x = 7$. (Round to the hundredths place.)

Use the following information to answer problems 21 to 23. A toy rocket is launched from a deck. The function $h = -16t^2 + 96t + 28$ defines the path of the rocket. (Note: h = feet above ground, t = seconds.)

21. Use a graphing calculator to draw a graph of the model.

22. Determine the height of the rocket at $t = 2$ seconds.

23. After how many seconds will the rocket hit the ground?

24. Suppose the population of a certain city is 125,000 with a projected growth rate of 3%. Estimate the population in 3 years.

Suggested Laboratory Exercises

Lab Exercise 1   Tax Bite 2013

The following information was gathered online from the Tax Foundation at www.taxfoundation.org. You may gather more current information of the same kind by visiting this site.

The major portion of a person's work year is spent earning money to pay taxes (federal, state, and local) and to pay for living expenses (food, clothing, and shelter). The Tax Foundation gives the following breakdown for an average calendar year. (See Table 4-23.)

(a) If you start work on January 1, 2013, and must work steady 8-hour work days (no weekends off) until you have worked enough days to pay all of your taxes (rows 1 and 2 of the table), on what calendar date would you have finished paying all of the taxes and started working for yourself?

(b) How much of your work year earns you your housing and food money?

(c) What percent of your work year is spent earning money that would be considered "disposable" income?

(d) If you earned $52,000 per year, approximately how much disposable income would you have?

(e) Wages have continued to increase (on the average) over the past few years, and yet disposable income has been decreasing. What might be some of the reasons for this sad state of affairs?

TABLE 4-23

Expense Category	Days
Federal taxes (include Social Security and Medicare)	79
State and local taxes	41
Housing	62
Health care	52
Food	30
Transportation	30
Recreation*	22
Clothing	13
Miscellaneous*	36
Total time =	365

*These items are not considered to be "essential" and may be called "disposable" income because you are not required to pay for any of these.

Lab Exercise 2 — The Growth of Subway Restaurants

The following information was taken from the www.subway.com website's historical time line data. Table 4-24 lists the number of Subway restaurants worldwide during the years indicated.

TABLE 4-24

Year	1985	1989	1990	1995	2006	2011
Restaurants	500	3000	5000	11,000	25,000	34,000

In the area of mathematics called "statistics," there is a process called linear regression analysis. Using data from Table 4-24 as (x, y) pairs (year, restaurants), it is possible to develop a linear equation that models the growth in the number of Subway restaurants. The equation obtained with this data is $y = 1301.7x - 2{,}585{,}093$, where y is the estimated number of restaurants, and x is the year number.

(a) Evaluate the linear regression equation for the years in Table 4-25 and fill in the estimated number of Subway restaurants rounded to the nearest whole number.

TABLE 4-25

Year	1995	2006	2011	2012
Restaurants				

(b) Is the model equation doing a good job of predicting the number of restaurants for known years (1995, 2006, and 2011)? For which year is it closest? For which year is it the worst predictor?

(c) Go to the Subway restaurant web page and see if the equation makes a good prediction for 2012.

Lab Exercise 3 — Using a Graphing Calculator for Linear Functions

Use a graphing calculator to analyze the following linear function. We will look at the equation, graph, and table in this exercise.

1. Jennifer makes a base pay of $250 per week. She also receives 5% of all sales that she makes during the week. Write a function equation to calculate her weekly paycheck.

2. Enter this function into your calculator using the $\boxed{Y=}$ button. After the equation is in the calculator, set the window as follows:

$$X_{min} = 0 \qquad Y_{min} = 0$$
$$X_{max} = 2500 \quad Y_{max} = 400$$
$$X_{scl} = 500 \qquad Y_{scl} = 50$$

Press the $\boxed{\text{GRAPH}}$ key and use the display to draw a sketch of this function equation. Be sure to label the axes.

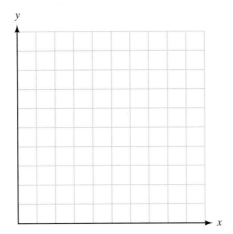

3. Then access the $\boxed{\text{TABLE}}$ function above the $\boxed{\text{GRAPH}}$ key and use this information to fill in the given table. You may need to access the $\boxed{\text{TBLSET}}$ function above the $\boxed{\text{WINDOW}}$ button in order to set up the table to match this chart. Set **TblStart = 0** and **ΔTbl = 500** because the change in sales (x) in the chart is in $500 increments.

Sales	Pay
$ 500	
$1000	
$1500	
$2000	
$2500	

4. Find $f(\$875)$ and explain its meaning for this problem. You can calculate this using the function equation or you can use the $\boxed{\text{CALCULATE}}$ feature over the $\boxed{\text{TRACE}}$ button to find the answer. Under $\boxed{\text{CALCULATE}}$, select **1:value** and press $\boxed{\text{ENTER}}$. Then you will see the graph and at the bottom of the screen, **x=**. Fill in 875 and press $\boxed{\text{ENTER}}$. The answer will be displayed and the location on the graph will be noted with a blinking cursor.

Lab Exercise 4 — Rebound Height and Variation

Get a golf, tennis, or hand ball. Drop (do not throw) the ball from some height. Does it rebound to the same height from which it is dropped? Does the height to which the ball rebounds change if the height from which it is dropped changes? Make some measurements and record the drop height and rebound height (approximately). Is the rebound height directly or inversely related to drop height? Is this relationship linear or not? Which variable is the dependent variable and which is independent? Is this relationship a function or not? Support your answers with data, graphs, and so forth.

Lab Exercise 5 — Name That Function!

Each of the following tables contains a list of words or numbers in the first column and a number in the second column. There is some functional relationship between the two columns in each case. Describe the function clearly in your own words. Complete the charts.

(a)
airplane	2
ship	1
automobile	4
boat	1
airliner	?
train	?

(b)
fossil	2
bones	2
dinosaur	4
paleontologist	4
raptor	?
pterodactyl	?

(c)
1	1
2	4
3	9
4	?
5	?

(d)
1	3
2	5
3	7
4	9
5	?

Lab Exercise 6 — Do You Want Pizza for Lunch?

Pizzas come in a bewildering assortment of sizes and with a plethora of topping choices and combinations. In this activity, you will attempt to clear up some of this confusion.

To begin, suppose that Luigi's Pizza Parlor serves the best-tasting pizza in town. Table 4-26 lists the sizes and prices of pizzas that are available from Luigi's.

1. To begin your research, complete Table 4-26 by filling in the last two columns (base price per square inch of pizza and price per square inch with three additional toppings). Remember, the formula for the area of a circle is $A = \pi r^2$. The base price is for a cheese pizza; all other toppings are at an additional cost. This additional cost is different for each size pizza.
2. Rank (from lowest to highest) the cost per square inch of all the sizes of cheese pizza. Which one is the "best" price? How did you determine this answer?
3. Rank the cost per square inch for pizzas with three extra toppings as you did in part 2. Answer the same questions.
4. Is there anything "odd" about Luigi's price structure?
5. Make up some questions (be reasonable) that you would like for the other groups in your class to answer about this activity. Be sure that you can answer them also.

6. Research the prices of pizzas in your local area. Check with as many restaurants as you can. Determine the "best" buy in your neighborhood.

TABLE 4-26

Size of Pizza	Diameter (all pizzas are round)	Base Price (cheese pizza)	Price per Extra Topping	Base Price per Square Inch	Price per Square Inch with Three Toppings
Small	9 in.	$6.00	$0.80		
Medium	12 in.	$9.75	$1.05		
Large	15 in.	$15.55	$1.75		
Extra Large	17.5 in.	$19.25	$2.50		

Lab Exercise 7

Transformations of Nonlinear Functions

The "basic" nonlinear functions such as $f(x) = x^2$, $f(x) = x^3$, $f(x) = \sqrt{x}$, and $f(x) = |x|$ are all "based" at the origin. These basic graphs can be shifted around the coordinate plane by changes in the equation. Use a graphing calculator to complete the following laboratory exercise.

1. Put the following functions into your graphing calculator in the y_1, y_2, and y_3 locations in the $y =$ list. Then complete the table. Sketch a graph on a piece of graph paper.

Function	y-intercepts	x-intercepts	Domain	Range
$f(x) = x^2$				
$f(x) = x^2 + 1$				
$f(x) = x^2 - 1$				

Answer the following questions:

(a) What do all three graphs have in common?
(b) Are the graphs all congruent (the same size and shape)?
(c) The graph of the general function $f(x) = x^2 + c$ is the graph of the function $f(x) = x^2$ moved $|c|$ units _____ if $c > 0$ and $|c|$ units _____ if $c < 0$.

2. Put the following functions into your graphing calculator in the y_1, y_2, and y_3 locations in the $y =$ list. Then complete the table. Sketch a graph on a piece of graph paper.

Function	y-intercepts	x-intercepts	Domain	Range
$f(x) = x^2$				
$f(x) = (x + 1)^2$				
$f(x) = (x - 1)^2$				

Answer the following questions:

(a) What do all three graphs have in common?
(b) Are the graphs all congruent?
(c) The graph of the general function $f(x) = (x + c)^2$ is the graph of the function

$f(x) = x^2$ moved $|c|$ units _____ if $c > 0$ and $|c|$ units _____ if $c < 0$.

3. Put the following functions into your graphing calculator in the y_1, y_2, and y_3 locations in the $y =$ list. Then complete the table. Sketch a graph on a piece of graph paper.

Function	y-intercepts	x-intercepts	Domain	Range
$f(x) = x^2$				
$f(x) = 2x^2$				
$f(x) = \frac{1}{2} x^2$				

Answer the following questions:

(a) What do all three graphs have in common?
(b) Are the graphs all congruent? How are they different?
(c) The graph of the general function $f(x) = ax^2$ is the graph of the function

$f(x) = x^2$, which is either narrower or wider

If $a > 1$, then the graph is _____ and if $0 < a < 1$, the graph is _____.

(d) Now, go to the $y =$ screen and insert negative signs in front of each equation. What happens to your graphs?

4. Without using your calculator, using the results from 1–3, sketch a graph of these functions:

(a) $f(x) = 2x^2 - 1$
(b) $f(x) = 2(x - 1)^2$
(c) $f(x) = \frac{1}{2}x^2 - 1$
(d) $f(x) = \frac{1}{2}(x - 1)^2$
(e) $f(x) = -2x^2 - 1$
(f) $f(x) = -\frac{1}{2}(x - 1)^2$

Now use your graphing calculator to verify your drawings.

5. Repeat these same steps for the functions $f(x) = x^3$, $f(x) = \sqrt{x}$, and $f(x) = |x|$.

Lab Exercise 8

Patterns in Exponential Functions

Your neighbor is going out of town for the month of July. It costs him $18 per day to board his two small dogs at the local vet. Since this will cost him $558 for the month, he asks you how much you will charge him to feed and walk the dogs for the entire 31 days in July. You tell him that you will work on July 1 for 2 cents, July 2 for 4 cents, July 3 for 8 cents, July 4 for 16 cents, and so on in the same pattern for every day of the month. He thinks this sounds like a real deal and says okay. At the end of the month, how much does he owe you?

1. Complete the table below for the first week of July.
2. Do you see a pattern in the pay amounts? How can you obtain the pay on a given day if you know the pay for the previous day?
3. Calculate the pay for days 8–14, 15–21, and 22–31. Then find the totals due to you for each of the last 3 weeks. What is the total amount in cents that your neighbor owes you at the end of the month? Convert this to dollars and cents.
4. The pay on any day can be written as a power of 2. Write each amount in the table in Problem 1 as a power of 2. For example, $2 = 2^1$, and $4 = 2^2$. Do you notice a pattern in the exponents?
5. Let n = the number of days worked. Using your pattern from problem 4, write an equation for the daily pay in cents as a function of days worked. Name it $P(n)$.
6. Use your function (allowing $n = 25$) to verify the pay for July 25 that you calculated in problem 3.
7. Put the function into your graphing calculator. Sketch it on a piece of graph paper. Is this a linear function? Explain your answer in terms of the graph and the values in the table.
8. Did your neighbor make a good deal? How are you going to spend your money?

Day in July	Pay in cents
1	
2	
3	
4	
5	
6	
7	
Total first week	

Lab Exercise 9

Examining Exponential Decay

When drugs are administered to a patient, the amount of medication in the bloodstream decreases continuously at a constant rate. This type of situation can be modeled with the function used for continuous exponential growth: $y = Ae^{rn}$, where r = rate of change, A = the initial amount of medication (in mg) injected into the bloodstream, and n = time periods. A particular medication is known to decrease at a rate of 6.5% per hour (-6.5% per hour). The common dose of this medication for adult patients is 5 mg per injection.

1. Use these numbers to write an exponential function that describes the decrease of this medication in a patient's bloodstream over time.

2. Now, use this function equation to fill in the missing values in the table.

Hours After Injection	Amount of Medication (in mg)
0	
1	
2	
3	
4	

3. Using these ordered pairs, sketch a graph of this function. What does the *x*-intercept of this function represent?
4. If a patient is given 5 mg of medication at 9 am, how much medication is still in his bloodstream at 11:30 am?
5. Using your exponential equation and a graphing calculator, define Y_1 by entering it into the calculator. If another dose is to be administered when there is approximately 1 mg of the drug present in the bloodstream, at approximately what time should you give the patient another injection? Use your TABLE function to help you answer this question.

5 Mathematical Models in Consumer Math

As we have seen in previous chapters, a function relates two variables to each other. A rule or relationship exists between two sets of data that connects values in one set with values in the other set. Many times, this rule is in the form of an algebraic equation or formula also known as a mathematical model. In this chapter, we will look at various mathematical models in making decisions about your personal finances. We will examine options for purchasing a home, purchasing an automobile, investments, and insurance.

"There is no branch of math, however abstract, which may not some day be applied to phenomena of the real world."
– Nikolai Lobatchevsky

In this chapter

5-1 Mathematical Models in the Business World

5-2 Mathematical Models in Banking

5-3 Mathematical Models in Consumer Credit

5-4 Mathematical Models in Purchasing an Automobile

5-5 Mathematical Models in Purchasing a Home

5-6 Mathematical Models in Insurance Options and Rates

5-7 Mathematical Models in Stocks, Mutual Funds, and Bonds

5-8 Mathematical Models in Personal Income

Chapter Summary

Chapter Review Problems

Chapter Test

Suggested Laboratory Exercises

Chapter 5 Mathematical Models in Consumer Math

Section 5-1 Mathematical Models in the Business World

There are numerous situations in the life of a business where mathematical models are necessary. Profit calculations, retail sales tax, depreciation, markup, and cost of production are a few of the areas where formulas must be developed and used. Some simple business situations will be presented in this section.

Sales Tax A large majority of state legislatures have approved a retail sales tax on purchases made by consumers. Some local governments also have a sales tax that helps fund local programs. Retail merchants are responsible for collecting the appropriate amount of sales tax from their retail sales and forwarding it to their state's Department of Revenue and/or local governments. Sales tax is given in the form of a percent that is applied to the total purchase price of goods.

IMPORTANT EQUATIONS

Sales tax: $t = rp$

where: t = sales tax r = tax rate p = amount of purchase

Since the calculation of sales tax usually involves rounding to the nearest penny, many states develop sales tax charts so cashiers can determine the tax owed without doing any calculating. Cash registers can now be programmed to enter the correct amount of sales tax owed based on items purchased. This is especially helpful when sales tax rates on some items differ from sales tax rates on others. For example, the sales tax on food items might be 2%, while the sales tax rate on nonfood items is 7%. The cash register can be programmed to distinguish between types of items based on the bar code on each item.

Example 1 Sales Tax Revenues

J. Bandera Company operates its business in a location where the state sales tax rate is 4% and the local sales tax rate is 1.5%. The company had $127,250 in retail sales in January. How much sales tax should be sent to the state? How much sales tax should be sent to the local government?

$$t = rp$$

(a) State sales tax = (0.04)($127,250) = $5090
(b) Local sales tax = (0.015)($127,250) = $1908.75

In a small business, the owner records the amount of money that he receives each day for purchases in his store. If the sales tax rate is 7% and he has only the total receipts for the day, he may need to calculate the amount of merchandise sold and the amount of sales tax he collected as separate figures. Look at Example 2 for a model of this problem.

Section 5-1 **Mathematical Models in the Business World** 197

Example 2

Calculating Original Sales from the Sales Total

Ari owns a small general merchandise store. At the end of the day on Saturday, he counted total receipts in his store of $6045.50. If the sales tax in his county is 7%, how much merchandise did Ari sell on Saturday? How much sales tax did he collect?

Let x = the amount of merchandise sold and $0.07x$ = sales tax collected.
  The model for the problem is:

$$\text{merchandise} + \text{tax} = \text{total receipts}$$

Now substitute the variables into the model and solve.

$$x + 0.07x = \$6045.50$$
$$1.07x = \$6045.50$$
$$x = \frac{\$6045.50}{1.07} = \$5650$$

$$\text{Sales tax} = 0.07(\$5650) = \$395.50$$

Therefore, he sold $5650 worth of merchandise and collected $395.50 in sales tax.

Depreciation The fixed assets or plant assets of a business are its buildings, machinery, equipment, land, and similar items that will be used for more than one year. **Depreciation** is the allowance made in bookkeeping for the decreases in value of property through wear, deterioration, or obsolescence. The **straight-line method of depreciation** is the simplest method used in bookkeeping. The **residual value** (also called scrap value or trade-in value) of an asset is its expected value at the end of its useful life. The total allowable depreciation (original cost − residual value) is divided evenly among the number of years of useful life of the property. Depreciation is considered to be an operating expense of a business and is deducted each year from the business profits when determining the year's taxable income.

IMPORTANT EQUATIONS

Straight-Line Method of Depreciation

$$\text{Annual depreciation} = \frac{\text{original value} - \text{residual value}}{\text{number of years}}$$

Example 3

Straight-Line Depreciation

A machine that engraves jewelry is purchased for $6500. It is expected to last 8 years and to have a residual or scrap value of $700. What amount of annual depreciation will be allowed using the straight-line method of depreciation?

Original cost	$6500
Scrap value	− $ 700
Allowable depreciation	$5800

$$\text{Annual depreciation} = \$5800 \div 8 = \$725$$

A yearly depreciation expense of $725 would be claimed each year for 8 years.

Using this depreciation amount, the value of this machine over its useful life can be modeled with a function. If x is the number of years of depreciation and V the value of the machine at the end of each year, then the function is:

$$V(x) = \$6500 - \$725x$$

This function is linear and would have a graph like that shown in Figure 5-1.

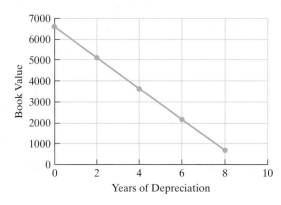

FIGURE 5-1
Straight-line depreciation.

Most businesses use the straight-line method of depreciation for reporting to their stockholders and creditors on their financial statements. Some companies keep a separate set of depreciation records for computing income taxes. They will then use a different method for calculating depreciation. This is done because it provides the most depreciation expense as quickly as possible, which decreases the amount of tax payments. The company can then apply the money saved to best fit their business needs. These alternate methods are more complicated and covered in all basic accounting courses.

Cost of Production and Profit For a manufacturing company to be profitable, the gross income from its sales must exceed the cost of the production of inventory. Fixed costs such as buildings, utilities, and wages must be considered as well as the cost of producing individual items. Models can be developed to show both the cost of manufacturing items and the income from sales. Total cost is a function of both fixed costs and per item costs. Gross income is the selling price of each item times the number of items sold to retailers. Net profit is the difference of these two numbers. In business production, if no items are produced in a month, the business must still pay its rent and other fixed costs. Therefore, if laborers go on strike, the business must find a way to continue production if at all possible. Labor unions have used this fact to help them negotiate for higher wages and better conditions in the workplace.

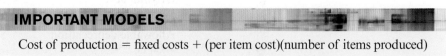

IMPORTANT MODELS

Cost of production = fixed costs + (per item cost)(number of items produced)

Gross income = (selling price)(number of items sold)

Net profit = gross income − cost of production

Section 5-1 **Mathematical Models in the Business World**

Example 4

Calculating Net Profit

The Hutchinson Corporation stamps popular designs on T-shirts and then sells them to retail stores. The fixed costs for this business during a month average $32,850. Shirts can be produced at a cost of $6.78 each. During March, the company produced and sold 12,750 T-shirts. Each shirt was sold for $12. Did the company make a profit in March?

(a) Cost of production = fixed costs + production cost of T-shirts

$$f(x) = \$32{,}850 + \$6.78x$$
$$f(12{,}750) = \$32{,}850 + (\$6.78)(12{,}750)$$
$$f(12{,}750) = \$32{,}850 + \$86{,}445$$
$$f(12{,}750) = \$119{,}295$$

(b) Gross income = (selling price)(number of items sold)

$$g(x) = \$12x$$
$$g(12{,}750) = (\$12)(12{,}750) = \$153{,}000$$

(c) Net profit = gross income − cost of production = $153,000 − $119,295 = $33,705

The company made a profit of $33,705 in March.

Markup A retail store buys its merchandise from a wholesaler or manufacturer and uses a markup rate to determine the selling price of each item. The **markup** on items is calculated by multiplying the markup rate times the price the retailer paid to acquire the item (wholesale price). The markup amount is then added to the cost, and this is the selling price or retail price of the item. Businesses such as furniture stores and jewelry stores may have as much as a 200% markup on their inventory. That is the reason that they can still make a profit even if they have a 50%-off sale!

IMPORTANT MODELS

Markup amount = markup rate × wholesale price

Example 5

Finding the Selling Price of an Item

(a) A manufacturer charges shoe stores $24 for a certain tennis shoe. Bellis Department Store uses a markup rate of 125% on its shoes. What will the selling price of this particular tennis shoe be?

Markup amount = (markup rate)(wholesale price) = (1.25)($24) = $30

Selling price = wholesale price + markup = $24 + $30 = $54

(b) If the suggested retail price of this type of shoe is $60, what is the suggested markup rate?

Markup amount = retail price − wholesale price = $60 − $24 = $36

Markup rate = markup ÷ wholesale price = $\frac{\$36}{\$24}$ = 1.5 = 150%

Practice Set 5-1

Write a mathematical model for each problem and use it to find the correct answer.

1. Sales tax in Reagan County is 7.5% on all purchases. Calculate the sales tax on the following purchase totals (round answers to the nearest cent):

    (a) $25.55   (b) $155.75   (c) $256.18

2. You buy a jet ski that costs $1450 plus 6% sales tax. Find the amount of the sales tax and the total bill.

3. Sales tax in San Patricio County is assessed at 8.25%, where 6.25% is designated for the state and 2% is designated for the county. If Johnson Mercantile has total retail sales of $25,615 during the month, how much sales tax will be owed to the state and how much will be owed to the county?

4. Some tourist areas assess a lodging tax to be paid to the local government to provide extra revenue for local projects such as local roads and utilities. If a family rents a beach cottage for $775 a week in the summer, how much will a 13% lodging tax add to the total cost of the cottage?

5. Myron owns a small sporting goods store. At the end of the week before Christmas, he counted total receipts in his store of $27,129.60. If the sales tax in his county is 8%, how much merchandise did Myron sell that week? How much sales tax did he collect?

6. Bart owns a neighborhood Kwik-E-Mart. At the end of the day on Wednesday, he counted total receipts in his store of $8399.70. If the sales tax on food items is 2%, how much merchandise did Bart sell on Wednesday? How much sales tax did he collect?

7. Selma bought a new television for a total cost of $213.95. If the sales tax at the store is 7%, what was the retail price of the television (price before tax)? (Round to the nearest penny.)

8. The total cost of a meal for a couple at a nice restaurant was $66.95. If the sales tax rate was 3%, what was the actual cost of the food?

9. A computer is purchased for a business at a cost of $2150. The computer is expected to be usable for 5 years and has a scrap value of $500. Using the straight-line method of depreciation, what amount of depreciation will be allowed each year?

10. Tate Realty buys new furniture for its waiting room. If the total cost of the furniture is $5475 and the usable life of the furniture is determined to be 10 years, what is the annual allowable depreciation if the residual value of the furniture is estimated at $850?

11. Dickinson's Law Firm buys new furniture for its waiting room. If the total cost of the furniture is $9450 and the usable life of the furniture is determined to be 7 years, using the straight-line method of depreciation, what is the annual allowable depreciation if the residual value of the furniture is estimated at $1000?

12. Motor Parts of Macon paid $20,000 for a motor vehicle. They expect to use it for 4 years before replacing it. They estimate that when they sell it in 4 years' time, they will get $8000 for it. Using the straight-line method, determine the annual amount of depreciation.

13. Many people prefer to buy used cars because a new car depreciates in value so quickly. Tyler buys a new car and pays $13,500 for it. A year later, he decides to sell it, and the retail price listed in car guides for a car similar to his is $11,880. What percent has the car depreciated in one year?

14. Computers depreciate very quickly because the technology is changing so rapidly. If a certain computer sells for $799 one year and the next year it is only worth $575, what percent has it depreciated in one year?

15. The Kerns Pen Company makes ballpoint pens. The fixed costs of operation for one month average $53,500. The cost of producing one pen is $0.75, and the pens are marketed to retailers for $1.05 each. Give the net profit or loss if the company manufactures zero pens in a month.

16. The Kerns Pen Company makes ballpoint pens. The fixed costs of operation for one month average $53,500. The cost of producing one pen is $0.75, and the pens are marketed to retailers

for $1.05 each. Give the net profit or loss if the company manufactures 150,000 pens in a month.

17. The Kerns Pen Company makes ballpoint pens. The fixed costs of operation for one month average $53,500. The cost of producing one pen is $0.75, and the pens are marketed to retailers for $1.05 each. Give the net profit or loss if the company manufactures 200,000 pens in a month.

18. The Kerns Pen Company makes ballpoint pens. The fixed costs of operation for one month average $53,500. The cost of producing one pen is $0.75, and the pens are marketed to retailers for $1.05 each. How many pens must be sold in order for the company to break even?

19. The inventor of a new game believes that the variable cost of producing the game is $1.95 per unit, and the fixed costs for a month are $8000. He sells the game to retailers for $5.00 each. Give the net profit or loss if the inventor makes zero games in a month.

20. The inventor of a new game believes that the variable cost of producing the game is $1.95 per unit, and the fixed costs for a month are $8000. He sells the game to retailers for $5.00 each. Give the net profit or loss if the inventor makes and sells 150,000 games in a month.

21. The inventor of a new game believes that the variable cost of producing the game is $1.95 per unit, and the fixed costs for a month are $8000. He sells the game to retailers for $5.00 each. Give the net profit or loss if the inventor makes and sells 200,000 games in a month.

22. The inventor of a new game believes that the variable cost of producing the game is $1.95 per unit, and the fixed costs for a month are $8000. He sells the game to retailers for $5.00 each. How many games must be sold in order for the inventor to break even?

23. Strayhorn Enterprises manufactures toy dogs. The fixed costs of operation are $3500 per month. A toy poodle costs $2.55 to manufacture, a toy cocker spaniel costs $3.05 to manufacture, and a toy collie costs $4.15 to manufacture. A poodle sells for $6.50, a cocker spaniel sells for $8.95, and a collie sells for $13.50. Give the net profit or loss if the company manufactures and sells 200 of each type of dog.

24. Strayhorn Enterprises manufactures toy dogs. The fixed costs of operation are $3500 per month. A toy poodle costs $2.55 to manufacture, a toy cocker spaniel costs $3.05 to manufacture, and a toy collie costs $4.15 to manufacture. A poodle sells for $6.50, a cocker spaniel sells for $8.95, and a collie sells for $13.50. Give the net profit or loss if the company manufactures and sells 150 poodles, 200 spaniels, and 300 collies.

25. Lloyd runs a copying service in his home. He paid $3500 for the copier and a lifetime service contract. Each sheet of paper he uses costs $0.01, and he gets paid $0.12 per copy he makes.

    (a) Write a model that can be used to calculate the cost of operations $C(x)$ for the first year. Let x be the number of copies he makes.
    (b) Calculate the total cost if he makes 20,500 copies this year.
    (c) If he is paid for all 20,500 copies, what will be his profit for the year?

26. The Dewhurst Company prints graduation announcements for the local high schools. The setup fee for each high school order is $150, and each announcement printed costs $0.05. The announcements cost the students $0.50 each.

    (a) Write a model that can be used to calculate the cost of printing announcements for East High School. Let x be the number of announcements ordered.
    (b) Calculate the total printing costs if the senior class at EHS orders 4500 announcements.
    (c) When all graduation announcements have been paid for by the students at EHS, what will the profit be for this printing job?

27. A sporting goods store uses a markup rate of 55% on all items. The cost of a golf club to the retailer is $35. What is the selling price of the golf club?

28. Jewelry has a 250% markup. If a diamond ring costs Frank's Jewelers $550, what will the retail price be on this ring?

29. A sweater costs a department store $26.95. The selling price of the sweater is $48.51. What is the markup rate of the sweater?

30. Ace Furniture Emporium charges $1425 for a sofa. If the sofa costs the store $475, what is the markup rate of the sofa?

Section 5-2: Mathematical Models in Banking

The borrowing and lending of money is a practice that dates far back into our history. In the past, many moneylenders charged exorbitant fees for loans. Eventually, regulations concerning lending practices were enacted, and today lending institutions are required to inform consumers of loan rates and methods used to calculate interest owed. Even with these regulations, the "fine print" of many loan agreements is difficult to understand. In this section, we will examine some of the simpler mathematical models that deal with the calculation of interest owed on personal loans and monthly payments.

When you borrow money from a bank or lending institution, you are actually renting the money for your own use. Just as you would expect to pay rent for a house you are living in, you must pay rent, or **interest,** for the money you borrow. The amount of money that you borrow is called the **principal** of a loan. A certain percentage of the loan is charged as interest, and the percent used to calculate this interest is called the **rate.** The rate is usually quoted on a yearly basis. If the interest rate is given as a yearly rate, then the time intervals used in formulas also should be expressed in terms of years.

Simple Interest One method of charging interest on loans is called **simple interest.** The interest here is calculated on the whole principal for the entire length of the loan. The amount due at the end of the loan, called **maturity value** (M), will be the sum of the original loan amount (P) and the total interest due (I), or $M = P + I$. This is the total amount the borrower will be required to repay the lender by the end of the loan period.

IMPORTANT EQUATIONS

Simple Interest Formulas

Simple interest:
$$I = Prt$$

Maturity value:
$$M = P + I$$

where I = interest, P = principal, t = time (in years), and r = rate

---

Example 1

Maturity Value of a Loan

A loan of $1200 is made by Citizens State Bank for 3 years at a simple interest rate of 9.5%. Calculate the total amount that must be repaid at the end of the loan period.

(a) Calculate interest owed:

$$I = Prt$$
$$I = (\$1200)(0.095)(3)$$
$$I = \$342$$

(b) Calculate the maturity value:

$$M = P + I$$
$$M = \$1200 + \$342$$
$$M = \$1542$$

At the end of the 3-year period, a total of $342 in interest has been charged, and the borrower had to repay a total of $1542 to the lender.

Example 2 — Borrowing Money Using a Short-Term Note

Latasha borrowed tuition money from the bank using a 6-month short-term note. The principal of the loan was $720 at an annual simple interest rate of 6.5%. What is the maturity value of this note?

(a) Calculate the interest owed. Since the variable t in the simple interest formula represents time in years and this loan is for 6 months, we must use the fraction $\frac{6}{12} = \frac{1}{2}$ in the formula.

$$I = Prt$$
$$I = (\$720)(0.065)\left(\frac{1}{2}\right)$$
$$I = \$23.40$$

(b) Calculate the maturity value of the note.

$$M = P + I$$
$$M = \$720 + \$23.40 = \$743.40$$

Therefore, at the end of 6 months, Latasha will repay the bank the sum of $743.40.

Some loans are made in the form of notes that are repaid in entirety at the end of the loan period. However, most loans require that the borrower repay the loan by making monthly payments. The monthly payment amount is calculated by dividing the number of payments to be made into the maturity value of the loan. Look at the following example.

Example 3 — Monthly Loan Payments for a Simple Interest Loan

A $2000 loan is made by Bank of the West at a simple interest rate of 8% for 2.5 years. What monthly payment will be required to completely repay the loan in 2.5 years?

(a) Calculate the interest owed:

$$I = Prt$$
$$I = (\$2000)(0.08)(2.5)$$
$$I = \$400$$

(b) Calculate the maturity value:

$$M = P + I$$
$$M = \$2000 + \$400$$
$$M = \$2400$$

(c) Calculate the monthly payments:

$$2.5 \text{ years} = (12)(2.5) = 30 \text{ payments}$$
$$\$2400 \div 30 = \$80.00 \text{ per month}$$

Compound Interest When an investor wishes to earn interest on his savings, he may also consult a bank or credit union. Money can be deposited into a savings account or a money market account. The depositor is paid interest by the bank on the amount of money in the account. Interest may be calculated on a daily, monthly, quarterly, or yearly basis and is credited to these accounts at appropriate intervals. This is called **compound interest,** and most banks and credit unions today pay interest in this manner. This is an advantage to the depositor because interest is paid both on the original deposit and on any interest credited to the account since the original deposit. The effect of paying "interest on interest" causes the account to grow in value more quickly than it would if interest were calculated using the simple interest formula.

To see how this works, let's look at an example. Suppose you deposit $1000 into a savings account that is paying 4.8% interest compounded monthly. This means that the 4.8% rate is divided by 12 (4.8% ÷ 12 = 0.4%) and that monthly rate is applied to your account balance each month. During the first month, your account would earn

$$I = Prt = (\$1000)(0.004/\text{month})(1 \text{ month}) = \$4.00$$

This interest is credited to your account to give you a new balance of $1004. At the end of the second month, the monthly interest rate is again applied to your balance.

$$I = Prt = (\$1004)(0.004/\text{month})(1 \text{ month}) = \$4.016 = \$4.02$$

Your new balance at the beginning of the third month is $1008.02. During the second month, you were paid interest on both the original deposit of $1000 and the interest that you earned during the first month. This is the principle of compounding interest.

Most compound interest is calculated with a table or computer program, but we will look at one formula that can be used to calculate compound interest.

IMPORTANT EQUATIONS

Compound Interest

$$M = P\left(1 + \frac{r}{n}\right)^{nt}$$

where M = maturity value, P = original principal, t = number of years, n = total number of compounding periods per year, and r = annual interest rate

By using the compound interest formula, we can calculate the future value of an investment given a certain interest rate and length of time for the account to grow. It is important to know if interest is paid monthly (12 pay periods per year), quarterly (4 pay periods per year), or daily (360 or 365 days a year). Note that some institutions use the **Banker's Rule** for calculating interest based on 30 days in a month, 360 days in a year. Others use **ordinary interest,** which is based on 365 days per year. If interest is compounded monthly and r represents the yearly interest rate, then we use $\dfrac{r}{12}$ in the formula and let the exponent be $12t$. If interest is compounded quarterly, we use $\dfrac{r}{4}$ and the exponent is $4t$. Using ordinary interest, if interest is compounded daily, we use $\dfrac{r}{365}$ and the exponent is $365t$. Do not round any numbers as you work the problem, but leave them in your calculator and only round the final answers. If you round before you have completed the calculations, your final answer will be incorrect.

IMPORTANT EQUATIONS
Quarterly/Monthly/Daily Compound Interest

Quarterly:
$$M = P\left(1 + \frac{r}{4}\right)^{4t}$$

Monthly:
$$M = P\left(1 + \frac{r}{12}\right)^{12t}$$

Daily:
$$M = P\left(1 + \frac{r}{365}\right)^{365t}$$

where r = annual rate and t = number of years.

Example 4 — Using the Compound Interest Formulas

Calculate the maturity value of a $1000 deposit into a savings account paying 5% after 10 years if it is compounded: (a) quarterly, (b) monthly, and (c) daily using ordinary interest.

(a) $M = P\left(1 + \dfrac{r}{4}\right)^{4t}$

$= 1000\left(1 + \dfrac{0.05}{4}\right)^{4 \cdot 10} = 1000(1 + 0.0125)^{40} = \$1643.62$

(b) $M = P\left(1 + \dfrac{r}{12}\right)^{12t}$

$= 1000\left(1 + \dfrac{0.05}{12}\right)^{12 \cdot 10} = 1000(1 + 0.004166\ldots)^{120} = \$1647.01$

Remember, do not round the number inside the parentheses when you are calculating the answer. Leave it in your calculator and round the final answer to the nearest penny.

(c) $M = P\left(1 + \dfrac{r}{365}\right)^{365t}$

$= 1000\left(1 + \dfrac{0.05}{365}\right)^{365 \cdot 10} = 1000(1 + 0.00013698\ldots)^{3650} = \$1648.66$

Notice that the value of the account compounded monthly is more than the account compounded quarterly and that the account compounded daily is larger than either of the other two. The more often the interest is compounded, the better the return will be.

For the sake of clarity, let us look at a comparison of the growth of a savings account. Suppose that you have $1000 to place in a savings account. You want to open an account, deposit your money, and then let the money just sit in the account for 10 years, untouched. The amount of money that would be in the account after 10 years would depend on the interest rate, of course, but also depends on the compounding rate. The value of $1000 at 5% annual interest for various compounding periods is shown in Table 5-1.

TABLE 5-1 Comparison of the Growth of a Savings Account at 5% Annual Interest

Number of Years	Simple Interest	Compounded Yearly	Compounded Quarterly	Compounded Monthly	Compounded Daily	Compounded Continuously
1	$1050.00	$1050.00	$1050.95	$1051.16	$1051.27	$1051.27
5	$1250.00	$1276.28	$1282.04	$1283.36	$1284.00	$1284.03
10	$1500.00	$1628.89	$1643.62	$1647.01	$1648.66	$1648.72

To emphasize the great difference that compounding can make, look at Table 5-2, which compares $1000 at 10% true simple interest (i.e., no compounding at all) for 100 years with 10% interest compounded yearly for 100 years.

Figure 5-2 shows graphically the difference if you were to compare the rate of growth at simple interest to the rate of growth at compound interest.

TABLE 5-2 Simple Interest versus Compound Interest

Number of Years	Compounded Yearly ($)	Simple Interest Rate ($)
0	1000.00	1000.00
1	1100.00	1100.00
2	1210.00	1200.00
5	1610.51	1500.00
10	2593.74	2000.00
20	6727.50	3000.00
50	117,390.85	6000.00
100	13,780,612.34	11,000.00

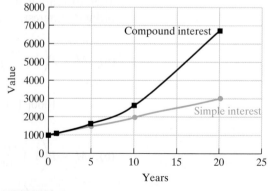

FIGURE 5-2
Simple interest versus compound interest.

It is easy to see from the figure that the line showing the growth in the value of your account at simple interest is a linear function (also called arithmetic growth). The curved line shows the function for the growth your account would have at compound interest. This curve shows what is sometimes called geometric or exponential growth.

CALCULATOR MINI-LESSON

Exponents

Any *standard scientific calculator* has a key that will allow you to find a base raised to any exponent. Look for a key labeled x^y or y^x. To test this key, find the value of 5^3 by entering 5 followed by y^x, 3 and $=$ button. You should have an answer of 125.

A *graphing calculator* (TI-84 Plus) has a key labeled $\wedge$, which can be used to raise any number or variable to any exponent. Calculate the value of 5^3 using this key. Press ENTER to obtain the correct answer of 125.

Example 5 — Saving for a Car

Debbie wants to buy a new car in two years. She sold a story to a national magazine and received a check for $5000 that she wants to save as the down payment on her new car. She plans to deposit it into an account at Guaranty Bank, where the interest rate is 6% compounded monthly. How much money will be in the account at the end of 2 years?

$$P = \$5000 \qquad r = 6\% = 0.06 \qquad t = 2$$

$$M = P\left(1 + \frac{r}{12}\right)^{12t}$$

$$M = \$5000\left(1 + \frac{0.06}{12}\right)^{12(2)}$$

$$M = 5000(1 + 0.005)^{24}$$

$$M = 5000(1.005)^{24}$$

$$M = \$5635.80$$

Debbie will have $5635.80 available for her down payment in 2 years.

The effect of compounding interest provides a higher maturity value in a long-term investment. When interest is paid into your account monthly or quarterly, you are then paid interest on that interest at the end of the next compounding period. Compounding can also occur daily and even continuously. As illustrated in Table 5-1, the more often that interest is compounded, the greater the return on your investment. Look at the next example, which compares the effect of compounding quarterly with compounding monthly.

Example 6 — Compounding Quarterly versus Monthly

Juan is interested in investing in a savings account that currently pays 4.8%. At First National Bank the interest is compounded quarterly, and at Community Bank the compounding is done monthly. If he plans to deposit $2500 for 5 years into the account, how much difference would there be in the maturity value of the investments at the two banks at the end of 5 years?

$$P = \$2500 \quad r = 4.8\% = 0.048 \quad t = 5 \text{ years}$$

(a) First National Bank: compounded quarterly

$$M = P\left(1 + \frac{r}{4}\right)^{4t}$$

$$M = \$2500\left(1 + \frac{0.048}{4}\right)^{4(5)}$$

$$M = \$2500(1 + 0.012)^{20}$$

$$M = \$3173.59$$

(b) Community Bank: compounded monthly

$$M = P\left(1 + \frac{r}{12}\right)^{12t}$$

$$M = \$2500\left(1 + \frac{0.048}{12}\right)^{12(5)}$$

$$M = \$2500(1 + 0.004)^{60}$$

$$M = \$3176.60$$

$3176.60 − $3173.59 = $3.01   (more interest at Community Bank)

By investing his money at Community Bank, Juan will earn more interest given the same interest rate and number of years because of the effect of monthly compounding. The difference ($3.01) is small in this example but increases as the size of the investment and the number of years in the account increase.

If a financial institution compounds interest for finite periods such as daily, quarterly, or monthly, the previous formulas apply. However, as the number of compounding periods per year increases (daily to hourly to . . .), we say that interest is **compounded continuously.**

IMPORTANT EQUATIONS

Continuous Compounding

$$A = Pe^{rt}$$

where P = principal dollars invested and A is the amount accumulated after t years at an annual rate, r.

The formula for continuous compounding is one type of the exponential growth function that was introduced in Chapter 4.

Example 7

Interest Compounded Continuously

Calculate the maturity value of a $5000 certificate of deposit (CD) paying 6.5% interest compounded continuously if it is held for 5 years. Use $A = Pe^{rt}$ where $P = \$5000$, $r = 0.065$, $t = 5$

$$A = 5000e^{(0.065)(5)}$$
$$A = 5000e^{0.325}$$
$$A = \$6920.15$$

The maturity value of this CD will be $6920.15.

Practice Set 5-2

For each problem, write a mathematical model and use it to find the correct answer.

Find the simple interest owed and maturity value of each loan given the following information.

	Principal	Rate	Time
1.	$ 500	9%	1 year
2.	$ 500	9%	6 months
3.	$1200	12%	18 months
4.	$2800	11.5%	5.5 years

5. What was the interest rate on a 2-year loan of $500 if the simple interest paid was $120?

6. What was the interest rate on a 3.5-year, $1000 loan if the simple interest paid was $297.50?

7. What interest will be charged on a $250 loan for 3 months if the rate is 18% simple interest?

8. What interest will be charged on a $1200 loan for 9 months if the rate is 12% simple interest?

Calculate the monthly payments necessary to retire the following loan amounts if simple interest is charged on the note.

	Loan Amount	Rate	Time
9.	$1200	12%	6 months
10.	$1500	8%	9 months
11.	$4500	10%	3 years
12.	$3750	11.5%	1.5 years

13. A refrigerator costs $1600 cash. It may be purchased for $100 down and 24 easy monthly payments of $80 each.
    (a) What are the finance charges (total interest)?
    (b) What is the total cost of this purchase if you finance the refrigerator?
    (c) What simple interest rate is being charged?

14. A man bought his girlfriend a new necklace for $259. Instead of paying cash, he paid $19 down, financed the balance at 12% simple interest, and agreed to make monthly payments for 12 months.
    (a) What is the amount of these monthly payments?
    (b) What will be the total cost of the necklace if he finances it under these terms?

15. Sara made 10 payments of $108.50 to the credit union. The original loan amount was $1000. After she had made all payments, how much interest did she pay? What simple interest rate was she charged by the credit union?

16. Andrew borrowed $1200 on a 6-month note. The maturity value of the note was $1284. What simple interest rate was he charged for this note?

Calculate the maturity value of the following certificate of deposits if the interest is compounded periodically.

	Principal	Rate	Time	Compounded
17.	$ 2000	7.2%	1 year	monthly
18.	$10,500	5.75%	15 years	monthly
19.	$25,000	6%	5 years	quarterly
20.	$ 3500	3.5%	4 years	daily

21. Find the maturity value and amount of interest earned on an investment of $900 invested for 3 years at 2%
    (a) compounded quarterly.
    (b) compounded monthly.

22. Millie invested $1000 at 3.5% interest, compounded quarterly for a term of 18 months.
    (a) What will be the maturity value of her investment at the end of this term?
    (b) How much interest did her investment earn?

23. Teresa is saving money to pay her daughter's college tuition. She hopes to have $6500 saved 5 years from now. If she buys a $5000 certificate of deposit that pays 4% interest compounded monthly, will she have reached her goal of $6500 five years from now?

24. A plaintiff in a lawsuit was awarded $20,000 by a jury. However, an appeal was filed, so the judge ordered that the money be held in an escrow account until the appeal was heard. Two years later, the plaintiff finally won the case and the money was paid out of the escrow account. If the escrow account had been paying interest at 4.5% compounded monthly, how much money did the plaintiff actually receive?

25. Sean plans to invest $10,000 in a savings account that pays 1.5% compounded quarterly. His friend suggests that he invest in an account paying the same interest rate but that pays interest compounded monthly. How much more interest will he earn at the end of 10 years if he follows his friend's advice?

26. Some certificates of deposit (CDs) pay simple interest that is credited to the deposit when the CD matures. CDs generally pay a higher interest rate because the depositor agrees to leave the money in the bank for a specific length of time. There are penalties for early withdrawal of funds from a CD. Compare the maturity value of a 3-year certificate of deposit for $5000 that pays 4% simple interest with the maturity value of $5000 in a savings account for 3 years that pays 2.5% interest compounded monthly.

27. Calculate the amount of money you will have in the bank after 5 years if you deposit $25,000 into a savings account that pays 3.0% interest compounded continuously.

28. Mary invests $3500 at 2.75% compounded continuously. How much money will be in the account after 6 years?

29. Inflationary estimates use the formula for continuous compounding when predicting future values of items. Use this formula to predict the future value (M) in 10 years of a house that is valued at $140,000 if the inflation rate is 2%.

30. Use the continuous compounding formula to predict the future value in 5 years of a gallon of milk if the current price is $4.45 per gallon and the inflation rate is 1.5%.

Section 5-3 Mathematical Models in Consumer Credit

When consumers borrow money to finance purchases such as cars or furniture, the money must be repaid in monthly payments. This method of purchasing items is called **installment buying.** There are generally two types of consumer credit available: **closed-end credit,** which involves the financing of a specific purchase amount for a set number of months; and **open-end credit,** where there are no fixed payments and the consumer continues to make payments until the balance is paid in full. Cars, furniture, and appliances are usually financed with closed-end credit (also known as fixed installment loans) while department store credit cards and bank charge cards such as Visa® or MasterCard® are examples of open-end credit.

Do you have a credit card or are you planning to get one soon? Credit limits are usually established when the credit card is issued. The required monthly payment is generally a percent of the balance due with a minimum payment required for low balances. You can save money if you know the interest rate, the annual fee,

Section 5-3 **Mathematical Models in Consumer Credit**

and the grace period (the interest-free period between purchases and billing given to consumers who pay off their balances entirely) on your cards. The interest rate may be fixed or variable (depending upon the amount you owe or some standard such as a prime rate, the rate banks charge their best customers). How can you save money? First, you can get a card with no annual fee, saving $5 to $50. Next, you can eliminate interest payments by paying off your entire balance each month.

If you cannot pay off the balance each month, credit card companies will charge interest on the balance due. Some companies still use the **unpaid balance method** to calculate the interest that you owe. The interest is calculated on the unpaid balance from the previous month, and any new purchases do not affect the interest charge calculation. The next example will show you how to find the new balance using this method.

Example 1 Calculating Interest Using the Unpaid Balance Method

Lynn received a statement from a department store where she has a charge card. Her previous balance was $380. This month, she made a $20 payment and charged an additional $31.15 to her account. The charge card has an interest rate of 18% per year. Find the finance charge and the new balance.

(a) First, calculate the interest owed using the simple interest formula.

$$I = Prt$$

If the annual interest rate is 18%, then the rate applied each month will be $18\% \div 12 = 1.5\%$.

Interest owed = (unpaid balance)(monthly rate)(time)

$$I = (\$380)(0.015)(1 \text{ month}) = \$5.70$$

(b) Second, calculate the new balance.

$$
\begin{array}{ll}
\$380.00 & \text{(unpaid balance)} \\
-\ \$\ 20.00 & \text{(payment made)} \\
+\ \$\ \ 5.70 & \text{(interest charged)} \\
+\ \$\ 31.15 & \text{(new purchase)} \\
\hline
\$396.85 & \text{(new balance)}
\end{array}
$$

Most credit card companies today use the **average daily balance method** to calculate interest charges. The companies apply the interest rate to the average balance that you carry on your card that month. Therefore, all payments and new purchases during the month are included in the interest calculation. The next example uses the same problem from Example 1 to illustrate this method.

Example 2 Calculating Interest Using the Average Daily Balance Method

Lynn received a statement from a department store where she has a charge card. Her previous balance was $380. She made a $20 payment and charged an additional $31.15 to her account. The charge card has an interest rate of 18% per year. Find the finance charge and the new balance.

June 3	Billing date	
June 12	Payment	$20.00
June 20	Clothes	$31.15
July 3	Next billing date	

(a) First, make a table that shows all transactions that have taken place during the month. Then compute a running balance for the month.

Date	Transaction	Balance
June 3	Opening balance	$380.00
June 12	Payment	$380.00 − $20.00 = $360.00
June 20	Purchase	$360.00 + $31.15 = $391.15

Next, calculate the balance due figures along with the number of days until the balance changed. Multiply each balance by the correct number of days and find the sum of these balances so that an "average daily balance" can be calculated.

Date	Balance Due	Number of Days Until Balance Changed	(Balance Due) × (Number of Days)
June 3	$380.00	9	$380.00 × 9 = $3420.00
June 12	$360.00	8	$360.00 × 8 = $2880.00
June 20	$391.15	13	$391.15 × 13 = $5084.95
	Total:	30 days	$11,384.95

$$\text{Average Daily Balance} = \frac{\text{sum of daily balances}}{\text{days in billing cycle}} = \frac{\$11,384.95}{30} = \$379.50$$

Lynn's finance charges will be calculated on this balance using the simple interest formula.

(b) The annual interest rate is 18%, so calculate the monthly rate as 18% ÷ 12 = 1.5%. Then calculate the finance charge for June as follows:

$$I = Prt = (\$379.50)(0.015)(1 \text{ month}) = \$5.69$$

The balance due on July 3 will be $391.15 + $5.69 = $396.84. ■

During the 1960s, Congress enacted several measures designed to protect the well-being of American consumers. One of these was the Consumer Protection Act or Truth-in-Lending Act that was enacted in 1969. An important feature of this law was the requirement that the total payment, the amount financed, and the finance charges be included in all credit contracts. This requirement was designed to assist consumers in comparing the different types of credit accounts available to them when making a purchase. All loan agreements must state the true annual interest rate known as the **annual percentage rate** or **APR**. Because the calculation of the APR is somewhat complicated, tables have been prepared so consumers can translate the finance charge per $100 to the APR. See Table 5-3. Look at the next example that illustrates how to find the true APR you are being charged using your loan amount and total interest paid.

Section 5-3 **Mathematical Models in Consumer Credit** 213

TABLE 5-3 APR Table (Finance Charge per $100 of Amount Financed)

Number of Payments	Annual Percentage Rate								
	4.0%	4.5%	5.0%	5.5%	6.0%	6.5%	7.0%	7.5%	8.0%
6	1.17	1.32	1.46	1.61	1.76	1.90	2.05	2.20	2.35
12	2.18	2.45	2.73	3.00	3.28	3.56	3.83	4.11	4.39
18	3.20	3.60	4.00	4.41	4.82	5.22	5.63	6.04	6.45
24	4.22	4.75	5.29	5.83	6.37	6.91	7.45	8.00	8.54
30	5.25	5.92	6.59	7.62	7.94	8.61	9.30	9.98	10.66
36	6.29	7.09	7.90	8.71	9.52	10.34	11.16	11.98	12.81
48	8.38	9.46	10.54	11.63	12.73	13.83	14.94	16.06	17.18
60	10.50	11.86	13.23	14.61	16.00	17.40	18.81	20.23	21.66

Number of Payments	Annual Percentage Rate								
	8.5%	9.0%	9.5%	10.0%	10.5%	11.0%	11.5%	12.0%	12.5%
6	2.49	2.64	2.79	2.93	3.08	3.23	3.38	3.53	3.68
12	4.66	4.94	5.22	5.50	5.78	6.06	6.34	6.62	6.90
18	6.86	7.28	7.69	8.10	8.52	8.93	9.35	9.77	10.19
24	9.09	9.64	10.19	10.75	11.30	11.86	12.42	12.98	13.54
30	11.35	12.04	12.74	13.43	14.13	14.83	15.54	16.24	16.95
36	13.64	14.48	15.32	16.16	17.01	17.86	18.71	19.57	20.43
48	18.31	19.45	20.59	21.74	22.90	24.06	25.23	26.40	27.58
60	23.10	24.55	26.01	27.48	28.96	30.45	31.96	33.47	34.99

Example 3 **Finding the APR on a Consumer Loan**

Latoya financed her new $1400 living room set that she bought at Dave's Discount Furniture World. She paid $200 down and financed the balance due for 30 months. Her payments were $44.26 per month. What was the APR for her purchase?

(a) Since the table gives us the finance charges per $100 of amount financed, we must first compute this amount.

$$\text{Payments } \$44.26 \times 30 \text{ months} = \$1327.80 \text{ paid to the store}$$
$$\text{Amount financed after down payment} = -\$1200.00$$
$$\text{Finance charge} = \$\ 127.80$$

$$\text{Finance charge per } \$100 = \frac{\text{finance charge}}{\text{amount financed}} \times \$100$$

$$\text{Finance charge per } \$100 = \frac{\$127.80}{\$1200} \times 100 = \$10.65$$

(b) Look in the APR table and read across the row for 30 monthly payments until you find the number closest to $10.65. This number is $10.66 and is in the column under the heading 8.0%. Therefore, the APR on this loan is approximately 8.0%.

Example 4

Calculating Monthly Payments Using the APR Table

Jeffrey plans to purchase a refrigerator for $1075. The store will finance his purchase at 8.0% APR for 24 months. Calculate the amount of Jeffrey's monthly payment if he makes a down payment on the refrigerator of $75.

(a) First, calculate the balance due that Jeffrey will finance.

$$\text{Cost} - \text{Down Payment} = \$1075 - \$75 = \$1000 \text{ to be financed}$$

(b) Use the APR table to calculate the total amount of interest that Jeffrey will pay on this loan. Go to the row for 24 monthly payments and find the amount in the column labeled 8.0%. This amount, $8.54, is the interest that he will be charged per $100 financed.

$$\text{Total interest due} = \left(\frac{\text{amount financed}}{100}\right) \times \text{amount in chart}$$

$$\text{Total interest due} = \left(\frac{\$1000}{100}\right) \times 8.54 = \$85.40$$

(c) Now, use the total interest due and balance due to calculate his monthly payments.

Balance due	$1000.00
Total interest due	+ $ 85.40
Total needed to pay off loan	$1085.40

$1085.40 ÷ 24 payments = $45.23 per month

The actual calculated amount is $45.225, but in order to ensure that the entire loan is paid, the company will round up when determining monthly payments and adjust the last payment to make the payoff exact.

Practice Set 5-3

Find the finance charge on each of the following open-end charge accounts using the unpaid balance method of calculating interest.

1. Unpaid balance: $350.75
   annual interest rate: 18%

2. Unpaid balance: $1230.40
   annual interest rate: 21.6%

3. Unpaid balance: $655.90
   annual interest rate: 19.8%

4. Unpaid balance: $540.35
   annual interest rate: 21%

Complete each of the following tables in problems 5 to 10 showing the unpaid balance at the end of each month. Use the unpaid balance method to calculate interest. Assume an 18% annual interest rate.

5.

Month	Unpaid Balance at Beginning of Month	Finance Charge	Purchases	Purchases	Payments	Unpaid Balance at End of Month
June	$350.10		$ 25.00	$36.75	$ 75.00	
July			$150.40	$ 0	$200.00	
August			$208.75	$55.40	$150.00	

6.

Month	Unpaid Balance at Beginning of Month	Finance Charge	Purchases	Purchases	Payments	Unpaid Balance at End of Month
January	$3071.10		$ 25.00	$ 0	$500.00	
February			$125.00	$13.85	$350.00	
March			$ 45.98	$16.79	$375.00	

7.

Month	Unpaid Balance at Beginning of Month	Finance Charge	Purchases	Purchases	Payments	Unpaid Balance at End of Month
October	$59.75		$325.88	$ 34.98	$ 50.00	
November			$ 59.06	$295.60	$150.00	
December			$155.67	$230.75	$ 50.00	

8.

Month	Unpaid Balance at Beginning of Month	Finance Charge	Purchases	Purchases	Payments	Unpaid Balance at End of Month
April	$312.88		$34.15	$ 0	$125.00	
May			$45.67	$165.98	$125.00	
June			$34.99	$ 90.45	$125.00	

9.

Month	Unpaid Balance at Beginning of Month	Finance Charge	Purchases	Purchases	Payments	Unpaid Balance at End of Month
June	$1295.00		$ 65.00	$ 0	$120.00	
July			$1325.00	$153.85	$350.00	
August			$ 415.98	$ 65.09	$575.00	

10.

Month	Unpaid Balance at Beginning of Month	Finance Charge	Purchases	Purchases	Payments	Unpaid Balance at End of Month
March	$0		$ 695.00	$ 0	$ 0	
April			$ 925.00	$ 58.85	$20.00	
May			$4816.92	$165.90	$75.00	

Find the average daily balance for each of the following credit card accounts (problems 11 to 16). Assume one month between billing dates and find the finance charge using an 18% annual rate and the average daily balance method.

11. Billing date: June 3 with a balance of $145.90
    June 17: Payment $ 25.00
    June 22: Groceries $121.95

12. Billing date: May 1 with a balance of $315.98
    May 5: Payment $125.00
    May 20: Birthday present $ 45.75

13. Billing date: March 1 with a balance of $ 35.95
    March 5: Payment $ 35.95
    March 12: Clothing $545.75

14. Billing date: July 1 with a balance of $485.52
    July 3: Fireworks $125.00
    July 10: Payment $100.00

15. Billing date: October 5 with a balance of $550.77
    October 5: Plane ticket $405.98
    October 12: Payment $350.00
    October 30: Halloween costume rental $ 45.00

16. Billing date: July 10 with a balance of $126.75
    July 15: Beach cottage rental $950.45
    July 25: Payment $500.00
    August 2: School clothes $245.68

Find the APR (true annual interest rate) to the nearest half percent for each of the following problems (17 to 22).

	Amount Financed	Finance Charge	Number of Monthly Payments
17.	$9500	$968.05	18
18.	$2000	$366.20	48
19.	$1000	$32	6
20.	$1500	$180	30
21.	$7500	$1841	60
22.	$5000	$850	36

23. Carlos financed a $1250 computer with 24 monthly payments of $56.82. Find the finance charge he paid and the APR he was charged.

24. Joaquin bought a wide-screen, flat-panel plasma TV for $4500. He paid $500 down and financed the balance for 36 months. His monthly payments were $127.20. Find the finance charge he paid and the APR he was charged.

25. Lenese is buying new furniture at Discount Furniture Mart because they are offering to finance her purchase at 5.5% APR for 24 months with no down payment. If her total bill is $3560, calculate the amount of interest she will pay by financing her purchase.

26. Fitness, Inc. is advertising a new exercise machine for $1999.99. They will finance the purchase at 6.0% APR for 24 months. How much interest will be paid if someone pays no money down and finances this equipment purchase?

27. Vanida is planning to buy a boat that costs $10,876 including taxes and tags. She will pay 10% down and finance the balance at 7.5% APR for 48 months. Calculate her monthly payments for this boat.

28. Appliance Central is having a special sale that offers customers the opportunity to buy a refrigerator with no down payment and with financing at 5.0% APR for 18 months. If the cost of a new refrigerator is $1895 including tax, how much will the monthly payments be if John-Paul buys this refrigerator?

29. Kimani's new jet ski cost him $4578 including tax and fees. His down payment was $500, and he financed the balance at 4.5% APR for 36 months. How much is his monthly payment?

30. Marisa is purchasing a new dining room set for $4295. She pays $250 down and finances the balance at 4.0% APR for 36 months. Calculate her monthly payments.

Section 5-4 Mathematical Models in Purchasing an Automobile

In the previous section on consumer credit, we discussed installment loans. The most common method for calculating interest on installment loans is by a method known as add-on interest. Add-on interest is an application of the simple interest formula. It is called add-on interest because the interest is added to the amount borrowed so that both the interest and the amount borrowed are paid for over the length of the loan. One of the most common applications of installment loans is in purchasing a car because the interest charged when purchasing a car is add-on interest.

IMPORTANT EQUATIONS
Installment Loan Formulas

Simple interest formula: $I = Prt$

Amount to be repaid: $A = P + I$

Number of payments: $N = 12t$

Amount of each payment: $m = \dfrac{A}{N}$

where P = amount to be financed, r = add-on interest rate, t = time (years) to repay the loan, I = amount of interest, A = amount to be repaid, m = amount of the monthly payment, and N = number of payments

Suppose that you have decided to purchase a new car. You will need to take several things into consideration. You will need to look at the sticker price of the car, the dealer's preparation charges, the tax rate, and the add-on interest rate. You will then need to make an offer for the car. If you are serious about getting the best price, find out the dealer's cost. The dealer's cost is the price the dealer paid for the car. You will need to do some research to find this out. You can consult a reporting service, an automobile association, a credit union, or a periodical like *Consumer Reports*. Most car dealers will accept an offer that is between 3% and 10% over what they actually paid for the car.

Example 1 Making an Offer for a Car

Allison wants to buy a new Honda Civic. The dealer's cost for the car is $12,750. She decides to offer the dealer 5% over the dealer's cost for this car. How much is Allison offering to pay for the car?

(a) Since she is offering 5% over the dealer's cost, we multiply 5% times $12,750.

$$\$12{,}750.00 \times 0.05 = \$637.50$$

To calculate the amount she will offer, add this to the dealer's cost.

$$\text{Amount of offer} = \$637.50 + \$12{,}750.00 = \$13{,}387.50$$

(b) An alternative calculation will help you arrive at the same answer. If we offer 5% over the dealer's cost, then we can multiply the dealer's cost by 105% (100% of dealer cost + 5% over) to calculate the amount that we should offer.

$$\text{Amount of offer} = 105\%(\$12{,}750) = 1.05(\$12{,}750) = \$13{,}387.50$$

Suppose that you have now decided to purchase a new car and you want to determine the monthly payment. Most car purchases require a down payment to be paid on the day of purchase. Then, the balance due on the car is financed for a period of 2 to 6 years, generally. Some dealerships provide on-site financing, but auto loans are available at banks, credit unions, and online lenders. Fees such as the cost of taxes and tags for the car must be included in the cost when you purchase a car. Example 2 demonstrates the process of calculating the balance due to be financed using the down payment, and the calculation of the monthly payment amount.

Example 2 — Purchasing a Car (Add-On Interest)

Sarah wants to buy a new car. The sticker price of the car is $16,000, the cost of tags is $50, and the tax rate is 7%. The dealership agrees to finance the car for 4 years at an interest rate of 5%. If Sarah pays the sticker price for the car and makes a $2500 down payment, how much will she need to finance? Calculate her monthly payment.

(a) We will first need to determine the amount that Sarah needs to finance.

$$\text{Sales tax} = 7\%(\$16{,}000) = \$1120$$

$16,000	Sticker price
$ 1120	Sales tax
+ $ 50	Tags
$17,170	Total cost

Amount to finance:	$17,170	Total cost
	− $ 2500	Down payment
	$14,670	

(b) To calculate her monthly payment, we begin by using the installment loan formula to calculate the interest she must pay.

$$I = Prt$$
$$I = \$14{,}670(0.05)(4) = \$2934.00$$

Next, calculate the total amount to be repaid.

$$A = P + I$$
$$A = \$14{,}670 + \$2934.00 = \$17{,}604.00$$

Finally, calculate the monthly payment.

$$m = \frac{A}{N} = \frac{\$17{,}604.00}{(4)(12)} = \frac{\$17{,}604.00}{48} = \$366.75$$

Therefore, Sarah's monthly payment will be $366.75.

Because a car purchase is a closed-end installment loan, another way to calculate monthly payments including finance charges is by using an APR table. This method of calculating monthly payments was introduced in the last section. Example 3 demonstrates the calculation of a monthly payment using the APR table (see Table 5-3).

Example 3

Purchasing a Car (APR Table)

Jeffrey plans to purchase a used car for $10,750. The dealership will finance his purchase at 8.0% APR for 36 months. Calculate the amount of Jeffrey's monthly payment if he makes a down payment on the car of $2000.

(a) First calculate the balance due that Jeffrey will finance.

Cost of car − down payment = $10,750 − $2000 = $8750 to be financed

(b) Use the APR table to calculate the total amount of interest that Jeffrey will pay on this loan. Go to the row for 36 payments and find the amount in the column labeled 8.0%. This amount, $12.81, is the interest that will be charged per $100 financed.

$$\text{Total interest due} = \left(\frac{\text{amount financed}}{100}\right) \times \text{amount in chart}$$

$$\text{Total interest due} = \left(\frac{\$8750}{100}\right) \times \$12.81 = \$1120.88$$

(c) Now use the total interest due and balance due on the car to calculate his monthly payments.

$8750.00   Balance due
+ $1120.88 Total interest due
$9870.88   Total needed to pay off loan

$9870.88 ÷ 36 payments = $274.20 per month

The actual calculated amount is $274.19111, but in order to ensure that the entire loan is paid, the company will round up when determining monthly payments and adjust the last payment to make the payoff exact.

Many people prefer to **lease** a car rather than buy a car. There are big differences between buying and leasing. Typically, if you were to purchase a new car, you would make a down payment and finance the remaining cost. At the end of the term, the car would be yours. Leasing is essentially renting, with your payment going toward the car's depreciation. If the lease includes a purchase option, you may buy it at the end of a specific time period.

There are short-term cost advantages to leasing. Most leases are for 3 years, so the repairs on the car are covered by factory warranty. The sales tax is also less because you only pay tax on the financed portion of the lease. The monthly payments and the down payment on a leased car are usually less than buying a car. You are only paying for the car's depreciation during the lease, plus the lease company's profit. For example, if the car is initially worth $28,000 but after 24 months is only worth $19,000, you pay only $9000 over 24 months. However, at the end of the lease term you have nothing. This is a major disadvantage in leasing a car. If you buy the car at the end of the lease, it will cost you a lot more than if you had just bought the car in the beginning.

Be sure to read the "fine print" on the lease agreement. For example, some banks charge an acquisition fee to initiate the lease. This adds to the initial cost of the car. Leases are also restrictive. If you exceed the yearly mileage limit, you will be charged an additional amount. The mileage surcharge may be 15¢ or more per mile. This can add up to a major expense very quickly. The car must also be returned in good shape. Any wear and tear on the car will be charged to you at the end of the lease.

Comparing lease offers can be very confusing, making it hard to know if you got a good deal. However, the Federal Reserve Board now requires a standardized federal disclosure form. The dealer is required to give details about specific items in the lease agreement. You must also remember that you will find it difficult to get out of your lease early if you want to—a problem if your driving needs or financial circumstances change.

At the end of the lease you have three options.

- Return the car to the dealer, where it will be checked for mileage and wear and tear. A disposition fee may be charged at the end of the lease to offset costs related to pick-up, reconditioning, and sale of the vehicle.
- Return the car and use it to trade on another lease or purchase. This is only possible if the wholesale value is more than the residual value of the car as stated in your contract.
- Purchase the car from the lease company. Most lease companies will allow you to purchase the car at the residual value (plus maybe a few hundred dollars more for profit for the leasing agent).

Example 4

Leasing a Car

Mr. Martinez plans to lease a red 2012 Ford Mustang 2-door convertible. The sales price of the car is $35,000. The lease agreement states the following:

Due at signing: $525.50 Mileage allowance: 15,000/year
Monthly payment: $525.50 Excess mileage fee: $0.17/mile
36 months Acquisition fee: $795

(a) If Martinez pays the acquisition fee, the amount due at signing, and the 36 monthly payments, how much has he paid to lease this car?

First calculate the total amount paid after he has made 36 lease payments. Then add this to his initial cost and acquisition fee.

$$\text{Monthly payments} = 36(\$525.50) = \$18{,}918.00$$

$ 525.50	Initial payment
$ 795.00	Acquisition fee
+ $18,918.00	Total of 36 payments
$20,238.50	

(b) If he drives 47,550 miles before he turns in the car, how much extra cost will he incur?

Martinez exceeded his mileage allowance for 3 years by 2550 miles (47,550 − 45,000). Therefore, we calculate the additional cost for this mileage by multiplying his excess mileage by $0.17/mile.

$$2550 \text{ miles}(\$0.17/\text{mile}) = \$433.50$$

Therefore, he will owe the company an additional $433.50 when he turns in the car.

One advantage of buying a car is that it is yours! You can customize it and drive it as hard and far as you want, penalty-free. Rather than having infinite payments, buying means you will eventually pay the car off. And if you want to sell it, you can do so at any time, as you are not locked into a contract. However, down payments on cars that you purchase can be substantial. Monthly payments are usually higher than payments for a leased car, and once your warranty expires, you will be responsible for the maintenance costs. When you want to sell the car (or trade it in), you will have to go through the hassle of doing so. And, as an investment, new cars depreciate rather than appreciate. Be sure to shop around and do your research before buying or leasing a car.

So which is better? That depends on your individual situation and needs. Some money advisers say that in general, if you buy a new car every 3 to 4 years, leasing can save you money. However, if you tend to hang on to your car for longer periods and pay it off, buying is a better deal. You will have to decide for yourself by analyzing the advantages and disadvantages of each.

Example 5 Comparing the Purchase of a Car with Leasing a Car

Emily is interested in a 2012 Jeep Wrangler 4WD vehicle and negotiates the purchase price to $22,000. She has $1200 to pay toward the purchase of this car. She wants to compare the cost of purchasing the vehicle with the cost of a lease agreement. The current rate for car loans is 8% APR, and she plans to finance the car for 5 years if she purchases it. The cost of a 36-month lease agreement is stated as follows:

Payment amount: $433.45
Purchase price: $22,000
Total money due at signing: $433.45
Residual value: $10,808
Current retail value: $22,700
Mileage allowance: 15,000/year

Excess mileage fee: $0.17/mile
Lease term: 36 months
Disposition fee: $395
Acquisition fee: $795
Purchase option: $11,307

(a) Purchasing the car by financing it at 8% APR for 5 years:

Emily can pay $1200 down for the car, and she will finance the balance. Therefore, she will finance $22,000 − $1200 = $20,800. Use the APR table to determine the total amount of interest she will pay for this purchase.

$$\text{Total interest due} = \left(\frac{\text{amount financed}}{100}\right) \times \text{amount in chart}$$

$$\text{Total interest due} = \left(\frac{\$20{,}800}{100}\right) \times \$21.66 = \$4505.28$$

The total amount due for the car: $20,800 + $4505.28 = $25,305.28
If we divide this by 60 (payments for 5 years), her monthly payment will be $421.76.

Total paid for the car: $25,305.28 + $1200 = $26,505.28

(b) Lease agreement:

When Emily takes the car home, she will be required to pay the $795 acquisition fee and $433.45 down payment, for a total of $1228.45. She will then make 36 payments of $433.45 for a total of $15,604.20. If she decides at that time to purchase the car, she will owe an additional $11,307.

If she does not finance this amount, the total she has paid for the car by leasing is $1228.45 + $15,604.20 + $11,307 = $28,139.65

Practice Set 5-4

1. What is add-on interest, and how is it calculated?
2. How do you find the monthly payment for add-on interest?
3. Name two advantages of leasing a car.
4. Name two disadvantages of leasing a car.
5. Name two advantages of buying a car.
6. Name two disadvantages of buying a car.

Determine the amount offered for the following cars.

7. Peter offers 5% over dealer cost on a Porsche that has a sticker price of $51,000 and a dealer cost of $46,090.
8. Molly offers 7% over dealer cost on a Volkswagen that has a sticker price of $26,475 and a dealer cost of $16,413.
9. Joseph offers 6% over dealer cost on a Chevrolet that has a sticker price of $32,515 and a dealer cost of $29,460.
10. Megan offers 5% over dealer cost on a Honda that has a sticker price of $24,480 and a dealer cost of $22,998.
11. Charles offers 7% over dealer cost on a BMW that has a sticker price of $63,000 and a dealer cost of $56,904.
12. Laura offers 5% over dealer cost on a Volvo that has a sticker price of $35,750 and a dealer cost of $34,682.

Determine the amount of interest and the monthly payment for each of the car loans in problems 13 to 18.

13. Peter purchases a car for $37,884 at 2.9% add-on rate for 5 years.
14. Molly purchases a car for $16,079 at 3.6% add-on rate for 4 years.
15. Joseph purchases a car for $15,747 at 5% add-on rate for 3 years.
16. Megan purchases a car for $34,631 at 4% add-on rate for 5 years.
17. Charles purchases a car for $50,187 at 1.9% add-on rate for 6 years.
18. Laura purchases a car for $29,108 at 8.5% add-on rate for 4 years.

19. A newspaper advertisement offers a $9000 car for nothing down and 36 easy monthly payments of $317.50. What is the total amount paid for the car, and what is the total finance charge?
20. A newspaper advertisement offers a $4000 used car for nothing down and 36 easy monthly payments of $141.62. What is the total amount paid for the car, and what is the total finance charge?
21. A car dealer will sell you a $16,450 car for $3290 down and payments of $339.97 per month for 48 months. What is the total amount paid for the car, and what is the total finance charge?
22. A car dealer will sell you a used car for $6798 with $798 down and payments of $168.51 per month for 48 months. What is the total amount paid for the car and what is the total finance charge?
23. A newspaper advertisement offers a $9000 car for nothing down and 36 easy monthly payments of $317.50. What is the simple interest rate?
24. A newspaper advertisement offers a $4000 used car for nothing down and 36 easy monthly payments of $141.62. What is the simple interest rate?
25. A car dealer will sell you a $16,450 car for $3290 down and payments of $339.97 per month for 48 months. What is the simple interest rate?
26. A car dealer will sell you a used car for $6798 with $798 down and payments of $168.51 per month for 48 months. What is the simple interest rate?
27. Phil bought a used car for $4500. He paid $500 down and financed the balance for 36 months. His monthly payments were $127.20. Find the finance charge that he paid and the APR that he was charged.
28. Vera is planning to buy a car that costs $17,795 including taxes and tags. She will pay 10% down and finance the balance at 5.0% APR for 60 months. Calculate her monthly payments.
29. Brandon's new motorcycle cost $8578 including taxes and tags. His down payment was $500, and he financed the balance at 4.5% APR for 36 months. How much was his monthly payment?
30. Molly is planning to buy a car that costs $24,095 including taxes and tags. She will pay $5000

down and finance the balance at 6.02% APR for 48 months. Calculate her monthly payments.

Use the following information to answer problems 31 to 33.

Matt plans to lease a 2012 Dodge Dakota pick-up truck 4WD. The sales price of the truck is $25,000. The lease agreement states the following:

Due at signing: $426.71   Mileage allowance: 15,000/year

Monthly payment: $426.71   Excess mileage fee: $0.18/mile

48 months   Acquisition fee: $895

31. If Matt pays the acquisition fee, the amount due at signing, and the 48 monthly payments, how much has he paid to lease this car?

32. If he drives 67,000 miles before he turns in the truck, how much extra cost will he incur?

33. If Matt buys the truck for $25,000 and pays 10% down, how much will he finance? If he finances this balance for 4 years at 2.9% add-on interest, how much will his monthly payments be? What is the total cost of the truck if he finances it?

Use this information for problems 34 to 36.

Mary Alice plans to lease a 2012 Mercedes C230 sport coupe. The sales price of the car is $35,000. The lease agreement states the following:

Due at signing: $615.95   Mileage allowance: 15,000/year

Monthly payment: $615.95   Excess mileage fee: $0.18/mile

48 months   Acquisition fee: $895

34. If Mary Alice pays the acquisition fee, the amount due at signing, and the 48 monthly payments, how much has she paid to lease this car?

35. If she drives 65,200 miles before she turns in the car, how much extra cost will she incur?

36. If Mary Alice buys the car for $35,000 and pays 20% down, how much will she finance? If she finances this balance for 6 years at the special rate of 0.9% add-on interest, how much will her monthly payments be? What is the total cost of the car if she finances it?

Section 5-5 Mathematical Models in Purchasing a Home

For most people, the single largest purchase of their life will be buying a house. This purchase will require many decisions in what may be unfamiliar areas. There are typically three steps in buying and financing a home. First, you will need to find a home you would like to buy and then reach an agreement with the seller on the price of the house. Second, you will need to find a lender to finance the loan. A loan specifically for buying a house is called a **mortgage.** You will need to shop around to find the lender that offers the best interest rates and terms. Finally, you will need to pay certain closing costs in a process called settlement or closing, where the deal is finalized. The term *closing costs* refers to money that must be paid above and beyond the down payment on the house. Attorney's fees, taxes, lender's administration fees, and appraisals are some of the items that may be included in closing costs.

The first step in buying a home is to find a house you can afford and come to an agreement with the seller. A realtor can help you find the type of house you want for the amount you can afford to pay for the house. The type of house you can afford depends greatly on your monthly income and the amount of down payment you can make. A down payment will generally be between 5% and 20% of the selling price of the house. The larger the down payment you make, the lower the interest rate may be, in many cases. If you make a down payment of less than 20% of the selling price, you will probably be required to pay mortgage insurance, which will add to your monthly payment amount. When determining the amount of the loan you need for a house, multiply your percent down payment times the selling price of the home and subtract that amount from the price. For example, a 15% down payment on a $150,000 house would be (0.15)($150,000) = $22,500, leaving a balance to be financed of $150,000 − $22,500 = $127,500.

A useful rule in determining the monthly payment you can afford is given here:

1. Subtract any monthly bills from your gross monthly income.
2. Multiply this amount by 36%.

Your monthly house payment should not exceed this amount.

Another way of deciding if you can afford a house is to multiply your annual salary by 4. The purchase price of the house should not exceed this amount. The method of multiplying 36% of your gross income is the more commonly used way of determining the maximum amount you can afford to pay monthly for a house. The following example illustrates this method.

Example 1

Maximum House Payment

Betty Menendez is looking for a new home in the Abilene area. Her gross monthly income is $4200, and her current monthly payments on bills total $515. What is the maximum amount she should plan to spend for house payments each month?

(a) The first step is to subtract her monthly bills from her gross monthly income.

$$\$4200 - \$515 = \$3685$$

(b) Multiply $3685 by 36%.

$$\$3685.00 \times 0.36 = \$1326.60$$

Her maximum house payment should be $1326.60 per month.

Once you have determined the maximum amount of monthly payment you can afford, you need to know the maximum amount you can borrow. This will help you decide what price range to look at in buying a house. An **amortization table** like the one in Table 5-4 can help determine the amount of money that you can afford to borrow. These tables are typically used by bankers and real estate agents to determine the monthly mortgage payments required at various rates on home loans.

TABLE 5-4 Amortization Table: Monthly Cost to Finance $1000

Rate of Interest	10 Years	15 Years	20 Years	25 Years	30 Years
3.0%	9.66	6.91	5.55	4.74	4.22
3.5%	9.89	7.15	5.80	5.01	4.49
4.0%	10.12	7.40	6.06	5.28	4.77
4.5%	10.36	7.65	6.33	5.56	5.07
5.0%	10.61	7.91	6.60	5.85	5.37
5.5%	10.85	8.17	6.88	6.14	5.68
6.0%	11.10	8.44	7.16	6.44	6.00
6.5%	11.35	8.71	7.46	6.75	6.32
7.0%	11.61	8.99	7.75	7.07	6.65
7.5%	11.87	9.27	8.06	7.39	6.99
8.0%	12.13	9.56	8.36	7.72	7.34
8.5%	12.40	9.85	8.68	8.05	7.69
9.0%	12.67	10.14	9.00	8.39	8.05
9.5%	12.94	10.44	9.32	8.74	8.41
10.0%	13.22	10.75	9.65	9.09	8.78
10.5%	13.49	11.05	9.98	9.44	9.15
11.0%	13.77	11.37	10.32	9.80	9.52

You can use the following formula to determine the amount of money you can afford to borrow using the amortization table.

IMPORTANT EQUATION

Maximum Amount of a Loan Formula

$$\text{Maximum Loan} = \frac{\text{monthly payment you can afford}}{\text{Table 5-4 entry}} \times 1000$$

Example 2 — Finding the Maximum Amount of a Loan

Charles can afford to make a maximum house payment of $900 per month. What is the maximum loan he should seek if he will finance his house over 30 years at 7.5% interest?

The maximum monthly payment is $900. Since the loan is being financed for 30 years at 7.5% interest, the table entry is 6.99. Substitute these values into the formula.

$$\text{Maximum Loan} = \frac{\$900}{6.99} \times 1000 = 128{,}755.36$$

Charles can afford to borrow $128,755.36.

Help is available from many sources to help you make financial decisions and to help you manage your money. One of these is www.fha.com. This site has a mortgage calculator and links to 11 types of calculators that give you information about mortgages. Suppose you wish to find out about what price of house you might be able to afford using this website. Look at the hypothetical situation in Example 3.

Example 3 — How Much Can I Borrow?

First go to the website www.fha.com, click on the "Mortgage Calculator" link, and then scroll down and choose the link called "How much can I borrow?" A simple spreadsheet will come up with slots for certain financial information to be inserted. Suppose you enter the following information (*Note*: Enter "0" in all empty slots on the form.):

Gross monthly income = $3200

Monthly payments: auto loan = $415

Hazard (renters) insurance = $10

Other monthly debts: credit cards, etc. = $125

Current bank interest rate available = 5.0%

Term in years = 30

Now click on the submit tab on the web page.

The results of the spreadsheet estimate is that a loan in the range of $112,000.00 would be possible for you with a monthly payment, including both principal and interest, of about $600.00. If you enter your real data and find that the amount that you could theoretically borrow is too low, you would need to see about reducing your other debts, car loan, credit cards, and so forth in order to get financing for a higher priced home.

Problems 40 and 41 give you some other hypothetical situations to use with this financial calculator.

When you have determined the amount of money that you need to borrow to purchase the house you have chosen, you can use the amortization table to calculate your actual monthly payment, or you can use a formula to calculate the monthly payment. The table and the formula are typically used for a **fixed-rate mortgage**. A fixed-rate mortgage is a loan in which the interest rate is guaranteed over the life of the loan. Fixed-rate loans usually range from 15 to 30 years.

First, we will use Table 5-4 to calculate the amount of the monthly payment for a fixed-rate loan. You will need to know the amount to be financed, the interest rate, and the length of time the loan is to be financed. The simple formula using the table is given in the following box.

IMPORTANT EQUATIONS

Fixed-Rate Mortgage Monthly Payment Formula (Using Amortization Table)

$$P = \frac{A}{1000} \times \text{Table 5-4 entry}$$

where P = monthly payment and A = amount borrowed

Example 4: Payments on a Fixed-Rate Mortgage Using the Amortization Table

The McIntyres are buying a $175,000 home. They plan to make a down payment of $67,000. If they get a fixed-rate mortgage for 30 years at 4.5%, what is their monthly payment? Use the amortization table to calculate their monthly payment.

(a) The amount borrowed (A) is the price of the house minus the down payment.

$$A = \$175{,}000 - \$67{,}000 = \$108{,}000$$

(b) Now use the formula $P = \frac{A}{1000} \times$ *Table 5-4 entry*, substituting the value $A = \$108{,}000$ and looking up the entry in Table 5-4. Since the rate is 4.5% and the loan is being financed for 30 years, the number from the table is 5.07.

$$P = \frac{\$108{,}000}{\$1000} \times 5.07$$

$$P = \$108 \times 5.07 = \$547.56$$

Therefore, the McIntyre family's monthly payment will be $547.56.

The formula to calculate the monthly payment directly is very complicated. However, it can be programmed into a computer, and by entering the required data, the monthly payment is easily calculated. Look at the formula and its use in calculating the monthly payments in Example 5.

IMPORTANT EQUATIONS

Fixed-Rate Mortgage Monthly Payment Formula

$$P = A \left[\frac{\frac{r}{12}\left(1 + \frac{r}{12}\right)^{12t}}{\left(1 + \frac{r}{12}\right)^{12t} - 1} \right]$$

where P = monthly payment, A = amount borrowed, r = annual rate, and t = number of years.

Example 5 — Payments on a Fixed-Rate Mortgage Using the Formula

The Hamiltons are buying a $156,000 home. They plan to make a down payment of $55,000. If they get a fixed-rate mortgage for 30 years at 3.0%, what is their monthly payment? Use the fixed-rate monthly payment formula to calculate their monthly payment.

(a) The amount borrowed (A) is the price of the house minus the down payment.

$$A = \$156{,}000 - \$55{,}000 = \$101{,}000$$

(b) Now use the formula, substituting the values $r = 0.030$, $t = 30$ and $A = \$101{,}000$. Remember not to round any numbers in the problem until you have the final answer.

$$P = 101{,}000 \left(\frac{\frac{0.030}{12}\left(1 + \frac{0.030}{12}\right)^{12 \cdot 30}}{\left(1 + \frac{0.030}{12}\right)^{12 \cdot 30} - 1} \right)$$

$$P = 101{,}000 \left(\frac{0.0025(1.0025)^{360}}{(1.0025)^{360} - 1} \right)$$

$$P = \$425.83$$

Therefore, the Hamilton family's monthly payment will be $425.83.

In addition to conventional fixed-rate loans, there are **adjustable-rate mortgages (ARMs)** available to consumers that generally start out with a lower interest rate than similar fixed-rate mortgages. However, even though the monthly mortgage payments remain the same for a 1-, 2-, or 5-year period, the interest rates change periodically, reflecting prevailing rates in the economy. After the introductory period of fixed mortgage payments, the payments are then adjusted to reflect the new interest rates. There are also additional monthly costs to consider when financing a house, including required homeowners insurance and property taxes. Closing costs when financing a home will also impact the final loan agreement. Some of these charges include a loan origination fee, lender document and underwriting fees, fees to the title company, title insurance fees, appraisal costs, and document recording fees.

Purchasing a house can be a complicated transaction. After you find your "dream house," be sure to shop for the lender offering the best interest rates and

Practice Set 5-5

Answer the following in your own words.

1. What factors are important when looking for a home loan?
2. What factors influence the amount of monthly payment for a home loan?
3. What is the general rule for determining the monthly payment that you can afford?
4. How do you determine the monthly payment for a home loan?
5. How do you determine the maximum amount of money that you can afford to borrow?

Determine the down payment and the amount to be financed for each home described in problems 6 to 11.

6. Cost of house: $148,700
   Down payment: 5% of cost of house

7. Cost of house: $153,250
   Down payment: 10% of cost of house

8. Cost of house: $195,000
   Down payment: 20% of cost of house

9. Cost of house: $179,900
   Down payment: 25% of cost of house

10. Cost of house: $273,450
    Down payment: 15% of cost of house

11. Cost of house: $315,500
    Down payment: 30% of cost of house

In problems 12 to 17 determine the maximum monthly payment you could afford for a house. Use the formula discussed in Example 1.

12. Gross monthly income: $1240
    Current monthly payments: $275

13. Gross monthly income: $985.65
    Current monthly payments: $135.50

14. Gross monthly income: $2750
    Current monthly payments: $560

15. Gross monthly income: $2200
    Current monthly payments: $550

16. Gross monthly income: $1680
    Current monthly payments: $520

17. Gross monthly income: $3700
    Current monthly payments: $650

Determine the monthly payment for each loan described in problems 18 to 23. Use the fixed-rate mortgage monthly payment formula.

18. Cost of house: $148,500
    Interest rate: 4.0%
    Term: 30 years

19. Cost of house: $185,000
    Interest rate: 3.5%
    Term: 25 years

20. Cost of house: $375,000
    Interest rate: 3.5%
    Term: 15 years

21. Cost of house: $110,500
    Interest rate: 4.5%
    Term: 30 years

22. Cost of house: $175,000
    Interest rate: 5%
    Term: 30 years

23. Cost of house: $113,650
    Interest rate: 5.85%
    Term: 20 years

Determine the monthly payment for each loan described in problems 24 to 29. Use Table 5-4 (amortization table formula).

24. Cost of house: $248,500
    Interest rate: 4.0%
    Term: 20 years

25. Cost of house: $85,000
    Interest rate: 3.5%
    Term: 25 years

26. Cost of house: $169,900
    Interest rate: 5.5%
    Term: 20 years

27. Cost of house: $112,000
    Interest rate: 4.5%
    Term: 15 years

28. Cost of house: $175,000
    Interest rate: 3.0%
    Term: 30 years

29. Cost of house: $265,000
    Interest rate: 5.0%
    Term: 30 years

In problems 30 to 35, determine the maximum amount of a loan that you could obtain.

30. Maximum monthly house payment: $1215.75
    Interest rate: 4.5%
    Term: 30 years

31. Maximum monthly house payment: $750
    Interest rate: 3.5%
    Term: 30 years

32. Maximum monthly house payment: $942
    Interest rate: 4.0%
    Term: 20 years

33. Maximum monthly house payment: $1850
    Interest rate: 3.5%
    Term: 25 years

34. Maximum monthly house payment: $1235
    Interest rate: 5.5%
    Term: 15 years

35. Maximum monthly house payment: $635
    Interest rate: 4.5%
    Term: 30 years

36. Charlotte wishes to buy a house that sells for $90,000. Her credit union requires her to make a 20% down payment. The mortgage rate is 5% on a 20-year loan. Determine the amount of the required down payment and find the monthly mortgage. (Use the amortization table.)

37. The Petersons are buying a $250,000 house and planning to make a 25% down payment. The bank is financing the balance at 5.75% for 30 years. Calculate the amount of their monthly payments. (Use the fixed-rate mortgage formula.)

38. Johnny and Eunice Harris are buying a house that sells for $104,000. The bank is requiring a minimum down payment of 20%. The current mortgage rate is 3.5% for 25 years. (Use the amortization table.)

    (a) Determine the amount required for the down payment and the amount of their monthly payment.
    (b) Using the monthly payment amount that you have calculated, determine the total amount that they will have paid for their house in 25 years.
    (c) Use the website www.fha.com or any other mortgage calculator online to verify your answers.

39. The Bergmanns are buying a new condo that is priced at $118,500. They will make a 15% down payment and finance the balance at 4.25% for 30 years. (Use the fixed-rate mortgage formula.)

    (a) Determine the amount required for the down payment and the amount of their monthly payment.
    (b) Using the monthly payment amount you have calculated, determine the total amount they will have paid for their condo in 30 years.
    (c) Use the website www.fha.com or any other mortgage calculator online to verify your answers.

40. Go to www.fha.com and click on the link "How Much Can I Afford?" Enter the following amounts: wages = $6500, student loan = $123, other payments = $40, interest rate = 5.0% for 30 years with 20% down payment, property tax estimate = $1550, and homeowners insurance = $500. What is the conservative estimate for the price of house you might be able to purchase? What is the aggressive estimate?

41. Repeat problem 40, but this time use a 15-year loan instead of a 30-year loan. Note the major changes. Go back now and enter your own data and see how much money you might be able to borrow.

Section 5-6   Mathematical Models in Insurance Options and Rates

In the previous sections, we have discussed mathematical models in buying a car and buying a home. In this section, we will look at mathematical models involved in purchasing insurance. Insurance is an economical way of helping an individual deal with a severe financial loss by pooling the risk over a large number of people. Many people purchase insurance protection, and their payments are pooled in order to provide funds with which to pay those who experience losses. This helps to lower the rates when many different people purchase the same protection and divide the risk. Rates would have to be much higher if only a few people chose to pay in enough money to provide reimbursement for a major loss.

The risk that an insurance company assumes by insuring a person influences how high the rates will be. The insurance company does not know who will suffer loss, but it knows from experience how many losses are expected and the cost of the claim. This is why car insurance is more expensive for inexperienced drivers. Teenagers will pay a much higher premium than an experienced adult driver will. This is determined by looking at the claims filed by young inexperienced drivers. Likewise, life insurance premiums are determined by the age of the client, health concerns, and the lifestyle of the client.

The payment made to purchase insurance coverage is called a premium. An insurance policy defines in detail the provisions and limitations of the coverage. The amount of insurance specified by the policy is called the face value of the policy. Insurance premiums are paid in advance, and the term of the policy is the time period for which the policy will remain in effect. The insurance company is referred to as the insurer or underwriter. The person that purchases the insurance is called the insured or policyholder. When an insured loss occurs, the payment that the insurance company makes to reimburse the policyholder is often called an indemnity.

Individual states have insurance commissions that regulate premium structure and insurance requirements (such as required automobile insurance) passed by their legislatures. This causes insurance practices and premiums to vary from state to state.

We will now look specifically at automobile insurance. Automobile insurance is usually written for a maximum term of one year. The main reason for a one-year policy is the high nationwide accident rate. These accidents increase rates for automobile insurance. Since insurance rates always reflect the risk taken by the insurance company, the driving record and the age of the driver are other major factors in determining the cost of automobile insurance. This is another reason for having a one-year policy, since a driver's record may change over a period of years.

Automobile insurance is divided into two basic categories: liability coverage and comprehensive/collision coverage. Liability insurance pays for bodily injury, property damage, and some additional expenses of other drivers, their passengers, and your passengers when you or a driver covered by your policy causes an accident. Most states require the owners of all vehicles registered in the state to purchase minimum amounts of liability insurance. Each state determines the minimum liability coverage for its drivers. Texas law requires proof of financial responsibility for anyone who drives an automobile in the state. The minimum liability insurance required by law is $20,000 bodily injury per person, $40,000 bodily injury per accident, and $15,000 for property damage per accident. This is commonly called "20/40/15" coverage. Liability insurance does not pay for damage to your own vehicle.

The following two tables show the base annual premiums for automobile liability insurance. Most insurance companies will reduce the total premium cost by a certain percent when two or more vehicles are covered under one insurance policy.

TABLE 5-5 Automobile Liability Insurance Base Annual Premium

Bodily Injury Coverage	Territory 1	Territory 2	Territory 3
20/40	$ 81	$ 91	$221
25/40	$ 83	$ 94	$115
25/50	$ 86	$ 97	$120
50/50	$ 88	$101	$125
50/100	$ 90	$103	$129
100/100	$ 91	$104	$131
100/200	$ 94	$108	$136
100/300	$ 95	$110	$139
200/300	$ 98	$112	$141
300/300	$100	$115	$144

TABLE 5-6 Automobile Liability Base Annual Premiums

Property Damage Coverage	Territory 1	Territory 2	Territory 3
$ 15,000	$83	$ 95	$100
$ 20,000	$85	$ 97	$103
$ 25,000	$86	$ 99	$104
$ 50,000	$87	$101	$107
$ 100,000	$90	$103	$108

Table 5-7 contains excerpts of the driver classifications used to determine the annual cost of liability insurance.

TABLE 5-7 Automobile Insurance Multiples of Base Annual Premium

		Pleasure; Less Than 3 Miles to Work Each Way	Drives to Work 3 to 9 Miles Each Way	Drives to Work 10 Miles or More Each Way	Used in Business
Mature drivers	All others	1.00	1.10	1.40	1.50
Young females	Age 16	1.40	1.50	1.80	1.90
	Age 20	1.05	1.15	1.45	1.55
Young males (married)	Age 16	1.60	1.70	2.00	2.10
	Age 20	1.45	1.55	1.85	1.95
	Age 21	1.40	1.50	1.80	1.90
	Age 24	1.10	1.20	1.50	1.60
Young males (unmarried not principal operator/owner)	Age 16	2.05	2.15	2.45	2.55
	Age 20	1.60	1.70	2.00	2.10
	Age 21	1.55	1.65	1.95	2.05
	Age 24	1.10	1.20	1.50	1.60
Young males (unmarried and owner)	Age 16	2.70	2.80	3.10	3.20
	Age 20	2.55	2.65	2.95	3.05
	Age 21	2.50	2.60	2.90	3.00
	Age 24	1.90	2.00	2.30	2.40
	Age 26	1.50	1.60	1.90	2.00
	Age 29	1.10	1.20	1.50	1.60

As you can see from the table, drivers under 21 are classified according to sex and age. Males under 25 are classified according to marital status and whether they are the owners or principal operators of the cars. Unmarried males may continue to be so classified until they reach age 30. Most mature drivers would be classified "all others." Since younger drivers are more inexperienced and tend to have more accidents, their liability insurance will be much higher.

Example 1 — Automobile Liability Insurance Premiums

Joshua Windsor is 16 years old, unmarried, and drives his car to work at the local grocery store each day, a one-way distance of 6 miles. He is not the principal operator of the car. The area in which he lives is classified Territory 1. Determine the cost of 20/40/15, or the minimum liability coverage required in the state of Texas. (Remember that 20/40/15 means $20,000 bodily injury per person, $40,000 bodily injury per accident, and $15,000 for property damage per accident.) Use Tables 5-5, 5-6, and 5-7 to determine his total annual premium.

First, look at Table 5-5 and determine the premium amount for $20,000 single and $40,000 total bodily injury, or 20/40. This amount is $81.00. Then look at Table 5-6 to determine the premium amount for $15,000 property damage. This amount is $83.00. Then add these two amounts together, $81.00 + $83.00 = $164.00 for the total base premium. Next, look at Table 5-7 to determine what you need to multiply the base amount by in order to get the annual premium amount. Since he is 16 years old, is unmarried, is not the principal operator, and drives 6 miles to work, multiply $164.00 × 2.15 = $352.60. Therefore, Joshua's total annual premium is $352.60.

Example 2 — Automobile Liability Insurance

Gerald Sluder of Sluder's Construction Company was driving a company truck when it hit an automobile, injuring a mother and her son. Sluder Construction was sued for $130,000 in personal injuries and $5500 property damage. The court awarded the victims a $70,000 personal injury judgment and the entire property damage suit. Sluder Construction carries 25/50/15 liability coverage. How much will the insurance company pay? How much of the award will Sluder Construction have to pay?

The insurance company will pay the $50,000 maximum total bodily injury coverage ($25,000 maximum per person) and the total award for property damage ($5500). Therefore, the insurance company will pay a total of $55,500.

Since the court awarded the claimants more than Sluder Construction's bodily injury coverage, the firm will be responsible for the excess:

	$70,000	Bodily injury awarded
−	$50,000	Liability coverage for bodily injury
	$20,000	Amount to be paid by Sluder Construction

The second category of automobile insurance is comprehensive and collision insurance. Comprehensive and collision insurance protect against damage to the policyholder's own automobile. Comprehensive insurance covers damage to the car resulting from fire, theft, vandalism, acts of nature, falling objects, and so on. Collision insurance covers damage resulting from one-car accidents. Collision insurance pays for repairs to the vehicle of the insured when the policyholder is responsible for an accident, when the insured's car was damaged by a hit-and-run driver, or when another driver was responsible for the collision but did not have liability insurance and was unable to pay for the property damage caused by the accident. Collision insurance does not pay for loss or damage to other vehicles or property. Collision policies are available with various deductible amounts. Both collision and comprehensive coverage become considerably less expensive as the policyholder pays a higher percentage of each repair cost. Comprehensive/collision rates also depend on the make, model, and age of the vehicle. For example, newer automobiles are more expensive to repair, so the insurance premiums will cost more than those for an older car.

Uninsured motorist insurance is also available. This offers financial protection for the policyholder's own bodily injuries or those of a family member when hit by a driver who did not carry bodily injury liability insurance. It does not cover the uninsured motorist.

Table 5-8 lists excerpts of base annual comprehensive and collision insurance rates for vehicles of various ages. Automobiles of the current model year are classified as "Age Group 1"; cars in Group 2 (the first preceding model year) and Group 3 (the second preceding model year) both pay the same rate; and models 3 or more years old are classified in Group 4. Each automobile model is assigned an identification letter from A to Z. (Typically, less expensive models are identified by letters near the beginning of the alphabet, and successive letters identify increasingly expensive models.) As in the case of liability insurance, the base annual comprehensive/collision premiums must be multiplied by a multiple reflecting the driver's classification.

TABLE 5-8 Comprehensive and Collision Insurance Base Annual Premiums

Model Class	Age Group	Comprehensive	$250 Deductible Collision	$500 Deductible Collision
A–G	1	$55	$ 82	$ 76
	2, 3	$52	$ 77	$ 73
	4	$49	$ 71	$ 67
J–K	1	$63	$111	$101
	2, 3	$59	$103	$ 95
	4	$54	$ 93	$ 86
L–M	1	$68	$133	$112
	2, 3	$64	$125	$104
	4	$57	$102	$ 94
N–O	1	$77	$140	$126
	2, 3	$70	$130	$117
	4	$62	$115	$105

Example 3

Comprehensive/Collision Insurance Premiums

Kevin Widderich, who is 26 years old and single, uses his car for business. His car is a Nissan Altima, which is considered a Model K, less than 1 year old. Find his annual premium for the following insurance: 50/50/20 liability insurance in Territory 3, with full comprehensive and $500-deductible collision.

Tables 5-5 and 5-6 are used to determine the base annual liability (bodily injury and property damage), as in the previous example. The comprehensive and collision base premiums are shown in Table 5-8. The total of all these base annual premiums must then be multiplied by the driver classification multiple from Table 5-7 to obtain the total annual premium.

```
   $ 125    50/50 Bodily injury
     103    $20,000 Property damage
      63    Comprehensive (Model K, Age Group 1)
+    101    $500-deductible collision (Model K, Age Group 1)
   $ 392    Base annual premium
×   2.00    Driver classification—unmarried male, age 26, business use
   $ 784    Total annual premium
```

Example 4

Comprehensive/Collision Insurance

Suzie Sharpe carries 10/20/5 liability insurance and $250-deductible collision insurance. Suzie was at fault in an accident that caused $900 damage to her own car and $1200 damage to the other vehicle. How much of this property damage will Suzie's insurance company pay? Suppose that a court suit results in a $12,000 award for personal injuries to the other driver. How much would the insurance company pay and how much is Suzie's responsibility?

The property damage to the other car is $1200, which will be paid by the insurance company since Suzie has a $5000 property damage policy. The property damage to Suzie's car is $900 − $250 (deductible) = $650. So the total property damage settlement is $1200 + $650 = $1850. Under Suzie's $10,000 single and $20,000 total bodily injury policy, the maximum amount the insurance company will pay to any one victim is $10,000. Suzie is personally liable for the remaining $2000 of the $12,000 settlement. Therefore, her total obligation is $2000 + $250 (deductible) = $2250.

As you can see, insurance is a complex topic. We have talked exclusively about purchasing automobile insurance in this section, but there are many other types of insurance that you will need to consider in your lifetime. Examples of other kinds of insurance include medical, home, renters, flood, and life insurance. Annuities are a type of retirement insurance. There are many factors to consider when purchasing insurance. You will want to shop around for the best rates and look at the reputation of the insurance company that you are considering.

Practice Set 5-6

1. What is insurance?
2. What is a premium?
3. What is face value?
4. Who is the insurer?
5. What is indemnity?
6. What is liability insurance?
7. What is the minimum amount of liability insurance in the state of Texas?
8. What is comprehensive insurance?
9. What is collision insurance?

Section 5-6 **Mathematical Models in Insurance Options and Rates** 235

Determine the total amount of liability insurance premiums for the following automobile insurance problems using Tables 5-5, 5-6, and 5-7.

10. Driver classification: Male, 47, drives 8 miles to work
    Liability coverage: 25/50/15
    Territory: 1

11. Driver classification: Female, 32, business use
    Liability coverage: 50/50/25
    Territory: 2

12. Driver classification: Unmarried male, 16, drives 3 miles to work, not principal operator
    Liability coverage: 50/100/25
    Territory: 3

13. Driver classification: Female, 25, drives 6 miles to work
    Liability coverage: 20/40/15
    Territory: 2

14. Driver classification: Female, 30, drives 12 miles to work
    Liability coverage: 50/50/25
    Territory: 1

15. Driver classification: Unmarried male, 21, drives 5 miles to work, principal operator
    Liability coverage: 100/200/50
    Territory: 3

Determine the total comprehensive/collision premiums for the following. Use Tables 5-7 and 5-8.

16. Model: G
    Age group: 2
    Deductible on collision: $250
    Driver classification: Female, 54, drives 1 mile to work

17. Model: L
    Age group: 4
    Deductible on collision: $500
    Driver classification: Male, 35, drives 11 miles to work

18. Model: N
    Age group: 1
    Deductible on collision: $500
    Driver classification: Unmarried male, 24, drives 6 miles to work, principal operator

19. Model: A
    Age group: 3
    Deductible on collision: $500
    Driver classification: Female, 47, drives 30 miles to work

20. Model: J
    Age group: 1
    Deductible on collision: $250
    Driver classification: Married male, 20, drives 12 miles to work, principal operator

21. Model: L
    Age group: 2
    Deductible on collision: $250
    Driver classification: Female, 65, drives 15 miles to work

22. Everett McCook, age 42, lives in Territory 3. Each day he drives 5 miles each way to the college where he teaches. His liability insurance includes $50,000 for single bodily injury, $100,000 for total bodily injury, and $15,000 for property damage. Determine his annual payment.

23. Kenny Sluder, age 26, is unmarried and owns a Volkswagen Jetta, considered a Model D car in Age Group 1. Each day he commutes 15 miles each way to work from his home in Territory 1. His auto insurance includes 25/50/15 liability coverage. He also has comprehensive and a $250 deductible on collision coverage. How much is his annual premium?

24. Erika Chavez was liable for an accident in which two people were injured. Her insurance included 50/100/25 liability protection. The court awarded one person $30,000 and the other $65,000 for personal bodily injuries. An award of $25,000 for damages to the other car was also part of the court settlement.

    (a) How much of the court settlement will the insurance company pay?
    (b) How much of the expense must Erika pay personally?

25. During a hurricane, Leslie Hull-Ryde's car was struck by a falling tree, which caused $8000 in damages. Leslie carries comprehensive and $250-deductible collision insurance.

    (a) How much of the damages will her insurance cover?
    (b) How much of the repair bill will Leslie have to pay?

Section 5-7: Mathematical Models in Stocks, Mutual Funds, and Bonds

Good financial planning includes the use of various types of investments as you examine your goals for the future. Planning for retirement at an early age may seem unimportant to you, but as you learned in an earlier section, the effect of compounding can lead to a significant amount of savings after 30 or more years. For example, if at age 16, you were to invest $2000 in a retirement account each year for 6 years ($12,000 invested by age 22), you would have more than $1 million by the time you are 65 if your account returns an average of 10% per year (the historical average for the stock market). In this section, we will examine different investment options that are available as you learn to manage your finances and plan for the future.

One type of very safe investment is a savings account. Though the interest rates may be relatively low, bank accounts up to $100,000 are insured by the federal government, so there is no risk of losing the principal you have invested. The interest rate of the account guarantees a certain increase in the value of your account. You have already examined these interest rates and the effects of compounding in Section 5-2.

Other types of investments are riskier but may offer better returns over a long period of time. These investments include stocks, bonds, and mutual funds.

Stocks A **stock** represents ownership in a particular company. When stocks are openly traded, they "go public," and each ownership unit is called a **share**. For example, if you buy 100 shares of stock in Wal-Mart, you are now a proud owner of that company (so you own some of the bricks in the local store!). Any investor who owns stock in a company is called a **shareholder.**

Most stocks are traded on the New York Stock Exchange (NYSE) or National Association of Securities Dealer Automated Quotations (NASDAQ). Individual investors most often use licensed stockbrokers or online brokerages to purchase shares of stock. Investors pay a brokerage fee for the purchase and sale of stocks. These costs vary based on the brokerage being used and sometimes on the number of shares being traded. Online brokerages may charge a flat fee, such as $9.99 per purchase for all stock purchases regardless of the size of the investment. However, they offer no investment advice or assistance with decisions about investment opportunities. Brokers who work directly with clients and their stock investments may charge a commission fee that is based on the number of shares bought or sold.

Example 1

Purchasing Shares of Stock

An investor wishes to purchase 5000 shares of Disney® stock (NYSE symbol DIS). The current price of Disney stock is $34.46 per share. His stockbroker charges 2% of the cost of the stock as her commission. Find the total cost of this purchase.

(a) First, find the cost of the stock.

$$5000 \text{ shares} \times \$34.46 \text{ per share} = \$172,300$$

(b) Now calculate the stockbroker's commission based on this purchase price.

$$\$172,300 \times 2\% = \$172,300 \times 0.02 = \$3446$$

(c) The total cost of this purchase is the cost of the stock plus the broker's fee.

$$\$172{,}300 + \$3446 = \$175{,}746$$

The value of a stock will fluctuate daily depending on the economy and the particular company. If a large number of people are trying to buy a particular stock when only a limited number of shares are available, the price of the stock will increase. Conversely, if many shareholders are trying to sell a stock and few people are purchasing, the prices will fall. Most large company stocks are considered to be a lower risk than the newer, small company stocks. There is some risk associated with all stocks.

Example 2 Calculating Profit from Stock Transactions

Suppose that you bought 750 shares of Dell stock (NASDAQ symbol DELL) at $35.15. The current price is $15.81, and you decide that you should sell this stock. You use an online brokerage firm that charges $9.99 per transaction. Calculate the amount of profit or loss you will have after you sell your stock.

(a) First, calculate the cost of the original stock purchase and the amount you will receive from the sale.

$$\text{Sale} = 750 \text{ shares} \times \$15.81 = \$11{,}857.50$$
$$\text{Cost} = 750 \text{ shares} \times \$35.15 = \$26{,}362.50$$
$$\text{Difference} = \$14{,}505 \text{ loss}$$

(b) The brokerage costs include a $9.99 fee for the original purchase and an additional $9.99 fee for the sale. Therefore, the total brokerage cost is $19.98.

$$\text{Net loss} = \text{loss} + \text{costs}$$
$$\text{Net loss} = \$14{,}505.00 + \$19.98 = \$14{,}524.98$$

If a company is profitable, it may pay owners all or part of its profits in the form of **dividends.** These dividends will be sent to you quarterly or can be used to automatically reinvest in the company by purchasing additional shares of stock for your portfolio. Newer companies often pay no dividends and choose to use the profits of the company to invest in the growth and expansion of the company. Currently, Wal-Mart pays an annual dividend of $1.46. Therefore, as a shareholder with 100 shares, you would receive a dividend of $\frac{1.46}{4} \times 100 = 36.50$ four times a year.

Over the last 75 years, the return on stocks has averaged more than 10% per year. However, it is important to view investments in stocks as long-term investments (more than 5 years). Monthly and yearly returns can be much more volatile than this average, and you can lose a great deal of money as a short-term investor.

Mutual Funds Most people do not have the time or expertise to choose good stock investments from the thousands of stocks that are available. For this reason, many small investors purchase **mutual funds** instead of individual stocks. A mutual fund is a group of stocks managed by a professional who invests your money in a portfolio of various investments. This portfolio may be a blend of various stocks or bonds or a combination of both. Shares of mutual funds are purchased in a manner similar to stock purchases. You will generally pay a fee to the manager, sometimes called the "load" fee. The money you invest is combined with the money of other investors, and the fund manager buys and sells stocks to try and obtain the best return possible. Some funds specialize in an area such as technology or health care, while others are more broad-based and include stocks from many types and sizes of businesses.

Bonds Another investment option available to the public is **bonds.** Bonds are traded like stocks, and their price is a function of the law of supply and demand. Companies and municipalities may issue bonds when they need to borrow money for a project. For example, if a county needs funds to build a new school, it may issue school bonds. People who buy these bonds are lending money to the county for its project. The county then agrees to pay the purchaser the face value (price paid at purchase) plus interest at a fixed rate. These are long-term investments with a typical term of 10 years or more. If a company goes bankrupt, the bondholders are the first to claim the assets of the company, even before the stockholders of the company. Bonds earn simple interest, which is typically paid periodically to the purchaser. Generally, investments in bonds are considered safer than investments in stocks, but the returns are usually lower. Bonds may also be included in mutual fund portfolios.

Reading a Stock Table Most daily newspapers and online finance websites include stock tables of the most widely traded stocks on the New York Stock Exchange, NASDAQ, and American Stock Exchange. They will also include a list of widely held mutual funds and their current values. There are many facts that can be included in a stock table, so we will examine a few of the most common by looking at Table 5-9.

TABLE 5-9	Reading a Stock Market Table			
NAME	DIV	VOL	LAST	NET CHG.
HOG	0.62	29573	38.90	−0.76

Table 5-9 refers to a recent listing for Harley Davidson (HOG) stock. The heading DIV refers to the dividends that the company pays per share to its stockholders annually. According to this table, Harley Davidson paid a $0.62 dividend per share to its shareholders during the last year. Therefore, if you own 2000 shares of this stock, you would have received $0.62 \times 2000 = \$1240$ during the last 12 months from the company in four quarterly payments. The heading VOL refers to the unofficial daily total of shares traded, expressed in hundreds. A listing of 29573 indicates that $29573 \times 100 = 2,957,300$ shares of Harley Davidson stock were traded yesterday. The heading LAST stands for the price at which shares traded when the market closed yesterday at 4:00 pm. The closing price for Harley was $38.90 per share at the close of trade yesterday. The NET CHG column represents the net change in value from the price at the close of the market two days ago. Since the value in this column is −.76, Harley Davidson's closing stock price of $38.90 was 76 cents lower than its closing price the day before.

Section 5-7 **Mathematical Models in Stocks, Mutual Funds, and Bonds** 239

Example 3 Reading a Stock Table

Use the stock table for Pepsi stock to answer the following questions.

(a) If you owned 2500 shares of Pepsi stock (PEP) last year, what annual dividend did you receive?

NAME	DIV	VOL	LAST	NET CHG.
PEP	2.06	34805	62.19	−0.60

The table lists an annual dividend of $ 2.06 for Pepsi. Therefore, someone with 2500 shares will receive 2500 × 2.06 = $5150.

(b) How many shares of Pepsi stock were traded yesterday?
The volume traded yesterday was 34805 × 100 = 3,480,500 shares.

(c) What was the price at which Pepsi stock closed yesterday?
The closing price of the stock yesterday was $62.19.

(d) What was the price at which Pepsi stock closed the day before yesterday?
Because the net change is −.60, that indicates that the closing price on the previous day was $62.19 + $0.60 = $62.79.

(e) What was the percent decrease in the value of the stock from two days ago until yesterday?
The percent decrease is calculated as follows:

$$\frac{\text{new value} - \text{original value}}{\text{original value}} \times 100 = \frac{\text{net change}}{\text{price before the change}} \times 100$$

$$= \frac{-.60}{62.79} \times 100 = -0.96\%$$

The use of computer spreadsheets is becoming more popular in today's society. People who own shares of stock, for example, can design a spreadsheet program to calculate the value of their portfolios. The mathematical models involved are fairly simple. Once a formula is programmed into the computer, the values of the variables can be changed easily, and the recalculation of the portfolio value is quickly completed.

In Table 5-10, mathematical models are used to calculate the values in cells F2, F3, F4, and F5. Formats will differ according to the specific spreadsheet program, but the models will use the following relationships:

$$F2 = B2*D2 \qquad F3 = B3*D3$$
$$F4 = B4*D4 \qquad F5 = \text{SUM}(F2{:}F4)$$

TABLE 5-10 Stock Portfolio

	A	B	C	D	E	F
1	Stock Name	# of Shares		Price Per Share		Current Value
2	LUB	150		4.63		$ 694.50
3	WEN	250		5.26		$1315.00
4	JNJ	120		64.09		$7690.80
					Total	$9700.30

As the daily price of each stock fluctuates, the price can be updated and the values in column F will automatically be recalculated using the same formulas or mathematical models that were previously programmed into the spreadsheet.

Practice Set 5-7

1. An investor purchases 1500 shares of Exxon Mobil stock (XOM) at $76.61 per share.
   (a) Find the cost of the stock.
   (b) If the broker charges 2% of the cost of the stock, find his commission on this purchase.

2. An investor purchases 500 shares of McDonald's stock (MCD) at $91.81 per share.
   (a) Find the cost of the stock.
   (b) If the broker charges 2.5% of the cost of the stock, find his commission on this purchase.

3. An investor purchases 50 shares of Intel stock (INTC) at $23.90 per share.
   (a) Find the cost of the stock.
   (b) If the broker charges 1.5% of the cost of the stock plus 10¢ per share, find his commission on this purchase.

4. An investor purchases 75 shares of Ryder System stock (R) at $50.81 per share.
   (a) Find the cost of the stock.
   (b) If the broker charges 1.75% of the cost of the stock plus 12.5¢ per share, find his commission on this purchase.

5. An investor purchases 1100 shares of eBay stock (EBAY) at $37.80 per share.
   (a) Find the cost of the stock.
   (b) If the broker charges 2.1% of the cost of the stock, find his commission on this purchase.

6. An investor purchases 700 shares of Microsoft stock (MSFT) at $25.43 per share.
   (a) Find the cost of the stock.
   (b) If the broker charges 2.5% of the cost of the stock, find his commission on this purchase.

7. An investor purchases 40 shares of Honda stock (HNDAF) at $30.02 per share.
   (a) Find the cost of the stock.
   (b) If the broker charges 2.5% of the cost of the stock plus 12.5¢ per share, find his commission on this purchase.

8. An investor purchases 85 shares of Nike stock (NKE) at $50.81 per share.
   (a) Find the cost of the stock.
   (b) If the broker charges 2.75% of the cost of the stock plus 7.5¢ per share, find his commission on this purchase.

9.

NAME	DIV	VOL	LAST	NET CHG.
FordM (F)	0.20	291200	11.11	−0.18

   (a) If you owned 500 shares of Ford stock last year, what annual dividend did you receive?
   (b) How many shares of Ford stock were traded yesterday?
   (c) What was the price at which Ford stock closed yesterday?
   (d) What was the price at which Ford stock closed the day before yesterday?
   (e) What was the percent decrease in the value of the stock from two days ago until yesterday?

10.

NAME	DIV	VOL	LAST	NET CHG.
GeneralElectric (GE)	0.68	253805	16.19	+0.17

   (a) If you owned 2000 shares of General Electric stock last year, what annual dividend did you receive?
   (b) How many shares of General Electric stock were traded yesterday?
   (c) What was the price at which General Electric stock closed yesterday?
   (d) What was the price at which General Electric stock closed the day before yesterday?
   (e) What was the percent increase in the value of the stock from two days ago until yesterday?

11.

NAME	DIV	VOL	LAST	NET CHG.
HewlettPackard (HPQ)	0.48	75750	25.97	+0.33

(a) If you owned 50 shares of Hewlett Packard stock last year, what annual dividend did you receive?

(b) How many shares of Hewlett Packard stock were traded yesterday?

(c) What was the price at which Hewlett Packard stock closed yesterday?

(d) What was the price at which Hewlett Packard stock closed the day before yesterday?

(e) What was the percent increase in the value of the stock from two days ago until yesterday?

12.

NAME	DIV	VOL	LAST	NET CHG.
Sprint (S)	NA	322805	2.80	+0.23

(a) If you owned 750 shares of Sprint stock last year, what annual dividend did you receive?

(b) How many shares of Sprint stock were traded yesterday?

(c) What was the price at which Sprint stock closed yesterday?

(d) What was the price at which Sprint stock closed the day before yesterday?

(e) What was the percent increase in the value of the stock from two days ago until yesterday?

13. Maria plans to invest $2155 in a mutual fund. James Large Cap Fund is currently selling for $8.62 per share. How many shares can she buy at this price?

14. Juan inherited $9280 from his grandmother and is planning to invest in a mutual fund called John Hancock Classic Value Fund. The current price per share of this fund is $23.20. How many shares of this fund can he purchase?

15. Georgio is saving for his son's college education by investing in mutual funds. He sells some land for $10,150 and plans to invest it in Texas Value Capital Funds. The current price of this fund is $29.00 per share. How many shares can he purchase?

16. Joseph is planning to buy shares in a mutual fund that invests in companies in China. He has $10,260 to invest, and his broker recommends Templeton China World A Fund, currently selling for $20.52 per share. How many shares of this fund can he buy?

17. What is a bond? Describe the difference between a bond and a stock.

18. What is a mutual fund? What advantage is there to investing in a mutual fund?

19. If you are a financial advisor, you assist your clients in making decisions about the amount of money they invest in each kind of investment—savings, stocks, bonds, and mutual funds. If you were advising a 25-year-old client who is interested in saving for retirement in 30 years, what advice would you give if he plans to invest at least some money in each type of account?

20. As a financial advisor, what investments would you recommend to a 50-year-old client who plans to retire at age 65? Why?

Section 5-8 Mathematical Models in Personal Income

Mathematics becomes a very powerful tool when it can be applied to our personal finances. Calculating a net salary increase after receiving a 5% raise translates into usable dollars for our own daily lives. The ability to correctly calculate the federal income tax owed may keep us from overpaying taxes. Creating a family budget helps us become good money managers. Although there may be general formulas available for some applications, many times we must modify general formulas and write mathematical models to fit specific circumstances.

Many jobs pay workers a specific salary amount per week or month. Other jobs pay an hourly rate to their employees. However, in some professions, sales jobs pay salaries based on the amount of sales someone generates during the pay period. A real estate salesman may net 4% of the total value of the houses he sells during a month. If he sells no houses, he receives no paycheck. The greater the value of his sales in a month, the larger the paycheck. Car sales and life insurance sales are other examples of professions that base income directly on sales. This method of payment is called **straight commission.** A mathematical model can be written to express this idea.

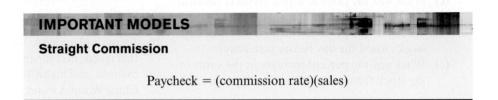

IMPORTANT MODELS

Straight Commission

Paycheck = (commission rate)(sales)

For the realtor making a 4% commission, the equation would be $P = 0.04s$, where P = paycheck and s = amount of sales.

Other types of sales jobs may pay the salesman a base salary and then a percentage of sales made during the pay period. This concept can be modeled with this equation.

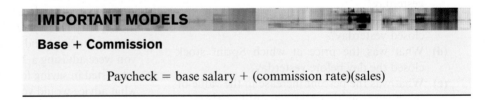

IMPORTANT MODELS

Base + Commission

Paycheck = base salary + (commission rate)(sales)

Waiters and waitresses are paid in a similar manner by earning both tips left by customers and a small base pay for coming to work. These salaries are all functions of the amount of money the customer spends. Many salary agreements have similar patterns, but a mathematical model or equation must be written to calculate the amount of pay earned for a specific job situation.

Example 1

Retail Store Sales

Coleen works in a retail store. She is paid a base salary of $120 per week plus a commission of 10% of the total weekly retail sales she makes.

(a) Write a mathematical model that can be used to calculate Coleen's weekly salary. Let P = weekly pay and s = total retail sales. Written in functional notation, the model would be:

$$P(s) = 120 + 0.10s$$

(b) Draw a graph to illustrate Coleen's potential salary. (See Figure 5-3.)

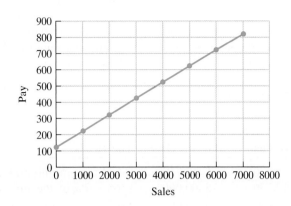

FIGURE 5-3
Retail store sales.

(c) Calculate Coleen's paycheck for the week if she makes $6200 in retail sales the week before Christmas.

$$P(s) = 120 + 0.10s$$
$$P(6200) = 120 + 0.10(6200)$$
$$= 120 + 620$$
$$= \$740 \text{ for the week}$$

You should verify this answer using the graph in Part b.

(d) Coleen needs to gross $500 each week in order to meet her expenses. How much in retail sales should she average each day of a 5-day working week in order to gross $500 for the week?

1. Let $P(s) = \$500$ so we can calculate the weekly sales (s) needed.

$$P(s) = 120 + 0.10s$$
$$500 = 120 + 0.10s$$
$$500 - 120 = 0.10s$$
$$380 = 0.10s$$
$$\$3800 = s$$

2. Since $3800 is the total amount of sales needed for the week, we divide by 5 to calculate the average amount of sales needed each day to reach this goal.

$$\$3800 \div 5 = \$760 \text{ of sales needed per day}$$

Tax laws and requirements seem to change on a yearly basis. The deductions allowed, exemptions allowed, and the tax rate may change as Congress passes new laws. However, the general model for calculating income tax can be used to write a specific equation for tax calculations. Deductions are subtracted from gross income to calculate net income or taxable income. Then, tax tables or tax rate schedules are used to find the amount of income tax owed by an individual.

Example 2

Calculating State Income Tax

Enrico is single and has a gross salary of $68,500 per year. After taking all legal exemptions and deductions, he calculates his taxable income to be $51,275. Enrico's W-2 form shows that his employer withheld $3150 in state taxes for the year. Use the tax schedule provided in Table 5-11 and calculate the amount of state tax that he owes. Will he file for a refund or will he owe an additional payment to the state?

TABLE 5-11 State Tax Rate Schedule

If Your Filing Status Is:	And the Amount on Line 11b Is More Than	But Not Over	The Tax Is:
Single	$ 0	$ 12,750	6% of the amount on line 11b
	$ 12,750	$ 60,000	$765 + 7% of the amount over $12,750
	$ 60,000	$120,000	$4072.50 + 7.75% of the amount over $60,000
	$120,000	—	$8722.50 + 8.25% of the amount over $120,000
Head of household	$ 0	$ 17,000	6% of the amount on line 11b
	$ 17,000	$ 80,000	$1020 + 7% of the amount over $17,000
	$ 80,000	$160,000	$5430 + 7.75% of the amount over $80,000
	$160,000	—	$11,630 + 8.25% of the amount over $160,000
Married filing jointly or qualifying widow(er)	$ 0	$ 21,250	6% of the amount on line 11b
	$ 21,250	$100,000	$1275 + 7% of the amount over $21,250
	$100,000	$200,000	$6787.50 + 7.75% of the amount over $100,000
	$200,000	—	$14,537.50 + 8.25% of the amount over $200,000
Married filing separately	$ 0	$ 10,625	6% of the amount on line 11b
	$ 10,625	$ 50,000	$637.50 + 7% of the amount over $10,625
	$ 50,000	$100,000	$3393.75 + 7.75% of the amount over $50,000
	$100,000	—	$7268.75 + 8.25% of the amount over $100,000

According to the State Tax Rate Schedule, a single taxpayer whose taxable income is between $12,750 and $60,000 must pay $765 plus 7% of the amount over $12,750. Written as a mathematical model:

$$\text{Tax} = \$765 + 0.07(I - \$12{,}750) \quad \text{where } I = \text{taxable income}$$

Completing the formula:

$$\text{Tax} = \$765 + 0.07(\$51{,}275 - \$12{,}750)$$
$$= \$765 + 0.07(\$38{,}525)$$
$$= \$765 + \$2696.75$$
$$\text{Tax} = \$3461.75$$

Since Enrico owes $3461.75 in state income tax and his employer has only withheld $3150 during the year, he will owe an additional tax payment of:

$$\$3461.75 - \$3150 = \$311.75$$

An important part of personal finances is the use of a family budget. An article on budgeting recently stated that most people spend 10% more than they make. This leads to increased use of credit, meaning a family can incur a great deal of debt. An effective budget puts you in control of your finances and expenditures.

Section 5-8 **Mathematical Models in Personal Income** 245

TABLE 5-12	Budget Form		
Category	Budget Items	Budgeted Amount	Actual Expense
Housing	Mortgage/Rent Electricity/Gas Water/Sewer Cable Telephone		
Food	Groceries Eating out		
Loans	Car loan Credit cards Other loans		
Miscellaneous	Clothing purchases Medical bills Charitable contributions Entertainment Gasoline/Car maintenance Savings		

Then you can avoid the question, "Where did all our money go?" You will already know the answer because you have a budget.

The first step in developing a budget is to examine your current financial information, including both monthly income and monthly expenditures. You can use a form like the one in Table 5-12 to help organize your figures. You should prepare a summary of your current spending habits and then adjust expenditures based on family goals such as saving for college tuition, a new car, or a down payment on a house.

The Consumer Credit Counseling Service recommends that you save at least 5% of your take-home earnings each month and that your credit obligations (not including your house payment) not exceed 15% of your take-home pay. A useful rule in determining the monthly house payment that you can afford is given here:

1. Subtract any monthly bills from your gross monthly income.
2. Multiply this amount by 36%.

Your monthly house payment should not exceed this amount.

Example 3

Budgeting Monthly Income

Joanna is single and has just started her first full-time job. Her monthly take-home pay after taxes have been withheld is $2074. She wants to save 7% of her take-home pay each month. Her monthly expenses are: apartment rent—$650, utilities (electricity and water)—$25, cell phone—$40, cable TV—$45, car payment—$225, credit card payments—$100, groceries—$350, entertainment—$250, clothing costs—$50, and gasoline—$120. Fill in a budget table and determine if she will be able to save 7% of her take-home pay.

The total amount that Joanna has budgeted at this point is $1855. If she wants to save 7% of her take-home salary, we must calculate that amount to determine if she will have enough left after expenses to save it.

$$0.07 \times \$2074 = \$145.18 \text{ or approximately } \$145$$

Category	Budget Items	Budget
Housing	Mortgage/Rent	$650
	Electricity/Gas	$ 25
	Water/Sewer	
	Cable	$ 45
	Telephone	$ 40
Food	Groceries	$350
	Eating Out	
Loans	Car Loan	$225
	Credit Cards	$100
	Other Loans	
Miscellaneous	Clothing Purchases	$ 50
	Medical Bills	
	Charitable Contributions	
	Entertainment	$250
	Gasoline/Car Maintenance	$120
	Savings	?

If we add her current expenses to this amount ($1855 + $145), the total is $2000. Therefore, she can save 7% of her take-home pay and still have enough money to meet her monthly expenses (with $74 left over for miscellaneous expenses!).

Creating a Budget

Marcus and his family plan to purchase a home in Little Rock. The family's gross monthly income is $6500. Their current monthly expenses include the following items: two car leases of $295 and $225, education loan—$75, cable—$105, telephone—$60, and credit card payment—$125.

(a) Determine the house payment that this family can afford. Round it to the nearest dollar.

(b) Complete a budget sheet for this family using the following percentages of the family gross income.

Fed/State Tax	28%	Entertainment	2%
Utilities	4.5%	Gasoline/Car Maintenance	4.75%
Food	7.5%	Miscellaneous	remaining amt.
Savings	5%		

Solution:

(a) To calculate the payment that this family can afford, subtract the monthly obligations from the gross income and then multiply the remaining amount by 36%. Round to the nearest dollar.

$6500 gross income
− 295 car lease
− 225 car lease

−	75	education loan			
−	105	cable bill			
−	60	telephone bill			
−	125	monthly credit card payment			
	$5615				

Now, 36% of that amount is 0.36 × $5615 = $2021.40 = $2021 (rounded)

(b) Fill in the budget sheet with the items included in most family budgets. Use the given data and specified percents to complete the table.

House Payment	$2021	Cable Bill	$105	Savings	5% × 6500 = $325
Car Lease	$295	Credit Card Payment	$125	Entertainment	2% × 6500 = $130
Car Lease	$225	Fed/State Tax	28% × 6500 = $1820	Gasoline/Car Maintenance	4.75% × 6500 = $308.75
Telephone Bill	$60	Utilities	4.5% × 6500 = $292.50	Miscellaneous	$230.25 remains for this category
Education Loan	$75	Food	7.5% × 6500 = $487.50		

To complete the budget, each category is calculated and the total amount designated to all categories is subtracted from $6500 to find the amount left in the Miscellaneous category.

Practice Set 5-8

For each problem, write a mathematical model and use it to find the correct answer.

1. Cassandra works for CJ Paper Products selling paper products to businesses and restaurants. Her monthly salary is computed based only on her sales per month. If she is paid 9.5% of her total monthly sales, calculate her gross pay this month if she sells $15,670 worth of paper products.

2. After Cassandra has been with the company for 12 months, she receives an increase of 0.75% in her commission percentage. If she sells $15,670 worth of paper products after she receives her raise, what will her gross pay be for the month?

3. Jerry sells cars at Cookie's Used Car Emporium. His monthly salary is calculated each month by adding a commission of 3.5% of his monthly sales to a base pay of $1245. If Jerry sells several cars worth a total of $65,000 during this month, calculate his gross pay.

4. Jerry needs to gross at least $2500 each month in order to pay his bills. Using your mathematical model from problem 3, calculate the amount of sales Jerry must make in order to be paid $2500.

5. Casey sells computer equipment. He is paid a base salary of $2200 per month and a commission of 2% of all sales over $7500. If Casey sells $12,350 worth of computer equipment this month, calculate his gross pay for the month.

6. Would Casey make a better salary if he were paid only a straight commission of 15% of total sales?

If Casey decides that he wants to be paid with a straight 15% commission, how much computer equipment would he need to sell to make the equivalent of the $2200 base salary he was previously earning?

7. Paolo makes a straight 2.5% commission on his sales of equipment. If he sells $125,000 worth of equipment this month, how much will he be paid this month?

8. Joan makes a base salary of $275 per week and a commission of 4% of sales over $1000. If she sells $1250 of merchandise this week, calculate the amount of her paycheck for the week.

9. A waitress generally expects to receive 15% of the cost of a meal as a tip from her customers. If the total bill for a table is $65.00, how much should be left as a tip for the waitress?

10. The standard tip that a customer should leave in a restaurant is 15%. How much money should a customer leave as a tip for the waitress if the meal costs $22.50?

11. A realtor sells a house for $150,000. The real estate agency that he works for will receive a commission on the sale in the amount of 6% of the selling price. The selling agent (realtor) will then receive 60% of that amount as his commission. How much will the agent earn from this sale?

12. The listing agent for a house receives 20% of the agency fee as commission when a sale occurs even if another realtor sells the house. If a house sells for $250,000 and the agency receives 6% of the selling price, how much will the listing agent be paid?

13. What is the commission rate for a sales employee who earned $425 straight commission on sales of $5000?

14. What is the commission rate for a car salesman who earned $1620 straight commission on a $36,000 car?

15. (a) Use the tax rate schedule from Example 2 to calculate the amount of state tax owed by a married couple filing jointly if the net taxable income is $48,745.
    (b) Calculate the amount of state tax owed by a single person with the same net taxable income.

16. Use the income tax table (Table 5-11) to calculate the tax that is owed by a taxpayer who is a Head of Household with net income of $257,000.

17. Proposals have been made that would implement a single-rate tax with few, if any, deductions allowed. Each person would pay the same fixed percent of his or her gross income. If Gilbert is a single person who grossed $31,000 last year, how much tax would he owe if the tax rate on gross income were 4.5%? If he takes all current allowable deductions, his taxable income would be $22,450. Using Table 5-11, would he owe more or less tax than he would using the 4.5% fixed rate?

18. Use Table 5-11 to determine if a married person with a taxable income of $55,000 will owe more tax if she files her return jointly or separately.

19. Ping is paid a salary based on a commission rate of 4.5% of his total sales in a month. Last year, he averaged sales of $50,000 per month throughout the year. Calculate his gross income for that year. After deductions, his net taxable income was $16,750. If he is filing as a single person, how much tax will he owe? (Use the tax rate schedule in Table 5-11.)

20. Mario is a car salesman, and he sold $1,248,000 worth of cars last year. He receives a 4.75% straight commission on his sales. Calculate his income for last year. After all deductions were taken, his net taxable income was $41,056. If he files as "married filing jointly," how much tax will he owe? (Use the tax rate schedule in Table 5-11.)

21. Suppose that you are a sales representative for the Rooty Tooty Trumpet Co. You earn a commission of 9.5% of your monthly sales made over your base pay of $800 per month.

    (a) Write an equation that models this situation.
    (b) What would the slope of the graph of this model be?
    (c) What are the units of this slope and what do they mean?
    (d) Is the relationship between your commission and sales a function?
    (e) If you were to graph this equation, which set of numbers (sales or commission) should be plotted on the y-axis and why?
    (f) Use your model to determine the amount of sales you would need to make in order to earn $2525 dollars in one month.

22. Ritzy Cars pays their salespersons on a graduated commission scale with no base pay. A salesperson earns 3% on the first $55,000 of a sale, 2% on the next $20,000, and 1% on any amount above $75,000 in sales.

   (a) Write a statement, instead of an equation, to model this pay scale. It should be written in such a way as to allow you to calculate the commission earned for any size sale.
   (b) Is the relationship between the amount of sales and the amount of commission earned a function in this situation?
   (c) What would be the commission paid on a sale of a used car at Ritzy's if the car cost $48,000?
   (d) What would the commission be on a "Yuppiemobile" that cost $128,000?

23. Josh is single and has just started his first full-time job. His take-home pay after taxes have been withheld is $1855. He wants to save 7% of his take-home pay each month. His monthly expenses are apartment rent—$590, utilities (electricity and water)—$25, cell phone—$45, cable TV—$45, car payment—$125, groceries—$320, entertainment—$150, and gasoline—$150. Fill in a budget table and determine if he will be able to save 7% of his take-home pay.

24. Marcus has just gotten married and wants to try and save 5% of his take-home pay each month for a down payment on a new house. His take-home pay after taxes have been withheld is $2350. His monthly expenses are apartment rent—$850, utilities (electricity and water)—$35, cell phone for two people—$75, cable TV—$45, car payment (his car)—$225, car payment (her car)—$200, credit card payments—$150, groceries—$550, entertainment—$250, clothing costs—$75, charitable contributions—$50, and gasoline—$250. Fill in a budget table and determine if he will be able to save 5% of his take-home pay.

25. The Consumer Credit Counseling Service recommends that your credit obligations (not including your house payment) should not exceed 15% of your take-home pay. If Ketan's take-home pay is $1978, what is the maximum monthly credit obligation that he should have?

26. The Consumer Credit Counseling Service recommends that your credit obligations (not including your house payment) should not exceed 15% of your take-home pay. If Olivia's take-home pay is $2206, what is the maximum monthly credit obligation that she should have?

27. Nedgelena plans to purchase a home in Columbia. Her gross monthly income is $4250. Her current monthly expenses include the following items: a car lease of $215, cable—$45, telephone—$55, and furniture payment—$45.

   (a) Determine the house payment that Nedgelena can afford. Round it to the nearest dollar.
   (b) Complete a budget sheet for her using the following percentages of her gross income.

Fed/State Tax	15%
Utilities	4.25%
Food	13.5%
Savings	5.5%
Entertainment	4%
Gasoline/Car Maintenance	5%
Health Insurance	2%
Miscellaneous	remaining amt.

28. George and his wife plan to purchase a home in Cramerton. Their gross monthly income is $5710. Their current monthly expenses include the following items: a car payment of $325, $55 payment for a refrigerator, cell phone plan—$65, and credit card payment—$75.

   (a) Determine the house payment that they can afford.
   (b) Complete a budget sheet for George and his wife using the following percentages of the family gross income.

Fed/State Tax	15%
Utilities	3.5%
Food	15.5%
Savings	7%
Entertainment	3%
Gasoline/Car Maintenance	4.5%
Health Insurance	1%
Miscellaneous	remaining amt.

29. Research your family's monthly expenses and fill out a chart of approximate costs for each category. In your opinion, are there places in the budget where adjustments in spending could be made?

30. Assume that you are single and earn $2600 per month. Federal and state taxes will be withheld from this total at an approximate rate of 28%. Calculate your take-home pay. Now, using this figure, research costs of apartments, gasoline, groceries, utility bills, cable TV, and other expenses in your local area. Create a reasonable budget, including at least 5% of your take-home pay for savings.

Chapter 5 Summary

Key Terms, Properties, and Formulas

adjustable-rate mortgage (ARM)
amortization table
annual percentage rate (APR)
average daily balance method
Banker's Rule
bonds
closed-end credit
closing costs
commission
compounded continuously
compound interest
depreciation
dividends
fixed-rate mortgage
installment buying
interest
lease
markup
maturity value
mortgage
mutual funds
open-end credit
ordinary interest
principal
rate
residual value
share
shareholders
simple interest
stocks
straight commission
straight-line method of depreciation
unpaid balance method

Formulas to Remember

Simple Interest:
$$I = Prt$$

Maturity Value:
$$M = P + I$$

Compound Interest Formulas

Quarterly:
$$M = P\left(1 + \frac{r}{4}\right)^{4t}$$

Monthly:
$$M = P\left(1 + \frac{r}{12}\right)^{12t}$$

Daily:
$$M = P\left(1 + \frac{r}{365}\right)^{365t}$$

where r = annual rate and t = number of years.

Fixed-Rate Mortgage Monthly Payment Formula

$$P = A \left[\frac{\frac{r}{12}\left(1 + \frac{r}{12}\right)^{12t}}{\left(1 + \frac{r}{12}\right)^{12t} - 1} \right]$$

where P = monthly payment, A = amount borrowed, r = annual rate, and t = number of years.

Chapter 5   Review Problems

1. The tax rate in Wake County is 7%. The county rate is 2.5% and the state rate is 4.5%. If Qwik-E-Mart sells $12,548 worth of merchandise this weekend, how much sales tax will the owner owe to the county and how much will he owe to the state?

2. TJ Sports had gross receipts of $4285.75 on Tuesday. This total included an 8.5% sales tax. What were the receipts from the merchandise (without tax) sold on Tuesday?

3. The local sales tax on food is 5% and the sales tax on other items is 6%. If Marie purchases $12.80 worth of food items and $26.50 worth of nonfood items at X Mart, what amount of sales tax will she be charged?

4. John gave a barber $15 for a service that costs $12.50. What is the tip rate?

5. You are depreciating a $25,000 van over 10 years. At the end of 10 years, the salvage value of the van is expected to be $1500. How much does the van depreciate each year?

6. The Jon-Jon Company makes small fishing boats. Each boat costs $125 to produce. The fixed costs of production are $1250 per month. These boats sell for $275 each. If the company produces and sells 75 boats this month, how much net profit will be earned?

7. The Ace Furniture Store uses a 200% markup on all furniture. If a sofa costs them $475, how much will the retail price be?

8. Julian has the choice of two salary plans as a salesman for TurboVac Company. He can earn his salary by taking a straight 25% commission on his monthly sales, or he can earn a fixed amount of $1000 per month and 5% of total monthly sales. If he sells $5000 in merchandise in July, which plan would yield the higher paycheck?

9. An amount of $5000 was deposited in a savings account paying simple interest. After 6 months, the account had earned $162.50 in interest. What is the annual interest rate on this account?

10. You borrow $15,000 on a 2-year note. The interest rate is 5.5% simple interest. What is the maturity value of the note? If monthly payments are made, how much will each payment be?

11. Find the interest on a $1000 bond paying 3% simple interest for 6 years.

12. James buys an $8000 certificate of deposit. It pays 5.8% compounded quarterly. If it matures after 2 years, what will the value of the investment be?

13. How much interest will be earned at the end of 5 years on a deposit of $2500 in an account paying 3.5% interest compounded monthly?

14. A person invests $10,000 at 4.0% for 20 years compounded continuously. How much will be in the account at the end of 20 years?

15. Find the finance charge for this month on an open-end charge account using the unpaid balance method of calculating interest if the unpaid balance is $3150.75 and the annual interest rate is 15%.

16. Find the average daily balance for the following credit card account. Assume one month between billing dates and find the finance charge using an 18% annual rate and the average daily balance method.

    June 1: Billing date with a balance of $1415.90
    June 15: Payment $ 125.00
    June 23: Birthday present $ 21.95

17. Marisa bought a new surround-sound stereo system for $2350. She paid $500 down and financed the balance for 36 months at 8%. Calculate the amount of her monthly payment. (Use the APR table found in Section 5-3.)

18. Michael plans to offer 5% over dealer cost on a Honda that has a sticker price of $32,480 and a dealer cost of $26,998. Calculate the amount of his offer.

19. Determine the amount of interest owed and the monthly payment if Pietro purchases a car for $27,885 at 0.9% add-on rate for 5 years.

20. A newspaper advertisement offers a $6500 used car for nothing down and 36 easy monthly payments of $196.80. What is the total amount paid for the car, and what is the total finance charge? Calculate the simple interest rate charged.

21. Anna plans to lease a new Mini Cooper 2-door coupe. The sales price of the car is $25,000. The lease agreement states the following:

    Due at signing: $395.59   Mileage allowance: 15,000/year
    Monthly payment: $395.59   Excess mileage fee: $0.15/mile
    48 months   Acquisition fee: $849

    If Anna pays the acquisition fee, the amount due at signing, and the 48 monthly payments, how much has she paid to lease this car?

22. If Anna drives 62,148 miles before she turns in the car, how much extra cost will she incur?

23. If Anna purchases the Mini Cooper for $25,000 at a 3.6% add-on rate for 5 years, how much will her monthly payments be?

24. Determine the maximum monthly payment you could afford for a house if your gross monthly income is $4988 and your current monthly payments are $500.30. Use the formula discussed in Section 5-5 or an online mortgage calculator.

25. The Dalton family is purchasing a $145,000 house. They plan to make a 20% down payment. The credit union will finance the balance for 30 years at a 6.25% fixed rate. Calculate the amount of their monthly payments.

26. Determine the monthly payment for this loan by using Table 5-4 (amortization table).

    Cost of house: $148,500   Interest rate: 5.5%
    Term: 30 years

27. Kristen Robinson is 16 years old and drives a Honda Accord, which is a Model L. The car is in Age Group 2. She drives the car for pleasure. Her parents own the car. Determine her annual premium for comprehensive and $500-deductible collision coverage.

28. Determine the annual premium for 50/100/100 liability insurance for a 45-year-old male who drives 25 miles to work and lives in Territory 3.

29. Janet's car was stolen while parked outside her home. The car had a market value of $20,000. Janet carries comprehensive and $500-deductible collision insurance. How much indemnity will the insurance company pay?

30. John bought 150 shares of Apple Computer (AAPL) stock and paid $532.35 per share. He also had to pay a $10.99 online broker fee to make the purchase. What was the total cost to him for this transaction?

31. An investor purchases 450 shares of Texas Instruments stock (TXN) at $27.28 per share.

    (a) Find the cost of the stock.
    (b) If the broker charges 2.5% of the cost of the stock, find her commission on this purchase.

32. Travis plans to invest $87,250 in a mutual fund. Century Shares Trust is currently selling for $34.90 per share. How many shares can he buy at this price?

33. What is the commission rate for an employee who earned $812.50 straight commission on sales of $12,500?

34. The weekly salary of an employee is $400 plus a 6.5% commission on the employee's sales over $1200. Find the employee's gross pay for a week when the sales amounted to $6000.

35. Use the income tax table in Section 5-8 to write a model that can be used to calculate the amount

of tax owed by a person who is married filing jointly if his income is between $21,250 and $100,000.

36. The Meltons had a net taxable income of $22,500 this year. How much income tax do they owe if they file jointly?

37. The Consumer Credit Counseling Service recommends that your credit obligations (not including your house payment) should not exceed 15% of your take-home pay. If Caitlyn's take-home pay is $2386, what is the maximum monthly credit obligation that she should have?

Chapter 5 Test

1. Jerry purchases a new car for $24,500. If the sales tax on this purchase is 3.5%, how much sales tax will the dealer add to his cost?

2. A retailer pays $12.40 for a coffee maker. He uses a 125% markup on small appliances. Find the retail price of this coffee maker.

3. A business buys an $18,000 car for its courier to use for delivering correspondence between branch offices. If the business office plans to depreciate the car using the straight-line method over a 10-year period, find the yearly depreciation allowance if the residual value of the car is $2500.

4. BMA Corporation manufactures beanies at a cost of $0.82 each. Their fixed costs per month are $3575. Each beanie sells for $2.20. If BMA produces and sells 14,500 beanies this month, calculate their net profit.

5. Mary makes a 12% commission on her weekly sales at Stereo Extravaganza. She also receives a base pay of $200 per week. If she sells $1275 worth of stereo equipment this week, what will her gross pay be?

6. One proposal in Congress has been to change the tax laws and have a "flat tax rate." This would require all citizens to pay the same percent of their gross pay to the federal government. If Jose's gross income this year is $55,250 and the flat tax rate is 18%, how much income tax would Jose owe the government?

7. Find the amount of money that you will have in the bank at the end of 15 years if you deposit $2500 into an account that pays 4.5% interest compounded quarterly.

8. Jeff earned $250 on his $5000 certificate of deposit last year. What simple interest rate did the bank pay him?

9. Jeremy deposits $7500 in a savings account that pays 5.5% compounded continuously. How much will be in this account in 6 years?

10. Find the finance charge for this month on an open-end charge account using the unpaid balance method of calculating interest if the unpaid balance is $5250.05 and the annual interest rate is 8.0%.

11. Find the average daily balance for the following credit card account. Assume one month between billing dates, and find the finance charge using an 18% annual rate and the average daily balance method.

    July 1: Billing date with a balance of $115.90
    July 15: Payment $ 25.00
    July 20: Clothes purchase $225.55

12. Robin bought a new washing machine for $1385. She paid $100 down and financed the balance for 24 months at 5%. Calculate the amount of her monthly payment. (Use the APR table.)

13. Determine the amount of interest owed and the monthly payment if Martin purchases a Lexus for $57,885 with no down payment at 0.9% add-on rate for 6 years.

14. Darrius plans to lease a new Jeep Grand Cherokee 4WD. The sales price of the car is $36,200. The lease agreement states the following:

Due at signing: $571	Mileage allowance: 15,000/year
Monthly payment: $571	Excess mileage fee: $0.17/mile
54 months	Acquisition fee: $795

    If Darrius pays the acquisition fee, the amount due at signing, and the 54 monthly payments, how much has he paid to lease this car?

15. If Darrius purchases the Jeep for $36,200, pays $600 down, and finances it at a 1.9% add-on rate for 6 years, how much will his monthly payment be?

16. The Rasmussen family is purchasing a $375,000 house. They plan to make a 25% down payment. The credit union will finance the balance for 30 years at a 5.5% fixed rate. Calculate the amount of their monthly payments using the amortization table.

17. Determine the annual liability insurance premium for 50/100/50 for a 16-year-old female who drives 4 miles one way to work in Territory 2.

18. Janet's car was stolen while parked outside her home. The car had a market value of $20,000. Janet carries comprehensive and $500-deductible collision insurance. How much out-of-pocket expense will Janet incur to replace the car, assuming that she can buy one for $20,000?

19. Larry purchased 175 shares of stock A at a price of $49.50 and 250 shares of stock B at $29.88. He also paid $9.99 in online broker's fees for each of the two purchases. What was the total cost of his purchases?

20. An investor purchases 50 shares of Wendy's stock (WEN) at $5.15 per share.

    (a) Find the cost of the stock.
    (b) If the broker charges 1.5% of the cost of the stock plus $0.15 per share, find his commission on this purchase.
    (c) What is the total cost of this stock purchase?

Suggested Laboratory Exercises

Lab Exercise 1   Saving for College Expenses

On Polly's 6th birthday her parents bought a $10,000 CD that pays at the rate, r, of 4% *compounded annually*. Use the simple interest formula to fill in the following table or set up a spreadsheet to calculate the values for you.

Polly's Age	Principal (P)	Interest Earned (I)	Total of Account ($P + I$)
7	$10,000.00	$400.00	$10,400.00
8	$10,400.00		
9			
10			
⋮			
18			

Now, use a calculator and the following formula to calculate the total value of Polly's CD on her 11th birthday. In this formula, A = the total value, P = the original principal at age 6, r = the interest rate, and t = the number of years that have elapsed since Polly's 6th birthday.

$$A = P(1 + r)^t$$

Now fill in the last line of the chart with Polly's 18th birthday and calculate the total value of her CD at that time.

Lab Exercise 2

Depreciation

Equipment purchased for use by business will decrease in value due to use and age as time goes by. The IRS recognizes this fact by allowing businesses to decrease the value of assets on their books for tax purposes. This is cleverly called *depreciation*.

Suppose that you own a company and you purchase a new widget maker for $50,000. This machine will depreciate in value so that after 5 years it will have a book value of $10,000. This depreciation is linear, that is, straight-line.

1. Write an equation for the book value, V, as a function of the number of years of use, x.
2. What is the slope of this line and what does it mean here?
3. What would the book value be after 4 years?
4. Set up a spreadsheet to do all of the calculations for you for all 5 years.

Lab Exercise 3

MACRS Depreciation Method

The IRS requires many businesses to depreciate equipment using a method called Modified Accelerated Cost Recovery System (MACRS). Business assets are grouped into categories specifying the time period that these assets can be written off or depreciated for tax purposes. This table lists some classes of assets.

Class of Property Items	Business Assets
3-year property	tractor units, racehorses over two years old, horses over 12 years old when placed into service
5-year property	automobiles, taxis, buses, trucks, computers and peripheral equipment, office machinery, dairy cattle
7-year property	office furniture, and any property not designated as belonging to a different class
10-year property	vessels, barges, tugs, single-purpose agricultural or horticultural structures, trees or vines bearing fruit or nuts

The depreciation method that you use is determined when the asset is put into use and that method is used for the life of that item. The table below is one of many MACRS tables. This is for a 7-year property using a method called the half-year convention. There are many tables and requirements that are too complicated for our use. In this problem, we will consider a simple example. (www.IRS.gov/publications/#946)

MACRS 7-Year Property

Years	Half-Year Convention
1	24.9%
2	17.49
3	12.49
4	8.93
5	8.92
6	8.93
7	4.46

Complete the following chart to show the depreciation amounts under MACRS for office furniture purchased in 2011 for $10,000. The amounts in the third column should be taken from the MACRS half-year convention table, which is the one most commonly used. Notice that the asset's tax basis does not change over the years—only the percentage used as a multiplier changes each year.

Year	Basis	Percentage	Deduction
2011	$10,000		
2012	$10,000		
2013	$10,000		
2014	$10,000		
2015	$10,000		
2016	$10,000		
2017	$10,000		

Now, complete this project by doing some research online to find a 5-year MACRS table. Calculate the depreciation table for $200,000 worth of computer equipment bought in 2011. (www.IRS.gov/publications/#946)

Lab Exercise 4 — Purchasing Furniture

1. Using the given diagram of the sample apartment, you must "purchase" furniture for the living/dining room and for the master bedroom. Use a website like www.thomasville.com or a major furniture chain website to shop for your furnishings. List the items you are buying and their costs below. You must purchase enough furniture to comfortably live, including a dining table and chairs, living room furniture, and a bedroom set. Don't forget to buy a mattress.

   Draw the furniture in the room where you plan to place it.

Item Description	Cost
Total cost of furnishings	

2. For the current promotion, you can purchase the furniture interest-free under the following condition:

   Monthly payments are based on
   20% down payment and 24 monthly payments.

Based on this plan, calculate the amount of your down payment: _____
Calculate the amount of each monthly payment: _____

3. The fine print!

   Until January next year, interest will accrue at 23.4%.
   If the balance is paid in full by end of deferral period,
   all finance charges will be credited, resulting in no interest to you.

   If you do not pay off this loan by the end of December of this year, then you will pay interest on the entire balance due at the time of purchase. Using the simple interest formula and the 24 months of financing available, how much interest would you pay if you weren't able to pay off the total balance due by the end of the year? _____

4. Use the Annual Percentage Rate (APR) Table (Table 5-3) to answer these questions.

   (a) You plan to pay 10% down on your purchase. How much do you pay at the time of purchase? _____ How much is left to finance? _____
   (b) If you finance the balance due at 7.5% for 36 months, how much interest will you owe? _____
   (c) What will your monthly payment be? _____

5. Using the installment loan table, calculate the monthly payments for your purchase if you decide to finance the purchase for 4 years at 8.5% APR.

   Total interest due _____
   Monthly payment _____

Lab Exercise 5 — Creating a Stock Portfolio

In Section 5-7 of this chapter, there was a brief discussion of the use of spreadsheets for tracking the value of a stock portfolio. Divide the class into groups of four and have each group choose eight stocks for a portfolio. Each group should research the companies chosen and write a brief paragraph for each describing the type of company chosen, the products it markets or produces, the location of the headquarters, and other interesting facts about the company. Each group should then "invest" $1000 in each company by "purchasing" the maximum number of whole shares possible with $1000. Students should use the Internet or newspaper to get the current per share cost of the chosen stocks. Using a spreadsheet, each group should first record the data for the stock market purchases that are made. Then each group should follow the closing prices of their stocks each day for two weeks. The ups and downs of these closing prices should be tracked and recorded on graph paper. At the end of two weeks, each group should calculate its portfolio's percent gain or loss. Which group made the most money?

Lab Exercise 6 — Examining Mortgage Options

The local bank is offering fixed-rate home loans at 4.25% for 30 years and 3.75% for 15 years. They will provide 100% financing of the appraised value of the house. The Holder family is purchasing a condo for $180,000. They currently have $20,000 in the bank that could be used as a down payment. Should they finance for 30 years or 15 years? Should they invest the $20,000 or use it as a down payment? Work with your group to determine the best decision for this family.

(a) Use an online website to help you calculate the monthly payment if they finance the entire cost of the house for 30 years. Then, calculate the total cost of the house after making payments for 30 years. Now, do the same calculations for a 15-year loan.

(b) Now use the same website to calculate the monthly payment if they pay $20,000 down and finance the balance for 30 years. Then calculate the total cost of the house after making payments for 30 years. (Don't forget to include the down payment in the total cost.) Now, do the same calculations for a 15-year loan.

(c) Assume that the Holders can invest the $20,000 in an investment that will return an average of 2.5% annually. How much would such an investment be worth after 15 years? After 30 years?

(d) Use all of the information that you have gathered to discuss with your group the best way to finance your home. Answer the questions above, writing your conclusion in a paragraph with clear reasons for your choice.

6 Modeling with Systems of Equations

Some problems or situations involve the interaction of two or more variables. Business profit, for example, is based on the cost of production and the income from selling a product. A break-even point can be determined using a system of equations so the business will be sure to make a profit. Several methods for solving systems of equations such as these will be explored in this chapter.

> *"Mathematicians do not deal in objects, but in relations between objects; thus, they are free to replace some objects by others so long as the relations remain unchanged."*
> — Poincaré

In this chapter

6-1 Solving Systems by Graphing
6-2 Solving Systems Algebraically
6-3 Applications of Linear Systems
6-4 Systems of Nonlinear Equations
Chapter Summary
Chapter Review Problems
Chapter Test
Suggested Laboratory Exercises

Section 6-1 Solving Systems by Graphing

Many problems involve more than one unknown quantity and involve using a **system of equations**. In these problems, there are usually two conditions that must be met. For example, you might be asked to make a specific volume of a solution having a certain concentration. The two requirements that must be satisfied are the volume of the solution and the strength of the solution. It is often easier to solve a problem like this with two equations that relate the unknown quantities. Then we look for the one solution that satisfies both sets of requirements—the one solution that satisfies both equations algebraically. This may be done by graphing, using algebraic techniques, or using a process with determinants called Cramer's rule. In this section, we will look at the graphing method used to solve systems.

A system of linear equations consists of two equations containing two related variables. For example, the following system involves the related variables x and y.

$$y = 2x - 8$$
$$3x + 4y = 12$$

Many ordered pairs (x, y) can be found that will satisfy the first equation. There are many ordered pairs that will satisfy the second equation as well. The **solution** of the system, however, is the single ordered pair that will satisfy *both* of the equations. The single ordered pair that will satisfy the system shown in the previous equations is $(4, 0)$. In this section, we will learn to find this solution graphically.

In Chapter 2, you learned to graph a linear equation using both a table of ordered-pair solutions and the slope-intercept form ($y = mx + b$) of the equation. We can use either of these techniques to graph each of the equations in the system. Both lines are graphed on the same set of axes. The point of intersection of the two lines is the solution of the system.

Example 1 Solving a System by Graphing

Find the solution of the following system of equations graphically.

$$x + 3y = 7$$
$$y = 3x - 1$$

TABLE 6-1
Example 1

x	y
0	
	0
1	

Use a table for each equation to find at least three points for each line. Although you may choose any values for x or y, we will let $x = 0$, $y = 0$, and $x = 1$. (See Table 6-1.)

Line 1: $x + 3y = 7$.

$$(0) + 3y = 7 \qquad x + 3(0) = 7$$
$$3y = 7 \qquad\qquad x = 7$$
$$y = 2\frac{1}{3}$$

$$1 + 3y = 7$$
$$3y = 7 - 1$$
$$3y = 6$$
$$y = 2$$

TABLE 6-2
Example 1, Line 1

x	y
0	$2\frac{1}{3}$
7	0
1	2

See Table 6-2.

Section 6-1 Solving Systems by Graphing

Now graph these three points on the grid to show the line representing the equation $x + 3y = 7$.

Line 2: $y = 3x - 1$.

$y = 3(0) - 1$ $0 = 3x - 1$
$y = -1$ $1 = 3x$
                    $\frac{1}{3} = x$

$y = 3(1) - 1$
$y = 3 - 1 = 2$

TABLE 6-3
Example 1, Line 2

x	y
0	-1
$\frac{1}{3}$	0
1	2

See Table 6-3.

Graph these three points representing the line $y = 3x - 1$ on the same grid and locate the point of intersection of the two lines. (See Figure 6-1.)

The two lines intersect at the point (1, 2), so this is the solution of this system of equations.

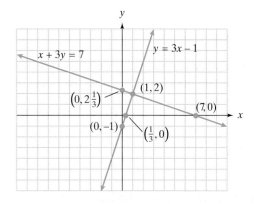

FIGURE 6-1
Example 1.

When the solution to a linear system has been found, it should be checked in the original equations. It should satisfy both equations.

Example 2 Checking a Solution

Show that the ordered pair (1, 2) is the correct solution of the system in Example 1:

$$x + 3y = 7$$
$$y = 3x - 1$$

(a) Substitute (1, 2) into the first equation.

$x + 3y = 7$
$1 + 3(2) = 7$
$1 + 6 = 7$
$7 = 7$ (true)

(b) Substitute (1, 2) into the second equation.

$y = 3x - 1$
$2 = 3(1) - 1$
$2 = 3 - 1$
$2 = 2$ (true)

Because the ordered pair (1, 2) checks in both equations, it is the solution of this system of equations.

When a system of two linear equations is graphed, three different outcomes are possible. The two lines may intersect in one point. This system is said to be **consistent**. The two lines may be parallel. In this case, there is no solution because the two lines do not intersect. These systems are said to be **inconsistent**. The third possible outcome results in two systems that coincide. In this case, all ordered pairs that satisfy the first equation also satisfy the second equation. Thus, infinitely many solutions are possible. These systems are said to be **dependent**. See Figure 6-2.

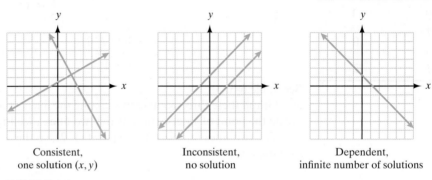

Consistent, one solution (x, y)

Inconsistent, no solution

Dependent, infinite number of solutions

FIGURE 6-2
Solutions of systems of equations.

Notice that when a system consists of parallel lines or lines that coincide, the directions of the two lines, and, therefore, the slopes of the two equations, are identical. On the other hand, if the lines intersect in one point, their directions must be different, so their slopes will also be different. If you graph the system using the slope-intercept form of the equation ($y = mx + b$), it is easy to determine the slope of each line. The slope numbers can give you important information about the solution of a system. Look at the following example.

Example 3 — Graphing a System with No Solution

Solve the following system graphically.

$$2x + 4y = 8$$
$$x + 2y = -2$$

(a) Put both equations into $y = mx + b$ form.

$$2x + 4y = 8$$
$$4y = -2x + 8 \quad \text{(add } -2x \text{ to both sides)}$$
$$y = -\frac{1}{2}x + 2 \quad \text{(divide by 4)}$$

In this case, the slope $= -\frac{1}{2}$ and the y-intercept $= 2$.

$$x + 2y = -2$$
$$2y = -x - 2 \quad \text{(add } -x \text{ to both sides)}$$
$$y = -\frac{1}{2}x - 1 \quad \text{(divide by 2)}$$

In this case, the slope $= -\frac{1}{2}$ and the y-intercept $= -1$.

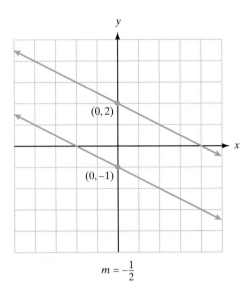

FIGURE 6-3
Example 3.

(b) Because both slopes are equal to $-\frac{1}{2}$, we know that the directions of the lines will be the same. Because the y-intercepts are different, the lines are not identical but must be parallel. (See Figure 6-3.)

Because the lines are parallel, this system is inconsistent, and the answer is "no solution."

Example 4 — Solving a System That Is Dependent

Solve the following system graphically.

$$3x - 2y = 8$$
$$6x - 4y = 16$$

(a) Put both equations into $y = mx + b$ form.

$$3x - 2y = 8$$
$$-2y = -3x + 8 \quad \text{(add } -3x \text{ to both sides)}$$
$$y = \frac{3}{2}x - 4 \quad \text{(divide by } -2\text{)}$$

In this case, the slope $= \frac{3}{2}$ and the y-intercept $= -4$.

$$6x - 4y = 16$$
$$-4y = -6x + 16 \quad \text{(add } -6x \text{ to both sides)}$$
$$y = \frac{3}{2}x - 4 \quad \text{(divide by } -4\text{)}$$

In this case, the slope $= \frac{3}{2}$ and the y-intercept $= -4$.

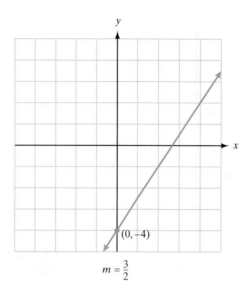

FIGURE 6-4
Example 4.

(b) Because the slopes and the *y*-intercepts are the same, the lines are identical. Therefore, any solution of the first equation also satisfies the second equation, and there are infinitely many solutions that will satisfy this system. (See Figure 6-4.)

A TI-83 Plus or TI-84 graphing calculator may be used to find the solution of a system of equations graphically. Solve each equation for *y* and enter the equations using the Y= key. When you press GRAPH, both lines will be displayed. Use the TRACE feature to move the cursor as close as possible to the point of intersection. The coordinates will be displayed at the bottom of the screen. They may not be exactly correct because the pixels on the screen cause slightly differing answers. There is a CALCULATE function located above the TRACE button; you can use it to calculate the point of intersection.

Example 5 — Solving a System Using a Graphing Calculator

Solve the following system using a graphing calculator.

$$4x + 3y = 24$$
$$2x - y = 2$$

(a) Solve each equation for *y*.

$$4x + 3y = 24 \qquad\qquad 2x - y = 2$$
$$3y = -4x + 24 \qquad\qquad -y = -2x + 2$$
$$y = -\frac{4}{3}x + 8 \qquad\qquad y = 2x - 2$$

(b) Enter the equations using the Y= function as

$$Y_1 = \left(-\frac{4}{3}\right)x + 8$$
$$Y_2 = 2x - 2$$

(c) Use the GRAPH button to display the lines. Use the TRACE function to move the cursor to the point of intersection. Use the CALCULATE feature to verify this point. (See Figure 6-5.)

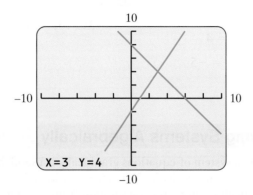

FIGURE 6-5
Example 5.

Practice Set 6-1

Which ordered pair is a solution of the system of equations?

1. $3x + 4y = 1$ $(7, 2)$ $(3, -2)$
   $x - y = 5$

2. $3x - 7y = -22$ $(2, 4)$ $(1, 3)$
   $x - y = -2$

3. $x - 4y = 1$ $(3, 1)$ $(-3, -1)$
   $5x - 3y = -12$

4. $2x - 3y = 6$ $(0, 3)$ $(3, 0)$
   $4x + 3y = 12$

5. $2x + y = 2$ $\left(1, \frac{1}{2}\right)$ $\left(\frac{1}{2}, 1\right)$
   $6x - 4y = -1$

6. $4x + 9y = 12$ $\left(-\frac{3}{2}, \frac{2}{3}\right)$ $\left(\frac{3}{2}, \frac{2}{3}\right)$
   $2x - 3y = 1$

Solve the following systems graphically. Check your solution.

7. $x + y = 3$
   $x - y = -1$

8. $x + 3y = 2$
   $2y = x + 3$

9. $2x - y = 4$
   $x + 2y = 2$

10. $3x - 7 = y$
    $y = 3x + 5$

11. $x = 4$
    $x + y = 5$

12. $4y - 3x = 8$
    $y = 5$

13. $2x - y = -1$
    $2x + y = 1$

14. $3x + 2y = 20$
    $x + y = 8$

15. $\frac{1}{3}x + 2y = 9$
    $x + 6y = 27$

16. $x + y = 9$
    $6x - 5y = 10$

17. $x - 10 = 5y$
    $x + 10 = 5y$

18. $x = 2$
    $y = 3$

19. $x + 2y = -10$
    $\frac{1}{2}x - 3 = y$

20. $3x = 2y + 5$
    $6x - 4y = 10$

21. $x - 4y = 1$
    $5x - 3y = -12$

22. $x + y = 13$
    $x - y = 3$

23. $x + y = 4$
    $3x + 3y = 12$

24. $3x - 2y = 6$
    $15x - 10y = -2$

25. $4x - 9y = 6$
    $\frac{1}{3}x - \frac{3}{4}y = 1$

Solve each system using a graphing calculator.

26. $y = \dfrac{3}{4}x + 2$
    $y = x + 1$

27. $2x + y = 4$
    $x - y = 2$

28. $2x - 5y = 10$
    $6x - 15y = 75$

29. $3x + 2y = 4$
    $-x + 3y = -5$

30. $x + 2y = -1$
    $2x - 3y = 12$

31. $2x - 3y = 1$
    $-6x + 9y = -3$

32. $x + y = -2$
    $-3x - 3y = 6$

Section 6-2 Solving Systems Algebraically

Solving a system of equations graphically is useful, but it is less accurate than solving a system algebraically. When the solution of a system is fractional, for example, it is difficult to give a precise solution from a graph.

The goal of solving systems algebraically is to reduce the system that has two equations in two variables to a single equation that has only one variable. This can be done using either of two methods: **substitution** or **linear combination**. In the substitution method, we substitute an expression for one variable to eliminate that variable, resulting in an equation containing only one unknown. In the linear combination method, the two equations are added together (based on the addition property of equality) in such a way that one of the two variables is eliminated from the new equation. Look at the following examples.

Example 1 Substitution Method

Solve:

$$4x - 3y = 16$$
$$y = 7x + 6$$

(a) In the second equation, y is defined to be $7x + 6$. Using substitution, we will replace the y in the first equation with the expression $7x + 6$ and then solve for x.

$$4x - 3y = 16$$
$$4x - 3(\mathbf{7x + 6}) = 16 \quad \text{(substitute)}$$
$$4x - 21x - 18 = 16 \quad \text{(distributive property)}$$
$$-17x - 18 = 16 \quad \text{(combine like terms)}$$
$$-17x = 34 \quad \text{(add 18 to both sides)}$$
$$x = -2 \quad \text{(divide by } -17\text{)}$$

(b) Now that we have solved for the value of x, we can easily calculate the value of y using the second equation.

$$y = 7x + 6$$
$$y = 7(-2) + 6 \quad \text{(substitute } -2 \text{ for } x\text{)}$$
$$y = -8 \quad \text{(simplify)}$$

Therefore, the solution of this system is the ordered pair $(-2, -8)$.
Check this solution by substituting these values into both of the original equations.

> **Rule** **The Substitution Method**
>
> 1. Solve one of the equations for one variable in terms of the other. Choose the variable with a coefficient of 1 to solve for, if possible.
> 2. Substitute the expression obtained in step 1 into the other equation and solve the resulting one-variable equation.
> 3. Solve for the remaining variable by substituting the value you obtained in step 2 into either of the original equations.
> 4. Check your answer to be sure that it satisfies both equations.

In the previous example, we did not have to do step 1 of this method because one of the equations was already in the correct form for substitution. Let us look at a problem where the first step must be done before substituting.

Example 2 **Substitution Method**

Solve the system:

$$2x + 6y = 16$$
$$6x + 5y = 9$$

We must first solve one of the equations for one variable in terms of the other. It appears to be easiest to solve for x in the first equation because all terms are divisible by 2.

$$2x + 6y = 16$$
$$2x = -6y + 16 \quad \text{(subtract } 6y \text{ from both sides)}$$
$$x = -3y + 8 \quad \text{(divide by 2)}$$

Now we take the expression $-3y + 8$ and substitute it in place of the x variable in the second equation.

$$6x + 5y = 9$$
$$6(-3y + 8) + 5y = 9 \quad \text{(substitute for } x\text{)}$$
$$-18y + 48 + 5y = 9 \quad \text{(distributive property)}$$
$$-13y + 48 = 9 \quad \text{(simplify)}$$
$$-13y = -39 \quad \text{(add } -48 \text{ to each side)}$$
$$y = 3 \quad \text{(solve for } y\text{)}$$

Now complete the solution by substituting 3 for y in one of the equations and solving for x.

$$x = -3y + 8$$
$$x = -3(3) + 8 \quad \text{(substitute 3 for } y\text{)}$$
$$x = -1 \quad \text{(simplify)}$$

The solution of this system is $(-1, 3)$. Check to verify the solution.

The second algebraic method for solving systems is the linear combination method. The key step is to obtain, for one of the variables, coefficients that differ only in sign so that the two equations can be combined or added together to eliminate that variable. This method is illustrated in the following example.

Example 3

Linear Combination Method

Solve the system:

$$2x + y = 4$$
$$x - y = 2$$

Notice that the y variables have coefficients of -1 and 1. Because these are opposites, if the two equations are added together, the y terms will be eliminated from the new equation.

$$2x + y = 4$$
$$\underline{x - y = 2}$$
$$3x = 6 \quad \text{(add the equations)}$$
$$x = 2 \quad \text{(divide both sides by 3)}$$

Now substitute 2 for x in either equation to solve for the value of y.

$$2x + y = 4$$
$$2(2) + y = 4 \quad \text{(substitute 2 for } x\text{)}$$
$$4 + y = 4 \quad \text{(simplify)}$$
$$y = 0 \quad \text{(add } -4 \text{ to both sides)}$$

Therefore, the solution of the system is (2, 0). Check this solution in both equations to verify the results.

Sometimes one or both equations will need to be modified so that the coefficients of one of the variables are opposites. This can be achieved by multiplying either or both of the equations by a constant as shown in the next example.

Example 4

Linear Combination Method

Solve the system:

$$5x + 2y = 7$$
$$3x - 6y = -3$$

You can obtain coefficients of y that are opposites by multiplying the first equation by 3.

$$5x + 2y = 7 \rightarrow \quad 15x + 6y = 21 \quad \text{(multiply by 3)}$$
$$3x - 6y = -3 \rightarrow \quad \underline{3x - 6y = -3} \quad \text{(same equation)}$$
$$18x \quad\quad = 18 \quad \text{(add equations)}$$
$$x \quad = 1 \quad \text{(solve for } x\text{)}$$

Now substitute $x = 1$ into either of the two original equations to find the value of y.

$$5x + 2y = 7$$
$$5(1) + 2y = 7 \quad \text{(substitute 1 for } x\text{)}$$
$$5 + 2y = 7 \quad \text{(simplify)}$$
$$2y = 2 \quad \text{(add } -5 \text{ to both sides)}$$
$$y = 1 \quad \text{(divide both sides by 2)}$$

Therefore, the solution of the system is (1, 1). Be sure to check your answer.

Rule — The Linear Combination Method

1. Obtain coefficients of either x or y that are opposites of each other. You may multiply one or both equations by a constant to achieve this goal.
2. Add the two equations together, producing a new equation containing only a single variable. Solve this equation for the value of that variable.
3. Solve for the remaining variable by substituting the value you obtained in step 2 into either of the original equations.
4. Check your answer to be sure that it satisfies both of the original equations.

Remember that a linear system of equations may intersect in one point and therefore have one solution (consistent), may be parallel and have no solution (inconsistent), or may be the same line and have an infinite set of solutions (dependent). It is easy to see these results graphically, but we need to examine how linear systems having no solution and those systems that are dependent behave algebraically.

Example 5 — Linear Systems with No Solution

Solve the system:

$$2x - 6y = 3$$
$$x - 3y = \frac{4}{3}$$

First, it will be simpler to work with the equations if there are no fractions in the problem. Therefore, we will multiply the second equation by 3 to clear the fractions.

$$3\left(x - 3y = \frac{4}{3}\right)$$
$$3(x) - 3(3y) = 3\left(\frac{4}{3}\right)$$
$$3x - 9y = 4$$

To obtain coefficients of x that are opposites, multiply the first equation by 3 and the second by -2.

$$2x - 6y = 3 \rightarrow \quad 6x - 18y = 9 \quad \text{(multiply by 3)}$$
$$3x - 9y = 4 \rightarrow \quad \underline{-6x + 18y = -8} \quad \text{(multiply by } -2\text{)}$$
$$ 0 = 1 \quad \text{(add the equations)}$$

Both the x and y variables have been eliminated at the same time. Because $0 = 1$ is a false statement, there is no ordered pair (x, y) that satisfies this system. Therefore, there is no solution for this system of equations. If the system were graphed, the lines would be parallel.

Example 6 — Linear Systems with an Infinite Number of Solutions

Solve the system:

$$2x = 4y - 8$$
$$x - 2y = -4$$

First, each equation needs to be written in standard form. Then the second equation can be multiplied by -2 to eliminate the x variable.

$$2x = 4y - 8 \rightarrow 2x - 4y = -8 \rightarrow \quad 2x - 4y = -8 \quad \text{(write in correct order)}$$
$$x - 2y = -4 \quad \rightarrow \quad x - 2y = -4 \rightarrow \underline{-2x + 4y = 8} \quad \text{(multiply by } -2\text{)}$$
$$ 0 = 0 \quad \text{(add equations)}$$

Both sides of the new equation are eliminated, leaving the identity $0 = 0$. This means that the two equations are identical. Such equations are called **dependent equations** and all solutions of the first equation are also solutions of the second equation. There are **infinitely many solutions** that satisfy both equations in this case. If the two equations were graphed, the same line would result from both equations.

Cramer's Rule

Another method used to solve linear systems is **Cramer's rule**, named after a Swiss mathematician who lived more than 200 years ago. Cramer's rule involves the use of an array of numbers called a determinant to calculate the values for the solution (x, y) of the system.

A **determinant** is a *square array of numbers* such as

$$\begin{vmatrix} a & b \\ c & d \end{vmatrix}$$

where the array has the same number of rows as columns and is enclosed between two vertical lines. The numbers a, b, c, and d are called **elements**. The value of a determinant is calculated as follows:

IMPORTANT EQUATIONS

Value of a Second-Order Determinant

$$\begin{vmatrix} a & b \\ c & d \end{vmatrix} = ad - bc$$

Look at the following example of this process.

Example 7 — Determining the Value of a Determinant

(a) Evaluate

$$\begin{vmatrix} 2 & 5 \\ 6 & 3 \end{vmatrix}$$

$$\begin{vmatrix} 2 & 5 \\ 6 & 3 \end{vmatrix} = (2)(3) - (5)(6) = 6 - 30 = -24$$

(b) Evaluate

$$\begin{vmatrix} -1 & -3 \\ 5 & 2 \end{vmatrix}$$

$$\begin{vmatrix} -1 & -3 \\ 5 & 2 \end{vmatrix} = (-1)(2) - (-3)(5) = -2 - (-15) = 13$$

We can use determinants to help us solve a system of equations by using the coefficients of the variables and the constants in each equation to form arrays. Cramer's rule can be written as follows:

Rule — Cramer's Rule

Given the system:

$$ax + by = c$$
$$dx + ey = f$$

$$x = \frac{\begin{vmatrix} c & b \\ f & e \end{vmatrix}}{\begin{vmatrix} a & b \\ d & e \end{vmatrix}} \qquad y = \frac{\begin{vmatrix} a & c \\ d & f \end{vmatrix}}{\begin{vmatrix} a & b \\ d & e \end{vmatrix}}$$

To use this rule, the equations must be written in standard form. If the determinant of the numerator is not 0 and the determinant of the denominator is 0, the system is inconsistent (the lines would be parallel). If the determinants of both the numerator and the denominator are 0, the system is dependent (same line). If the determinant of the denominator is a nonzero number, then the lines are consistent and there is a solution (x, y) for the system. The following examples use Cramer's rule to solve the given systems of equations.

Example 8 — Solving a System Using Determinants

Solve the system:

$$3x + 5y = -1$$
$$2x - 3y = 12$$

(a) Let $a = 3$, $b = 5$, $c = -1$, $d = 2$, $e = -3$, and $f = 12$.

(b) Substitute values into Cramer's rule:

$$x = \frac{\begin{vmatrix} c & b \\ f & e \end{vmatrix}}{\begin{vmatrix} a & b \\ d & e \end{vmatrix}} = \frac{\begin{vmatrix} -1 & 5 \\ 12 & -3 \end{vmatrix}}{\begin{vmatrix} 3 & 5 \\ 2 & -3 \end{vmatrix}} = \frac{(-1)(-3) - (5)(12)}{(3)(-3) - (5)(2)} = \frac{-57}{-19} = 3$$

$$y = \frac{\begin{vmatrix} a & c \\ d & f \end{vmatrix}}{\begin{vmatrix} a & b \\ d & e \end{vmatrix}} = \frac{\begin{vmatrix} 3 & -1 \\ 2 & 12 \end{vmatrix}}{\begin{vmatrix} 3 & 5 \\ 2 & -3 \end{vmatrix}} = \frac{(3)(12) - (-1)(2)}{(3)(-3) - (5)(2)} = \frac{38}{-19} = -2$$

(c) Therefore, this system is consistent and has a solution $(3, -2)$. Check this solution by substituting it into the original equations as previously discussed.

$$3x + 5y = -1 \qquad\qquad 2x - 3y = 12$$
$$3(3) + 5(-2) = -1 \qquad\qquad 2(3) - 3(-2) = 12$$
$$9 + (-10) = -1 \qquad\qquad 6 - (-6) = 12$$
$$-1 = -1 \quad \text{(Yes)} \qquad\qquad 12 = 12 \quad \text{(Yes)}$$

Example 9 — Solving a System Using Determinants

Solve the system:

$$2x = 8 - y$$
$$4x + 2y = 3$$

(a) To use Cramer's rule, we must first put our equations in standard $Ax + By = C$ order.

$$2x = 8 - y \rightarrow 2x + y = 8$$
$$4x + 2y = 3$$

Now let $a = 2$, $b = 1$, $c = 8$, $d = 4$, $e = 2$, and $f = 3$.

(b) Substitute values into Cramer's rule:

$$x = \frac{\begin{vmatrix} c & b \\ f & e \end{vmatrix}}{\begin{vmatrix} a & b \\ d & e \end{vmatrix}} = \frac{\begin{vmatrix} 8 & 1 \\ 3 & 2 \end{vmatrix}}{\begin{vmatrix} 2 & 1 \\ 4 & 2 \end{vmatrix}} = \frac{(8)(2) - (1)(3)}{(2)(2) - (1)(4)} = \frac{13}{0} = \text{undefined}$$

(c) Because the denominator has a value of 0 and the numerator is a nonzero number, this system is inconsistent and has no solution. The lines, if graphed, would be parallel.

In Example 9, the denominator of the fraction is 0 and the numerator is a nonzero number. That result indicates that there is no solution because the lines are parallel. If both the numerator and the denominator have a value of 0, this indicates that the lines are dependent and that there are an infinite number of solutions.

Practice Set 6-2

Solve the following systems using the substitution method.

1. $x = 2y$
   $x + y = 18$

2. $x - 2y = 0$
   $3x = y$

3. $5x + 3y = 11$
   $x = 5y + 5$

4. $3x + 2y = 0$
   $x + 2y = 4$

5. $3x - y = 4$
   $9x - 5y = -10$

6. $x - 2y = 5$
   $6y - 3x = -15$

Solve the following systems using the linear combination method.

7. $2x - 3y = 7$
   $4x + 3y = 5$

8. $x + 3y = 2$
   $x - 2y = -3$

9. $x + 7y = 12$
   $3x - 5y = 10$

10. $5x + 2y = 7$
    $3x - 6y = -3$

11. $x = y + 3$
    $3y = x + 2$

12. $6x - 5y = 3$
    $10y = 5 + 12x$

13. $3x - 2y = 6$
    $-6x + 4y = -12$

14. $3x + 3y = 7$
    $2x + 5y = 3$

Evaluate the following determinants.

15. $\begin{vmatrix} 1 & 2 \\ 2 & 5 \end{vmatrix}$

16. $\begin{vmatrix} 2 & 3 \\ 4 & 5 \end{vmatrix}$

17. $\begin{vmatrix} 3 & 2 \\ 6 & 4 \end{vmatrix}$

18. $\begin{vmatrix} -2 & 2 \\ 3 & 5 \end{vmatrix}$

19. $\begin{vmatrix} 3 & 5 \\ -2 & 4 \end{vmatrix}$

20. $\begin{vmatrix} 7 & 5 \\ -3 & -2 \end{vmatrix}$

Solve each system using Cramer's rule.

21. $x - 4y = 1$
    $2x + 3y = 13$

22. $3x - 8y = -43$
    $-9x - 6y = -21$

23. $6x - 2y = 36$
    $5x + 4y = 47$

24. $15x - 6y = -15$
    $9x + 12y = 4$

25. $6x - 7y = 28$
    $-4x + 5y = -20$

26. $3x - 4y = 6$
    $15x - 20y = 36$

27. $2x - y = -10$
    $x + y = 4$

28. $2x + 3y = 7$
    $x - 6y = 11$

29. $3x - 2y = 13$
    $-6x + 4y = -26$

30. $2x = 3 - 7y$
    $10x + 3y = -1$

31. $y = 3x$
    $2x - 15y = 86$

32. $2x - 3y = 6$
    $-4x + 6y = 8$

33. $2x - 3y = 10$
    $2x + 5y = 2$

34. $x + y = -25$
    $2x = 3y + 70$

35. $x = -4y$
    $3x = 5y + 17$

36. $x + 2y = 4$
    $3x - y = 1$

37. $3y + 2x = -36$
    $3x = 5y + 22$

38. $0.3x + y = 0.1$
    $0.1x + 0.2y = -0.1$

39. $0.2x - 0.4y = 2.2$
    $0.5x - y = 1.6$

40. $0.5x - 0.75y = -0.5$
    $0.125x - 0.75y = 2.375$

41. $2x + 3y = 1$
    $3x + 2y = 2$

42. $2x + y = 1$
    $x - 7y = -2$

43. $\dfrac{2}{3}x - \dfrac{2}{5}y = 1$
    $\dfrac{5}{2}x - \dfrac{3}{2}y = 2$

44. $\dfrac{1}{2}x + \dfrac{3}{4}y = \dfrac{7}{4}$
    $\dfrac{1}{3}x - \dfrac{1}{6}y = \dfrac{1}{2}$

Solve the following systems using any method.

45. $2x - y = 2$
    $4x + 3y = 24$

46. $x = y + 1$
    $2x - 2y = -5$

47. $y = 2x - 1$
    $y = x + 1$

48. $6x + 21y = 132$
    $6x - 4y = 32$

49. $5x - 2y = 14$
    $6x - 5y = 9$

50. $x - 2y = 7$
    $3x + y = \dfrac{7}{2}$

51. $x - y = 8$
    $3y = x + 2$

52. $2x - 8y = 4$
    $\dfrac{5}{2}x + y = 5$

53. $14x - 6 = -2y$
    $7x + y = 3$

54. $8y + 8x = 8$
    $y = 2x + 4$

55. $7x + 3y = -8$
    $5x - 4y = -61$

56. $3x - 4y = 6$
    $5x - 3y = -12$

57. Is (2, 3) a solution of the following system?
    $$3x + 5 = y$$
    $$2y + x = 8$$

58. Is $\left(-1, \dfrac{2}{5}\right)$ a solution of the following system?
    $$x + 5y = 1$$
    $$x + y = -\dfrac{2}{5}$$

59. The sum of two numbers is 20 and their difference is 2. Write a system of equations that will model this problem. Then solve the system using any method you choose.

60. The sum of two numbers is 63. Their difference is 19. Write a linear system to model this problem and use it to find the two numbers.

61. The sum of two numbers is 30. If twice one of them is 3 times the other, what are the numbers?

62. The difference between two numbers is 18. Find the two numbers if their sum is 0.

63. One-half the sum of two numbers equals their difference. What are the numbers if their sum is 64?

64. The difference between two numbers is 2. If 3 times the larger plus 5 times the smaller is 94, what are the numbers?

65. Three times one number is 4 times another. If the sum of the two numbers is 49, what are the numbers?

66. Find two numbers whose sum is 15 if one of the numbers subtracted from twice the other is also 15.

Section 6-3 Applications of Linear Systems

There are many applied problems in science, business, and health services for which the correct mathematical model is a system of linear equations. These models involve situations requiring two or more variables and two or more conditions on those variables. An appropriate system should be constructed and then solved using any of the methods previously discussed. Look at some of the following examples of applications involving linear systems.

Example 1 Problems of Quantity and Cost

The newly renovated Morris Theater will seat 1500 people. Reserved seats will cost $8 each, and all other seats will cost $5 each. How many of each type of ticket must be sold for the opening performance if the new owners want to generate a total of $10,500 in sales?

Let
$$x = \text{reserved seats @ \$8}$$
$$y = \text{general seats @ \$5}$$

(a) There are two conditions present in this problem. One involves the total number of tickets available (1500). The other condition involves the amount of money from sales that needs to be generated ($10,500). Write two equations that reflect these two conditions.

Total ticket sales: $x + y = 1500$

Revenue:

Reserved seats	x	·	$8	=	$8x$
General seats	y	·	$5	=	$5y$

Total Revenue: $8x + 5y = \$10{,}500$

(b) Solve the system using any of the methods we have studied. Here we will use linear combination.

$$x + y = 1500 \quad \rightarrow \quad -5x - 5y = -7500 \quad \text{(multiply by } -5\text{)}$$
$$8x + 5y = 10{,}500 \quad \rightarrow \quad \underline{8x + 5y = 10{,}500}$$
$$3x = 3000 \quad \text{(add the equations)}$$
$$x = 1000 \quad \text{(solve for } x\text{)}$$

(c) Now substitute $x = 1000$ into either of the original equations to find y.

$$x + y = 1500$$
$$1000 + y = 1500 \quad \text{(substitute } x = 1000\text{)}$$
$$y = 500 \quad \text{(add } -1000 \text{ to both sides)}$$

Therefore, 1000 seats at $8 and 500 seats at $5 should be sold.

(d) Check to see if this satisfies the original requirements of the problem.

Total seats should be $1500 = 1000 \ @ \ \$8 + 500 \ @ \ \$5$ (true)

Total revenue is $\$10{,}500 = (1000)(\$8) + (500)(\$5)$
$$= \$8000 + \$2500$$
$$= \$10{,}500 \quad \text{(true)}$$

Example 2   Problems of Quantity and Cost

A service organization is selling hot dogs and hamburgers at the local Fall Harvest Festival. Three hot dogs and two hamburgers cost $9.75, while one hot dog and three hamburgers cost only $8.50. Find the cost of a hamburger and the cost of a hot dog. Let x = the cost of a hot dog and y = the cost of a hamburger.

(a) There are two orders given in the problem. Write two equations to reflect these orders.

	Quantity		Item		Cost
Order #1	3	·	x	=	$3x$
	2	·	y	=	$2y$
Total					$9.75
Order #2	1	·	x	=	x
	3	·	y	=	$3y$
Total					$8.50

Order #1: $3x + 2y = \$9.75$
Order #2: $x + 3y = \$8.50$

(b) Solve the system using any of the methods we have studied. Here we will use linear combination.

$$
\begin{aligned}
3x + 2y = 9.75 &\rightarrow \quad 3x + 2y = 9.75 \\
x + 3y = 8.50 &\rightarrow \quad \underline{-3x - 9y = -25.50} \quad \text{(multiply by } -3\text{)} \\
& \qquad\qquad -7y = -15.75 \quad \text{(add the equations)} \\
& \qquad\qquad\quad y = 2.25 \quad \text{(solve for } y\text{)}
\end{aligned}
$$

(c) Now substitute $y = 2.25$ into either of the original equations to find x.

$$
\begin{aligned}
x + 3y &= 8.50 \\
x + 3(2.25) &= 8.50 \quad \text{(substitute } x = 2.25\text{)} \\
x + 6.75 &= 8.50 \quad \text{(simplify)} \\
x &= 1.75 \quad \text{(add } -6.75 \text{ to both sides)}
\end{aligned}
$$

Therefore, the cost of a hot dog is $1.75 and the cost of a hamburger is $2.25.

The next examples involve mixture problems. These types of problems occur in chemistry labs, where chemists mix solutions of differing concentrations, in metallurgy and the production of metal alloys, and even in the production of food products such as mixed nuts. Concentrations of items being mixed may be given in the form of a percent. Remember to convert the percent to its equivalent decimal form before using it in an equation.

Example 3 Mixture Problems

How many milliliters of a 10% acid solution should be mixed with a 30% acid solution to produce 500 mL of a 15% solution?

There are two conditions that must be satisfied in this problem. The total volume of the mixture must be 500 mL and the concentration of the new mixture must be 15%.

(a) Write two equations to reflect these conditions.
Let $x =$ amount of 10% solution and $y =$ amount of 30% solution.

	Volume of Solution		Concentration of Acid		Volume of Acid
10% solution	x	·	0.1	=	$0.1x$
30% solution	y	·	0.3	=	$0.3y$
15% solution	500	·	0.15	=	75

Concentration: $0.1x + 0.3y = 75$
Total Volume: $\qquad x + y = 500$

(b) In this problem, we will use Cramer's rule to find the solution.

$$x + y = 500$$
$$0.1x + 0.3x = 75$$

$$x = \frac{\begin{vmatrix} c & b \\ f & e \end{vmatrix}}{\begin{vmatrix} a & b \\ d & e \end{vmatrix}} = \frac{\begin{vmatrix} 500 & 1 \\ 75 & 0.3 \end{vmatrix}}{\begin{vmatrix} 1 & 1 \\ 0.1 & 0.3 \end{vmatrix}} = \frac{(500)(0.3) - (1)(75)}{(1)(0.3) - (1)(0.1)} = \frac{150 - 75}{0.3 - 0.1} = \frac{75}{0.2} = 375$$

$$y = \frac{\begin{vmatrix} a & c \\ d & f \end{vmatrix}}{\begin{vmatrix} a & b \\ d & e \end{vmatrix}} = \frac{\begin{vmatrix} 1 & 500 \\ 0.1 & 75 \end{vmatrix}}{\begin{vmatrix} 1 & 1 \\ 0.1 & 0.3 \end{vmatrix}} = \frac{(1)(75) - (0.1)(500)}{(1)(0.3) - (1)(0.1)} = \frac{75 - 50}{0.3 - 0.1} = \frac{25}{0.2} = 125$$

This means that 375 mL of 10% acid solution, and 125 mL of 30% acid solution will produce 500 mL of a 15% acid solution.

Check these solutions to see if they meet the original requirements.

Example 4

Value Mixture Problem

A local organic grocery store is making 25 pounds of a trail mix consisting of walnuts and raisins. The raisins cost $2.87 per pound, and the walnuts cost $7.06 per pound. How many pounds of each item are in the mix if the total value of the trail mix is $134.60?

There are two conditions that must be satisfied in this problem. The total amount of the mixture is 25 pounds, and the total value of the new mixture is $134.60.

(a) Write two equations to reflect these conditions.
Let x = the number of pounds of walnuts needed and let y = the number of pounds of raisins needed.

	Amount of Each Item		Cost of Each Item		Total Value of the Items
walnuts	x	·	$7.06	=	$7.06x
raisins	y	·	$2.87	=	$2.87y
mixture	25				$134.60

Amount: $x + y = 25$
Value: $7.06x + 2.87y = 134.60$

(b) In this problem, we will use the substitution method to solve.

(1) Solve the amount equation for y. $x + y = 25$
                                                       $y = 25 - x$

(2) $7.06x + 2.87(25 - x) = 134.60$ (substitute this expression for y)
       $7.06x + 71.75 - 2.87x = 134.60$ (distributive property)
             $4.19x + 71.75 = 134.60$ (simplify)
                    $4.19x = 62.85$ (add -71.75 to each side)
                          $x = 15$ (solve for x)

(3) Now substitute $x = 15$ into either of the original equations to find y.

$$x + y = 25$$
$$15 + y = 25 \quad \text{(substitute } x = 15\text{)}$$
$$y = 10 \quad \text{(add } -15 \text{ to both sides)}$$

Therefore, in order to make a mixture that has a value of $134.60, the grocer should mix 15 pounds of walnuts with 10 pound of raisins.

Check these solutions to see if they meet the original conditions of the problem.

Supply and demand are two important concepts in economics. **Supply** is the number of items manufacturers are willing to produce. **Demand** is the number of items consumers are willing to purchase. The equations or functions used to define supply and demand for a particular item are usually based on the selling price of the item. Each is a function of the selling price. If demand is greater than supply, then stores will continually have empty shelves and unhappy customers asking for a particular product. If supply is greater than demand, then the manufacturer will have warehouses filled with a product that consumers will not buy. Businesses strive for the **equilibrium point**, where supply equals demand. This point would be the intersection point of the supply and demand equations for a business.

Example 5 Supply and Demand Equations

The supply S and demand D (in thousands) for a certain product are given by the equations $S = f(p) = 3p - 15$ and $D = h(p) = 30 - 2p$, where p represents the price of the product in dollars. Find the value of p at the equilibrium point.

At the equilibrium point, supply equals demand, so we set the two equations equal to each other. (See Figure 6-6.)

$$3p - 15 = 30 - 2p$$
$$3p + 2p - 15 = 30 - 2p + 2p$$
$$5p - 15 = 30$$
$$5p - 15 + 15 = 30 + 15$$
$$5p = 45$$
$$p = 9$$

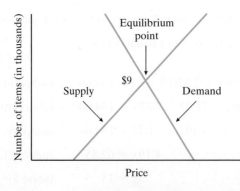

FIGURE 6-6
Example 5. Supply and demand equations.

Therefore, the equilibrium point where supply equals demand is $9. Notice that the demand exceeds the supply to the left of the equilibrium point (when the price is below $9) and supply exceeds demand to the right of the equilibrium point (when the price is above $9).

If we wish to also calculate the number of items that should be produced at this price, we substitute $9 into either equation to calculate the answer.

$$S = f(p) = 3p - 15$$
$$f(9) = 3(9) - 15 = 27 - 15 = 12 \text{ (or 12,000) items}$$

Some manufacturers use a technique called **break-even analysis** to help them determine the number of items they must produce and sell to cover expenses. All income after the break-even point is profit for the company. To determine the break-even point, we let the equation that defines total cost of production equal the equation that defines total revenue from sales. Cost of production and total revenue are both functions of the number of items produced and sold. The graph shown in Figure 6-7 illustrates this idea.

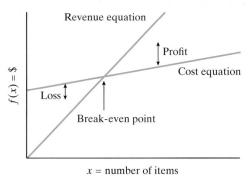

FIGURE 6-7
Break-even analysis.

Example 6 Break-Even Point

DSC Printing Company prints paperback books for authors who wish to self-publish their works. They charge $15 per book plus a setup fee of $1200. If an author uses this publishing company and sells each book for $25, how many books must be sold in order for the author to break even?

First, we will write mathematical models to represent the cost equation and the revenue equation. Then we will calculate the break-even point by setting the cost equation equal to the revenue equation.

(a) Let b be the number of books printed.

$$\text{Cost} = C(b) = \$1200 + 15b$$
$$\text{Revenue} = R(b) = \$25b$$

(b) Set the equations equal to each other.

$$R(b) = C(b)$$
$$25b = 1200 + 15b$$
$$25b - 15b = 1200$$
$$10b = 1200$$
$$b = 120$$

Therefore, an author will break even when 120 books have been printed and sold. If fewer than 120 books are printed sold at this price, the author will lose money, and if more than 120 books are printed and sold at this price, the author will make a profit.

Example 7 — Geometry Applications

The fence surrounding a backyard playground has a length 3 times its width. The perimeter of this rectangular playground is 296 ft. Find the dimensions of the playground.

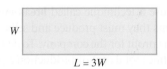

There are two conditions that must be satisfied in the problem. The length of the longer side is three times the width of the shorter side, and the total distance around the playground is 296 ft.

(a) Write two equations to reflect these conditions. Let L = length of the playground and W = width of the playground.

$$L = 3W$$
$$2L + 2W = 296$$

(b) In this problem, we will use substitution to find the solution.

$$2L + 2W = 296$$
$$2(3W) + 2W = 296 \quad \text{(substitute for } L\text{)}$$
$$6W + 2W = 296$$
$$8W = 296 \quad \text{(simplify)}$$
$$W = 37 \quad \text{(solve for } W\text{)}$$

(c) Now complete the solution by substituting 37 for W in one of the equations.

$$L = 3W$$
$$L = 3(37) = 111$$

Therefore, the playground is 111 ft long and 37 ft wide.

Practice Set 6-3

Write a system of equations for each problem and then solve using any method we have studied.

1. The McCook apartment building contains 20 rental units. One-bedroom apartments rent for $650 per month, and two-bedroom apartments rent for $825 per month. The total monthly income from all apartments is $14,400. How many one-bedroom and two-bedroom apartments are in the apartment building?

2. Tickets to a band concert were $5.00 for balcony and $10.00 for orchestra seats. If attendance at one show was 800, and if total receipts for that show were $7000, how many people bought orchestra seats?

3. A hotel charges $90 per day for a double room and $80 per day for a single. If there are 80 rooms that are occupied for a total revenue of $6930, how many rooms of each kind are there?

4. Marianne works two jobs for a total of 30 hours per week. One job pays $8 per hour, and the second job pays $10.50 per hour. If her weekly combined pay from the two jobs is $285, find the number of hours that she works at each job.

5. A total of $12,000 is invested in two funds paying 5% and 4.5% simple interest. If the yearly interest is $560, how much of the $12,000 is invested at each rate?

6. A total of $150,000 is invested in two funds paying 6.25% and 6% simple interest. If the yearly interest is $9212.50, how much of the $150,000 is invested at each rate?

7. Maribeth bought three leotards and four pairs of tights at the Dancer's Studio for $185.00, while Beth bought two leotards and three pairs of tights for $127.50. What are the price of a leotard and the price of a pair of tights?

8. Joan bought 3 heads of lettuce and 2 lb of tomatoes at Ingles for $4.05. Betty bought 2 heads of lettuce and 3 lb of tomatoes for $4.35 at the same store. What are the price of a head of lettuce and the price of a pound of tomatoes at Ingles?

9. The Math Club and the Physics Brain Bunch both meet at the local pizza parlor once a month. The math club purchases three large pepperoni pizzas and four pitchers of soda for a total cost of $61. The physics club has two large pepperoni pizzas and three pitchers of soda for a total cost of $42. Find the cost of a large pepperoni pizza and the cost of a pitcher of soda.

10. At Roasters Café, the cost of 3 cups of coffee and 6 donuts is $15, while the cost of 5 cups of coffee and a dozen donuts is $28. Use these totals to find the cost of one cup of coffee and the cost of one donut.

11. Cathy purchased 16 shares of Coca-Cola® stock and 10 shares of Pepsi® stock for $1097.40. Nancy bought 48 shares of Coca-Cola and 12 shares of Pepsi for $2509.20. What are the price of a share of Coca-Cola and the price of a share of Pepsi?

12. At an office supply store, Mary buys seven packages of notebook paper and four pens for $6.40. At the same store, Sue buys two packages of notebook paper and 19 pens for $5.40. What was the cost of each of these items?

13. A chemist wants to mix a solution containing 10% acid with another solution containing 25% acid. She needs to produce a total of 30 mL, and the concentration of acid in the final solution should be 18%. How much of each acid should she use?

14. Everett wishes to clean his boat with a 15% soap solution. How much of a 10% soap solution and a 20% soap solution should he add to a quart of pure water to make 30 qt of a 15% solution?

15. A silversmith uses two silver alloys. The first alloy is 60% silver and the second is 40% silver. How many grams of each should be mixed together to produce 20 g of an alloy that is 52% silver?

16. Annette wishes to mix a tomato sauce that is 17% sugar with a sauce that is 30% sugar to obtain 2.6 L of a tomato sauce that is 24% sugar. How much of each should she mix?

17. Juan has available a 10% acetic acid solution and a 60% acetic acid solution. Find how many milliliters of each solution he should mix to make 50 mL of a 40% acetic acid solution.

18. How many ounces of pure water and how many ounces of a 16% butterfat solution should Charlotte mix to obtain 32 oz of a 10% butterfat solution?

19. A grocer is making 26 pounds of a batch of mixed nuts containing cashews and peanuts. The peanuts cost $2.95 per pound, and the cashews cost $9.95 per pound. How many pounds of each item are in the mix if the total value of the mixed nuts is $132.70?

20. A coffee house sells a blended coffee that is made using a $6-per-pound grade and a $4.50-per-pound grade. If 5 pounds of this blend cost $27, how many pounds of each type of coffee are in this mixture?

21. A grocer plans to mix candy that sells for $3.40 per pound with candy that sells for $4.60 per pound. He wants to make 10 lb of a mixture that will sell for $4 per pound. How much of each type of candy should he use?

22. A pet store is mixing black oil sunflower seeds with white millet to produce a mixture that sells for $1.60 per pound. If sunflower seeds cost $1.85 per pound and white millet costs $1.45 per pound, how many pounds of each are needed to produce 80 pounds of this mixture?

23. Use the following supply and demand equations to determine the equilibrium price. Supply: $S(p) = 1200 - 2.8p$. Demand: $D(p) = 800 + 3.2p$. Explain what your answer means.

24. Use the following supply and demand equations to determine the equilibrium price. Supply: $S(p) = 3p$. Demand: $D(p) = 600 - 2p$. Explain what your answer means.

25. The supply and demand functions for a popular new DVD are reported by the company to be $D(p) = 9500 - 100p$ and $S(p) = 5000 + 200p$, where p is the price of the DVD in dollars, $D(p)$ is the demand, and $S(p)$ is the supply. Determine the equilibrium price for this DVD.

26. The supply and demand functions for a popular new compact disc are reported by the company to be $D(p) = 1500 - 20p$ and $S(p) = 1125 + 10p$, where p is the price of the compact disc in dollars, $D(p)$ is the demand, and $S(p)$ is the supply. Determine the equilibrium price for this compact disc.

27. Timjon Company produces children's cartoon watches. It has determined that the following equations describe the supply and demand curves for the watches. Demand: $D(p) = 30 - 3p$. Supply: $S(p) = 4p - 5$, where p = the price of a watch. Determine the equilibrium price for these supply and demand functions. Graph these functions to show the equilibrium price.

28. Windsor Company produces video games. It has determined that the following equations describe the supply and demand curves for the games. Demand: $D(p) = 60 - 2p$. Supply: $S(p) = 4p - 6$, where p = the price of a video game. Determine the equilibrium price for these supply and demand functions. Graph these functions to show the equilibrium price.

29. A company produces backpacks and charges $32 per pack. The monthly cost of manufacturing the backpacks includes a fixed cost of $25,500 and $15 per backpack. Find the number of backpacks that need to be sold in a month for the company to break even. Draw a sketch of these functions.

30. A company produces toaster ovens at a cost of $18 each and has a fixed production cost of $5500 per month. Each toaster oven is sold to stores for $38. Draw a sketch of the graph of these two equations and label the break-evens point. How many toaster ovens must be sold before the company begins to make a profit?

31. John's Rent-a-Car Agency charges $26 a day plus 15¢ per mile for a compact car. The same type of car can be rented from Cookie's Rental Agency for $18 a day plus 20¢ per mile. How many miles would you need to drive in one day for the cost of the cars to be the same? If you plan to drive 200 miles, which rental agency should you use?

32. Martel is paid $200 plus a 5% commission on sales per week. He is considering a new salary plan that would pay him a straight 15% commission on sales. What amount of sales would he need to make for the two salaries to be equal? If he normally averages $1500 per week in sales, which plan should he choose?

33. Industrial Paper Products pays their salespeople a straight 12% commission on all monthly sales. The Paper Factory pays salespeople 4% of all monthly sales plus a $1500 per month base salary. What amount of sales would a salesperson need to make for the two monthly salaries to be equal? For a salesperson averaging sales of $5000 per week, which company would provide a higher monthly salary?

34. Sonya needs to purchase a new refrigerator. Model A costs $950 to purchase and $32 in electricity per year to operate. Model B costs $1275 to purchase but costs only $22 in electricity per year to operate. After how many years will the total cost of both units be equal? If she expects to keep the refrigerator for about 15 years, which model is the better buy?

35. The length of a rectangle is 3 ft greater than its width. If its perimeter is 14 ft, find the length and width of the rectangle.

36. The perimeter of a rectangular window is 200 cm. The length of one side is 20 cm less than 11 times the other. What are the dimensions of the window?

37. The perimeter of a rectangular painting is 60 in. Its length is 6 inches more than its width. What are the dimensions of a frame for this painting?

38. The perimeter of Kevin's lawn is 54 m. It is in the shape of a rectangle with the length of one side 3 m less than twice the width. What are the dimensions of the lawn?

39. When not at the lake, Cassie spends her time in a rectangular dog pen that her master constructed with 28 yd of fencing. If the length of the pen is 2 yd longer than twice the width, what are its dimensions?

40. The perimeter of an isosceles triangle is 48 in. If the length of the shortest side is 3 in. less than either of the other two sides, what are the lengths of the sides of the triangle?

Section 6-4 Systems of Nonlinear Equations

Not all problems or relationships can be described using linear equations. We have previously discussed graphing nonlinear equations using a graphing calculator. In this section, we will look at the solutions of **nonlinear systems of equations**. In a nonlinear system of equations, at least one of the two equations contains a product of variables (e.g., xy) or a variable with an exponent other than 1, and its graph will be a curve.

Example 1 Solving a Nonlinear System of Equations with a Graphing Calculator

Solve the following system of equations.

$$x^2 - y = 0$$
$$2x - y = 0$$

The first equation is a nonlinear equation because the x variable has an exponent of 2. We will use the graphing calculator to show us the intersection points of these two equations. Unlike linear systems, nonlinear systems may have more than one intersection point.

(a) Solve each equation for y and use the $\boxed{Y=}$ function to enter the equations into the calculator. Then use the $\boxed{\text{GRAPH}}$ button to display the graphs. Use the following values in the $\boxed{\text{WINDOW}}$ setting:

$$X_{\min} = -5 \quad Y_{\min} = -5$$
$$X_{\max} = 5 \quad Y_{\max} = 5$$
$$X_{\text{scl}} = 1 \quad Y_{\text{scl}} = 1$$

See Figure 6-8.

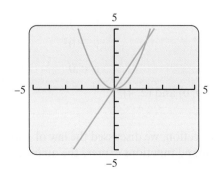

FIGURE 6-8
Example 1.

(b) Notice that the line $(y = 2x)$ intersects the parabola $(y = x^2)$ at two points. Use the TRACE function on the calculator to get the approximate values of these solutions. Use the CALCULATE function to get the exact values. The solution set is $\{(0, 0), (2, 4)\}$.

You can see from this example that the graphing calculator will help us visualize the problems and the number of intersection points in the solution set. However, using the TRACE function does not always give us an exact answer. We can use an algebraic method to solve some systems and calculate an exact answer.

Example 2 **Solving a Nonlinear System of Equations Algebraically**

Solve the system:
$$x^2 - y = 0$$
$$2x - y = 0$$

(a) First solve each equation for y.
$$y = x^2$$
$$y = 2x$$

(b) Because both x^2 and $2x$ equal y, they must be equal to each other. Now, using the method of substitution, we have a quadratic equation in one variable that can be solved by factoring (or with the quadratic formula).

$$\left. \begin{array}{l} y = x^2 \\ y = 2x \end{array} \right\} \rightarrow \quad x^2 = 2x$$

$$x^2 - 2x = 0 \quad \text{(set equation equal to 0)}$$
$$x(x - 2) = 0 \quad \text{(factor)}$$
$$x = 0 \quad x - 2 = 0 \quad \text{(set each factor equal to 0)}$$
$$ x = 2 \quad \text{(solve for } x\text{)}$$

(c) Calculate the y values for each of the x values by substituting $x = 0$ and $x = 2$ into either equation.

$$\begin{array}{ll} \text{If } x = 0 & \text{If } x = 2 \\ y = x^2 & y = x^2 \\ y = 0^2 & y = 2^2 \\ y = 0 & y = 4 \end{array}$$

Solution: $(0, 0)$ and $(2, 4)$.

In the last section, we discussed the law of supply and demand. Remember that supply and demand functions express the quantities of items supplied or demanded

Example 3

Supply and Demand Functions

The demand function for a specific toy is given by the function $p(x) = -2x^2 + 90$. The corresponding supply function is given by the function $p(x) = 9x + 34$. In both equations, p is the price in dollars and x is in thousands of units. Find the equilibrium price and number of units produced at this price.

(a) At market equilibrium, the price given by the demand function will equal the price given by the supply function.

Because $p(x) = -2x^2 + 90$ and $p(x) = 9x + 34$, we let

$$9x + 34 = -2x^2 + 90 \quad \text{(and solve it by factoring)}$$
$$2x^2 + 9x - 56 = 0 \quad \text{(set equation equal to 0)}$$
$$(2x - 7)(x + 8) = 0 \quad \text{(factor)}$$
$$2x - 7 = 0 \quad x + 8 = 0$$
$$2x = 7 \quad x = -8 \quad \text{(solve each factor for } x\text{)}$$
$$x = 3.5$$

Because x represents the number of toys in thousands, the answer cannot be negative. Therefore, $x = -8$ is an extraneous root and is not a possible answer.

The correct number of toys is 3.5 thousand or 3500.

(b) Now we need to calculate the equilibrium price. Use $x = 3.5$ and substitute into either equation to find p.

$$p(x) = 9x + 34$$
$$p(x) = 9(3.5) + 34$$
$$p(x) = 31.5 + 34 = 65.5 = \$65.50$$

Therefore, the equilibrium price for these toys is $65.50.

(c) Use your graphing calculator to sketch these functions and verify the answer you have calculated. Use the window setting shown in Figure 6-9.

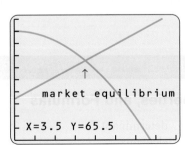

FIGURE 6-9
Example 3.

Practice Set 6-4

Use a graphing calculator to sketch each system of equations. Find the solution of the system algebraically and verify with the calculator.

1. $y = x^2$
   $3x + y = 10$

2. $y = x$
   $y = 2x^2 + 1$

3. $y = x$
   $y = -2x^2 + 1$

4. $y = x^2 + 2$
   $y = -x^2 + 4$

5. $y = x^2 - 5x + 6$
   $y = 2x$

6. $y = x^2 - 2$
   $y = -2x^2 + 6x + 7$

7. $y = x^2 - 4$
   $y = 2x - 1$

8. $y = x^2 + 1$
   $x + y = 3$

9. $2x + 3y^2 = 2$
   $x - y = 1$

10. $x + xy = -1$
    $xy = -6$

11. $x^2 + y^2 = 13$
    $x^2 - y^2 = 5$

12. $x^2 + y^2 = 5$
    $x + 2y = 3$

13. The sum of two numbers is 7, and the sum of their squares is 25. What are the numbers?

14. Find two numbers whose sum is 19 if the sum of their squares is 185.

15. The sum of the squares of two numbers is 58, and their difference is 10. Find one pair of such numbers.

16. The sum of two numbers is 12. Their difference is 2. Find the numbers.

17. The product of two positive numbers is 60, and their difference is 7. Let x be the smaller number and let y be the larger number. Write a system of equations that models this situation. Find the two numbers.

18. The product of two numbers is 4, and the sum of their squares is 8. Find the numbers.

19. The demand function for a certain type of sunglasses is given by the equation $p = -0.1x^2 - 2x + 90$, and the corresponding supply function is given by the function $p = 0.1x^2 - x + 30$, where p is in dollars and x is in thousands of pairs of sunglasses. Find the equilibrium price and the number of sunglasses produced at that price.

20. The supply and demand functions for a new product are as follows:

    Demand: $p(x) = 7 - 0.5x$
    Supply: $p(x) = 1 + 0.5x^2$

    where p is the price in dollars and x is the number of units in thousands. Find the number of units produced and the equilibrium price for this product.

21. The area of a room is 143 ft², and its perimeter is 48 ft. Let x be the width and let y be the length of the room. Write a system of equations that models this situation. Determine the length and the width of the room.

22. The area of the desktop is 1000 in.², and its perimeter is 130 in. Let x be the width and let y be the length of the desktop. Write a system of equations that models this situation. Find the length and the width of the desktop.

Chapter 6 Summary

Key Terms, Properties, and Formulas

break-even analysis
consistent
Cramer's rule
demand
dependent
dependent equation

determinant
elements
equilibrium point
inconsistent
infinitely many solutions
linear combination

nonlinear system of equations
solution
substitution
supply
system of equations

Points to Remember

Solving a system of equations graphically: Graph both lines on the same set of axes and locate their point of intersection. If they are parallel, the systems are inconsistent and there is no solution. If the two equations are for the same line, the system is dependent and there are infinitely many solutions.

Solving a system algebraically: Use substitution or linear combination. The substitution method involves replacing one of the variables in one equation with its equivalent from the other equation, resulting in a single equation with one variable. In linear combination, the two equations are added together in such a way as to produce a new equation with only one variable.

Cramer's rule: Use determinants to find the solution of a system of linear equations. Given the system:

$$ax + by = c$$
$$dx + ey = f$$

$$x = \frac{\begin{vmatrix} c & b \\ f & e \end{vmatrix}}{\begin{vmatrix} a & b \\ d & e \end{vmatrix}} \qquad y = \frac{\begin{vmatrix} a & c \\ d & f \end{vmatrix}}{\begin{vmatrix} a & b \\ d & e \end{vmatrix}}$$

A nonlinear system of equations includes at least one equation that is not linear. These systems may have several solutions, one solution, or no solution. A graphing calculator can help you visualize the system and its solutions.

Many application problems involve the use of systems of equations. Some of the more common types of problems include problems of quantity and cost, mixture problems, supply and demand problems, and break-even analysis problems.

Chapter 6 Review Problems

Decide if the given ordered pair is a solution of the given system.

1. $x - 2y = 4$ $(0, -2)$
   $2x + y = -2$

2. $3x - y = -13$ $(-5, -2)$
   $x - 2y = -1$

3. $x + y = 4$ $(1, 3)$
   $2x - y = 5$

4. $3x + y = 7$ $(2, 1)$
   $x = 2y$

Solve each system of equations by graphing.

5. $2x - y = 13$
   $x + y = 8$

6. $2x + y = 3$
   $4x + 2y = 1$

7. $5x - 3y = 9$
   $x + 2y = 7$

8. $x + y = 4$
   $2x - y = -1$

9. $2x - 3y = -6$
   $y = -3x + 2$

Solve each system of equations algebraically.

10. $3x - y = 7$
    $2x + y = 3$

11. $8x - 5y = 32$
    $4x + 5y = 4$

12. $8x + 2y = 2$
    $3x - y = 6$

13. $3x + 5y = 69$
    $y = 4x$

14. $6x + 3y = 0$
    $18x + 9y = 0$

15. $6x - 8y = 6$
    $-3x + 2y = -2$

16. $2x + 10y = 3$
    $x = 1 - 5y$

17. $4x + 5y = 44$
    $x + 2 = 2y$

18. $2x + 3y = -5$
    $3x + 4y = -8$

Use Cramer's rule to solve the following systems of equations.

19. $2x - y = 4$
    $3x + y = 21$

20. $4x + 2y = 2$
    $5x + 4y = 7$

21. $4x + 5y = 2$
    $-8x - 10y = 1$

22. $2x + y = -4$
    $x = y + 7$

23. $0.8x - 0.1y = 0.3$
    $0.5x - 0.2y = -0.5$

24. Solve the following system of equations using all methods—graphing, algebraic process, and Cramer's rule.

    $2x + y = 8$
    $5x - 2y = -16$

Solve the following nonlinear systems of equations. Use a graphing calculator to verify your solutions.

25. $x^2 + y = 0$
    $y = 2x - 3$

26. $y = 3x + 6$
    $y = 2x^2 + 1$

Write a system of equations for each problem and then solve the system.

27. For the school fundraiser, Tyler sold donuts at $4 per box and stadium cushions for $6.50 each. He sold 62 items and collected a total of $318. How many of each item did he sell?

28. A 40% acid solution is to be mixed with a 70% acid solution. The chemist needs to produce 120 mL of a 50% acid solution. How much of each acid will be needed to produce the correct mixture?

29. A company manufactures and sells digital clocks. The fixed costs of production are $20,000, and the variable costs per unit produced are $25. The selling price per clock is $125. How many clocks must the company sell to break even?

30. The demand function for a particular lamp is $D(p) = 120 - p$, and the supply function for the same lamp is $S(p) = 2p - 117$. Find the equilibrium price, where supply equals demand.

31. Cheryl has a total of 60 coins in her piggy bank. If the coins consist of only dimes and nickels and the total value of the coins in her bank is $5.75, how many of each coin does she have?

32. The cost of 10 lb of potatoes and 4 lb of bananas is $6.16. Four pounds of the same potatoes and 8 lb of bananas cost $6.88. What are the cost per pound of potatoes and the cost per pound of bananas?

33. A printer recently purchased $3000 worth of new equipment to offer personalized stationery. The cost of producing a package of stationery is $3, and it is sold for $5.50 per package. Find the number of packages that must be printed for him to break even.

34. The perimeter of a rectangular bedroom is 84 feet. The length of the room is 6 feet more than the width. Find the dimensions of the bedroom.

35. Write a word problem that is modeled by the following linear system:

    $a + s = 40$
    $4.50a + 2.00s = 100$

Chapter 6 Test

1. Is the ordered pair $(-3, 4)$ a solution of the following system?

    $2x + 3y = 6$
    $2x + y = -2$

2. Find the value of the determinant $\begin{vmatrix} 2 & -5 \\ -1 & 6 \end{vmatrix}$.

Solve each of the following systems of equations by graphing.

3. $x - y = 3$
   $x + y = 5$

4. $x = 2$
   $3x + 2y = 4$

5. $2x - y = 4$
   $6x = 3y + 15$

Solve each of the following systems of equations algebraically.

6. $y = 4 - 3x$
   $3x + y = 5$

7. $x - 2y = 4$
   $3x + 4y = 2$

8. $4x + 3y = 15$
   $2x - 5y = 1$

9. $2x + 4y = 7$
   $5x - 3y = -2$

Solve each of the following systems of equations using Cramer's rule.

10. $3x - 7y = 13$
    $6x + 5y = 7$

11. $0.5x - 1.2y = 0.3$
    $0.2x + y = 1.6$

12. Solve the following nonlinear system of equations. Use a graphing calculator to verify your solutions.
$$2x^2 - y = 0$$
$$3x - y = -2$$

13. Solve the following nonlinear system of equations. Use a graphing calculator to verify your solutions.
$$x + 2y = 4$$
$$y = x^2 + 3x + 2$$

Write a system of equations for each problem and solve the system.

14. A company manufactures and sells calculators. The fixed costs are $10,000, and each unit costs $5 to manufacture. The selling price of each calculator is $25. Find the number of calculators that must be sold for the company to break even.

15. The rooms at a motel are priced at $80 for a single and $100 for a double. If there are 65 rooms occupied on Friday night, and the total revenue for the night is $6400, how many double rooms have been rented?

16. A chemist is preparing 200 mL of an alcohol solution. He is mixing a 12% alcohol solution with an 8% alcohol solution in order to produce a 9% solution. How much of each solution will he need to produce 200 mL of 9% solution?

17. A combined total of $12,000 is invested in two bonds that pay 10.5% and 12% simple interest. The annual interest is $1380. How much is invested in each bond?

18. The perimeter of a rectangle is 76 in. The length of the rectangle is 5 in. more than twice the width. Find the length.

19. Find the equilibrium price for the following demand and supply functions: $D(p) = 89 - 7p$ and $S(p) = 29 + 3p$.

20. The Turning Pointe is planning a new line of leotards. For the first year, the fixed costs for setting up the production line are $17,000. The per item cost for each leotard is $6.00, and each leotard will be sold to a retailer for $10.00. Find the break-even point.

Suggested Laboratory Exercises

Lab Exercise 1

Which Salary Option?

James is applying to be a salesperson for Broadview Used Cars. He has a choice to make that may greatly affect his income. He may choose a base salary of $25,000 per year plus a commission of 10% on sales that he makes, or he may choose a base salary of $35,000 per year plus 7.5% commission on all sales he makes.

(a) Letting $x =$ the number of dollars in sales that James may make, and y_1 and $y_2 =$ the first and second annual earnings options, respectively, write two equations (a system) as described by the options given to James.

(b) Solve this system of equations. Graph this system.

(c) What is the meaning of this solution as applied to this situation?

(d) For what level of annual sales will each of the two methods yield the same gross annual income for James?
(e) For what range of sales should James choose option 1? Option 2?

Lab Exercise 2

Catering Business (A System of Inequalities)

A local caterer has kept good records over the years and has a good idea of just how long it takes to prepare food and set up to serve the food for several different types of possible events. Normally, the food is prepared on the day before the event (if possible), and then tables and decorations are set up and the food is laid out on the day of the event. Table 6-4 shows the estimated time required to prepare food and set up to serve for banquets and for receptions. Of course, there is a limit to the total number of hours that the caterer works during any given week.

TABLE 6-4 Lab Exercise 2: Catering Business

	Hours to Prepare the Food	Time to Set Up
Banquet	12	7
Reception	4	5
Maximum work hours available	38	40

(a) Write a system of inequalities for this situation.
(b) Graph this system and shade the region of the graph showing the number of banquets and receptions that may be catered in one week.
(c) Can this caterer handle three banquets and two receptions in one week?
(d) How about two banquets and four receptions?
(e) What is the maximum number of receptions that may be catered if no banquets are catered?
(f) What is the maximum number of banquets that may be catered if no receptions are catered?

Lab Exercise 3

Marketing Decisions (A System of Inequalities)

A local political campaign is planning its budget for newspaper advertisements during the month prior to the election. The campaign can afford to spend at most $10,200 on these advertisements, which cost $150 per ad during the week and $500 per ad on Sunday. The campaign plans to run at most 54 ads.

(a) Write a system of inequalities for this problem.
(b) Using your graphing calculator, draw a graph that illustrates this problem with the icon set to "shade below." Set the window to view only the first quadrant. Use the intersect option to find the vertices of the quadrilateral whose area makes up the pool of answers (the overlapping shaded region).
(c) If all ads are run on a weekday and none are run on Sundays, what is the largest number of ads that the campaign can run?
(d) If the campaign spends exactly $10,200 buying ads on weekdays and Sundays, how many ads will they run on weekdays and how many ads will they run on Sunday?
(e) If the campaign manager decides to run 8 ads on Sunday, what is the maximum number of ads they can run during weekdays and stay within the budget?

Lab Exercise 4

Phone 'Em Up

One of the many long-distance telephone companies (call this company X) charges a base fee of $2.70 per month plus 10 cents per minute of use. A new company (call this company Y) decides to charge 12 cents per minute of use and no base fee.

1. Write an equation to model the charges of company X.
2. Write an equation to model the charges of company Y.
3. Write an equation to model the number of minutes of use where the charges of company X will equal the charges of company Y.
4. Sketch these two equations. Based on your graph, which plan is cheaper if you talk an average of 200 minutes each month?

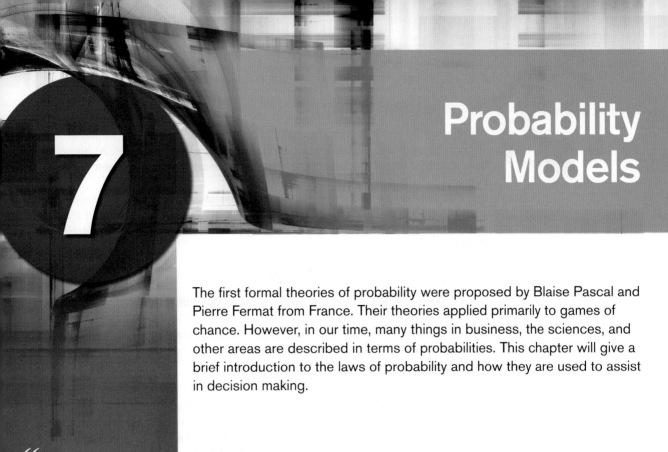

Probability Models

The first formal theories of probability were proposed by Blaise Pascal and Pierre Fermat from France. Their theories applied primarily to games of chance. However, in our time, many things in business, the sciences, and other areas are described in terms of probabilities. This chapter will give a brief introduction to the laws of probability and how they are used to assist in decision making.

> *It is a truth very certain that, when it is not in our power to determine what is true, we ought to follow what is most probable.*
> – René Descartes

In this chapter

- **7-1** Sets and Set Theory
- **7-2** What Is Probability?
- **7-3** Theoretical Probability
- **7-4** Odds
- **7-5** Tree Diagrams
- **7-6** *Or* Problems
- **7-7** *And* Problems
- **7-8** The Counting Principle, Permutations, and Combinations

    Chapter Summary

    Chapter Review Problems

    Chapter Test

    Suggested Laboratory Exercises

Section 7-1 Sets and Set Theory

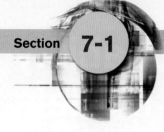

A fundamental concept in mathematics is the sorting of objects into similar groups. We use the word **set** in mathematics to refer to a collection of objects. At home you have a set of dishes. Of course, in algebra, we are usually speaking about groups of numbers.

The numbers or items in a set are called **elements**, or members of the set. Often a letter is used as a *name* for a set. The members of the set are enclosed in braces, { }. For example, to write the set of all counting numbers less than 6 in **set notation** we would write:

$$N = \{1, 2, 3, 4, 5\}.$$

A set is **finite** if either it contains no elements or the number of elements in the set is a natural number. The set of counting numbers less than 25 is a finite set because the number of elements in the set is 24, and 24 is a natural number. Written in set notation, this set would be something like this:

$$S = \{1, 2, 3, 4, \ldots, 24\}.$$

An **infinite** set is a set that is not finite. The set of counting numbers is an example of an infinite set. The set of counting numbers written in set notation would be $N = \{1, 2, 3, 4, 5, \ldots\}$. The three dots (. . .) at the end of the list indicate that the set of numbers continues endlessly and is infinite.

It is often better to write sets that are descriptions rather than lists. This is particularly true if the set is very large or if the set contains elements that are not numbers. To do this we use **set-builder notation**. Writing set S in this notation we have:

$$S = \{x \mid x \text{ is a natural number less than 25}\}$$

↑ ↑ ↑
the set of numbers   such that   conditions for being an element of this set

This notation reads as follows: "S is the set of numbers x such that x is a natural number less than 25." The number 25 is not a member of this set because 25 is not less than 25.

If we wish to compare individual numbers with a set, some other symbols may be used. For example, the number 5 is a member of the set S described above. This can be written in mathematical *shorthand* using the symbol $\in$. The statement $5 \in S$ is read as "5 is an element of set S." If a number is not a member of a particular set, then the symbol $\notin$ is used. For example, $40 \notin S$ is read "40 is not an element of set S."

Two or more sets are said to be **equal** to each other only if they contain exactly the same members, no more and no less. The usual symbols, $=$ and $\neq$, are used. For example, {dog, cat, bird} = {cat, bird, dog} and {dog, cat, bird} $\neq$ {dog, cat, fish}.

A useful way to depict relationships among sets is to use a picture diagram called a **Venn diagram**. Venn diagrams use a rectangle, which represents what is called the universal set. The **universal set**, denoted by U, contains all the elements under consideration in a given discussion. Within the rectangle are circular regions that represent the sets under consideration.

Suppose that you have two sets, A and B. These two sets can be combined to form a new set called the **union** of sets A and B. This new set would contain all of the elements in both sets. It is the "uniting" of two sets into a new set containing the elements from both sets. For example, if we merge two math classes into one single class, we have created a new set of students which is the union of the original two sets of students. The symbol ∪ is used to stand for the union of two sets.

The Venn diagram to illustrate $A \cup B$ is shown in Figure 7-1. The union of the two sets is noted by the shaded regions in the Venn diagram.

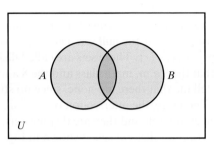

FIGURE 7-1

Example 1   Finding the Union of Sets

$$\text{Let } U = \{1, 2, 3, 4, 5, 6, 7, 8\}$$
$$A = \{1, 3, 5, 7\}$$
$$B = \{2, 4, 6\}$$
$$C = \{4, 6\}$$

Find the following:

(a) $A \cup B$
(b) $A \cup C$
(c) $B \cup C$

(a) The union of A and B is the set consisting of all elements in A or in B or in both. So, $A \cup B = \{1, 2, 3, 4, 5, 6, 7\}$
(b) $A \cup C = \{1, 3, 4, 5, 6, 7\}$
(c) $B \cup C = \{2, 4, 6\}$ Notice that even though 4 and 6 occur in both sets, we only list them once.

Often, we are interested in what various sets may have in common with each other. This is called the **intersection** of sets. Remember that the intersection of two roads is the pavement that they have in common. In two sets, the elements that are present in both sets are used to create a new set of elements called the intersection of the sets. For example, the students that are in the same 8 a.m. math class and 9 a.m. history class would be a new set of students that is the intersection of the two classes. The symbol ∩ is used to stand for the intersection of two sets.

The Venn diagram for $A \cap B$ is shown in Figure 7-2. The intersection of the two sets is noted by the shaded region in the Venn diagram.

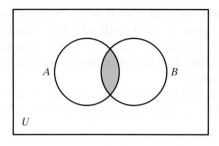

FIGURE 7-2

If two sets have no common elements then the intersection is a set with no elements called the **empty set** or the **null set**. The empty, or null, set is denoted by the symbol ∅ or empty braces { }. These sets are called **disjoint sets**. If I ask you to tell me the students that the 8 a.m. math class and the 8 a.m. history class have in common, you would tell me that there are none, since no student can be in both classes at the same time. Therefore, the intersection of the 8 a.m. math class and the 8 a.m. history class is the empty set, and they are disjoint sets.

The Venn diagram for disjoint sets is shown in Figure 7-3. There are no shaded regions, since the sets are disjoint and have nothing in common.

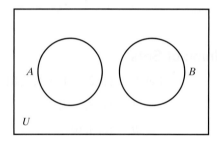

FIGURE 7-3

Example 2 — Finding the Intersection of Sets

Let $U = \{1, 2, 3, 4, 5, 6, 7, 8, 9, 10\}$

$A = \{1, 2, 3, 4, 5\}$

$B = \{2, 4, 6\}$

$C = \{6, 8, 10\}$

Find the following:

(a) $A \cap B$
(b) $B \cap C$
(c) $A \cap C$

(a) The intersection of A and B is the set consisting of elements in both A and B. So $A \cap B = \{2, 4\}$.
(b) $B \cap C = \{6\}$
(c) $A \cap C = \emptyset$. Since A and C have no elements in common, the intersection is the empty set. Remember that these are called disjoint sets.

If every member of set A is also a member of set B, then set A is said to be a **subset** of B. The members of your math class are a subset of the student body of the college. The symbol for subset is $\subseteq$. For example, if set $A = \{1, 2, 3, 4, 5\}$ and set $B = \{2, 3\}$, then we can say that set $B \subseteq A$ since all of the elements in B are included in A. The Venn diagram to illustrate this example is shown in Figure 7-4.

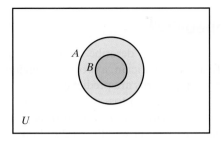

FIGURE 7-4

The remainder of this chapter discusses probability theory. The topics discussed in this section will be helpful in the discussion of probability and understanding some of the more difficult concepts in probability theory such as computing the probability of X "or" Y occurring (an example of union of possible outcomes) and the probability of X "and" Y occurring (an example of the intersection of possible outcomes).

Practice Set 7-1

Let set $A = \{1, 2, 3, 4, 5\}$ and let set $B = \{x \mid x$ is a natural number greater than 4$\}$.

Determine whether each of the following is true or false. (problems 1 to 12)

1. $2 \in A$
2. $10 \in B$
3. $8 \in A$
4. $6 \notin A$
5. $4 \in B$
6. $3 \notin A$
7. $B = \{3.5, 4, 4.5, 5, 5.5, 6, 6.5, \ldots\}$
8. $A \cup \varnothing = \varnothing$
9. $\{\varnothing\} = \varnothing$
10. $A \cap B = \varnothing$
11. $\{2, 3\} \subseteq A$
12. Set A is a finite set.
13. Write the set $A = \{1, 2, 3, 4, 5\}$ in set builder notation.
14. Write, in set notation, the set of natural numbers between 2 and 13.
15. Write, in set notation, the set of even natural numbers less than 50.
16. Write the set {January, February, March, ..., December} in set-builder notation.

Label each of the following as a finite set or an infinite set. (problems 17 to 20)

17. $\{1, 2, 3\}$
18. $\{x \mid x$ is an integer such that $5 \leq x \leq 11\}$
19. $\{x \mid x$ is a number such that $0 \leq x \leq 8\}$
20. $\{\ldots, -3, -2, -1, 0, 1, 2, 3, \ldots\}$

Let set $A = \{1, 2, 3, 4, 5, 6\}$, set $B = \{2, 4, 6, 8\}$, set $C = \{x \mid x$ is an even natural number$\}$, and set $D = \{x \mid x$ is an odd natural number$\}$. (problems 21 to 24)

21. Write $A \cup B$
22. Write $A \cap B$
23. Write $B \cap D$
24. Which set is a subset of one of the others?

Given $M = \{1, 2, 3, 4, 5\}$; $N = \{1, 4, 7\}$; and $P = \{2, 3, 5, 6, 8\}$. Perform each of the following operations and draw a Venn diagram to illustrate your answers. (problems 25 to 30)

25. $M \cap N$
26. $P \cap N$
27. $M \cup P$
28. $P \cup N$
29. $M \cap P$
30. $M \cup N$

Let $U = \{1, 2, 3, 4, 5, 6, 7, 8, 9, 10\}$; $A = \{1, 2, 3, 4, 5\}$; $B = \{1, 2, 5, 7\}$; and $C = \{3, 5, 7, 9\}$. (problems 31 to 40)

31. $A \cup B$
32. $A \cup C$
33. $B \cup C$
34. $A \cap B$
35. $A \cap C$
36. $B \cap C$
37. $A \cap \varnothing$
38. $A \cup \varnothing$
39. $A \cap (B \cap C)$
40. $(B \cap C) \cap A$

298 Chapter 7 **Probability Models**

Section 7-2

What Is Probability?

Probability theory attempts to describe the predictable long-run patterns of random processes. To a mathematician, a process is **random** if individual outcomes are uncertain but long-run patterns are predictable.

For example, if you were to toss a coin into the air, would it land so that the "head" is up or would it land "tails" up? You may not be able to correctly guess what any particular toss will show, but you should be able to guess that, for a large number of tosses, approximately half the time heads will show and half the time tails will show. Thus, tossing a coin is a random process.

To a mathematician, probability is not a guessing game. It is a matter of determining the relationship between various possible outcomes. For tossing a coin, there are limited results possible. For events that occur in our daily lives, the situation is very complicated. We will begin by looking at some familiar objects with limited possible outcomes.

As with any new topic, there are some terms and expressions that are commonly used in probability problems. These terms need to be defined before beginning a discussion of the particulars of probability.

Definition: Experiment

An experiment in probability is a controlled operation that yields a set of results.

If we set out to gather information in a systematic way so that a reasoned decision may be made, then an **experiment** has been done. An experiment in probability will not necessarily be like an experiment in physics or chemistry. An experiment can be as simple as flipping a coin and recording whether the result was "heads" or "tails." These possible results are called **outcomes**. It is usually important to know the number of possible outcomes to a given experiment.

Definition: Outcomes

The possible results of an experiment are called outcomes. It is also important to know the number of possible outcomes. This is denoted by N in formulas related to probability.

An **event** is a subset of the outcomes of an experiment. For example, when a die is rolled the *event* of rolling a number less than 6 can be satisfied by any one of five outcomes: 1, 2, 3, 4, or 5. The event of rolling a 3 can be satisfied by only one outcome: the 3 itself. The event of rolling an odd number can be satisfied by any of three outcomes: 1, 3, or 5.

Definition: Event

A subset of the outcomes of an experiment is called an event. This is denoted by E in formulas related to probability.

There are two classifications of probability: **empirical probability**, which is experimental, and **theoretical probability**, which is mathematical. Empirical probability is the relative frequency of the occurrence of an event and is determined by *actual* observations of an experiment. We will indicate the probability of an event E by $P(E)$. Theoretical probability is determined through a study of the possible outcomes that *can* occur for the given experiment. The following formula is used for computing the empirical probability:

IMPORTANT FORMULAS

Empirical Probability (Relative Frequency)

$$P(E) = \frac{\text{number of times event } E \text{ has occurred}}{\text{total number of times the experiment has been performed}}$$

The probability of an event, whether it is empirical or theoretical, will always be a number between 0 and 1, inclusive, and may be expressed as a fraction, decimal, or percent. An empirical probability of 0 means that the event *never* occurred, and an empirical probability of 1 means that the event *always* occurred.

Example 1 Calculating an Empirical Probability

A coin is tossed 100 times and lands heads up 56 of those times. Find the empirical probability of the coin landing heads up.

Let E be the event that the coin lands heads up.

$$P(E) = \frac{56}{100} = 0.56 \text{ or } 56\%$$

The coin lands on heads 56% of the time.

Example 2 Calculating an Empirical Probability

A pharmaceutical company is testing a new drug that is supposed to reduce high blood pressure. The drug is given to 600 individuals with the following outcomes.

$$\text{Blood pressure reduced} = 402$$

$$\text{Blood pressure increased} = 48$$

$$\text{Blood pressure remained the same} = 150$$

If this drug is given to an individual, find the empirical probability that

(a) the blood pressure is reduced.
(b) the blood pressure is increased.
(c) the blood pressure remains the same.

(a) Let E be the event that the blood pressure is reduced.

$$P(E) = \frac{402}{600} = 0.67 \text{ or } 67\%$$

(b) Let E be the event that the blood pressure is increased.

$$P(E) = \frac{48}{600} = 0.08 \text{ or } 8\%$$

(c) Let E be the event that the blood pressure remains the same.

$$P(E) = \frac{150}{600} = 0.25 \text{ or } 25\%$$

Empirical probability is used when probabilities cannot be theoretically calculated. Insurance companies use empirical probability to determine the chances of a 16-year-old male driver having an automobile accident based on data collected about past automobile accidents. This information is then used to determine the cost of car insurance for all 16-year-old male drivers.

By surveying human characteristics such as blood type or eye color, we can use these data to calculate the percent of the population that has certain characteristics. In addition, these numbers can be used to find the probability of any randomly chosen person having a particular characteristic such as type O+ blood or blue eyes. Look at the illustration in Example 3.

Example 3 An Application to Human Characteristics

Assume that in the general population 55% of persons have eyes that would be in the brown-color category, 15% blue, 10% green, and 20% other color variations.

(a) Based on these percentages, if you are in a classroom with 25 persons (including yourself), how many should have brown eyes?
(b) What is the probability that the next person entering the room will have blue eyes?

(a) Because 55% of the general population has brown eyes, we would expect that about 55% of persons in any general group would have brown eyes.

$$55\% \text{ of } 25 = 0.55(25) = 13.75$$

This means that about 14 of the 25 persons would be expected to have brown eyes.

(b) Because 15% of the general population has blue eyes,

$$P(\text{blue eyes}) = 15\% = 0.15$$

or, as a fraction, $P(\text{blue eyes}) = \dfrac{15}{100} = \dfrac{3}{20}$.

Most of us accept the fact that if a fair coin is tossed many times, it will land heads up approximately half of the time. We can guess that the probability a fair coin will land heads up is $\frac{1}{2}$. Does this mean that if a coin is tossed 2 times, it will land heads up exactly once and tails up exactly once? Obviously, this is not the case. The probability that a coin will land heads up is determined by looking at the relative frequency over the long run. In other words, the more times an experiment is performed, the more accurately we can predict the probability. This is referred to as the **law of large numbers**.

Theorem The Law of Large Numbers

The law of large numbers states that probability statements apply in practice to large numbers of trials—not to a single trial. It is the relative frequency over the long run that is accurately predictable, not individual events or precise totals.

What does it mean to say that the probability of rolling a 3 on a die is $\frac{1}{6}$? It means that over the long run, on the average, one of every six rolls will result in a 3. A related topic to empirical probability is **conditional probability**. Suppose that you are interested in the probability of two events occurring and you already know what will happen in the first event. In other words, you want to know the probability that a particular *second event* will occur *given* that you already know the results of the first event. This is called conditional probability. Conditional probabilities are written in a shorthand as $P(A|B)$. This is read as "the probability that A will occur given that B has already occurred."

In some instances, the occurrence of one event does not affect the outcome of a subsequent event. For example, if the first child born into a family is a boy, this outcome does not affect the probability that the next child will also be a boy. These are **independent events**. However, if we draw an ace out of a deck of cards and do not return it to the deck, this event does affect the probability that the next card we draw will be an ace. These events are classified as **dependent events** because the outcome of the first event affected the outcome of the second. Look at the following example of conditional probability.

Example 4 Finding a Conditional Probability

You have a standard deck of 52 playing cards. (See Figure 7-5.) Find the probability of drawing a card that is a club (♣) given each of the following conditions. Let C = drawing a club. Whenever two cards are drawn in this example, the first card drawn is not replaced in the deck before drawing the second card.

(a) $P(C)$
(b) $P(C\,|\,$ the card is a black card$)$
(c) $P(C\,|\,$ a club was not drawn on the first draw$)$

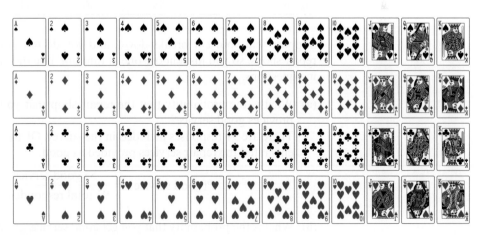

FIGURE 7-5
Standard deck of 52 playing cards.

(a) This is a simple probability of drawing one club from the deck. Because there are 13 clubs in the deck out of 52 cards,

$$P(C) = \frac{13}{52} = \frac{1}{4}$$

(b) We know that a black card has been drawn from the deck, and there are 13 clubs in the 26 black cards in a deck.

$$P(C|\text{the card is a black card}) = \frac{13}{26} = \frac{1}{2}$$

(c) On the first draw, a club was not chosen, so we know that all 13 clubs are still in the 51 remaining cards.

$$P(C|\text{a club was not drawn on the first draw}) = \frac{13}{51}$$

Data collection is often presented in tabular form. We can use these tables to calculate the empirical probability of certain events happening. We can restrict the data set to a particular characteristic by using conditional probability as demonstrated in Example 4.

Example 5

Probabilities from a Table

The following table is based on some recent customer surveys done at two automobile repair shops. One repair shop is at an automobile dealership while the other is an independent repair shop. Customers were asked to rate the service that they received on their cars from the shop.

	Good Service	Poor Service	Total
Car dealer shop	20	9	29
Independent shop	35	27	62
Total	55	36	91

Suppose that your car breaks down, and you do not know anything about these repair shops, so you just choose one at random. Use the table to answer the following questions.

(a) What is the probability that the repair shop that you choose will be one that provides good service?
There are 91 responses included in the survey. Of those, a total of 55 respondents rated the repair services provided as good.

$$P(\text{good service}) = \frac{55 \text{ good service}}{91 \text{ total respondents}} = 60.4\%$$

(b) What is the probability that you get good service, given that you chose an independent shop?
There were 62 people who chose the independent shop and 35 reported that they got good service.

$$P(\text{good service}|\text{independent shop}) = \frac{35 \text{ good service}}{62 \text{ responses about independent shops}} = 56.5\%$$

(c) What is the probability that you will get poor service, given that you chose to have your repairs done at the car dealer's shop?

There were 29 people who had their car repaired at the car dealership, and 9 of them reported poor service.

$$P(\text{poor service}|\text{car dealer's shop}) = \frac{9 \text{ poor service}}{29 \text{ responses about car dealer's shop}} = 31\%$$

Conditional probability will be reviewed again in Section 7-7 when we look at calculating the probability of a series of events occurring together.

Practice Set 7-2

Answer problems 1 to 4 in complete sentences.

1. What is an experiment? What are outcomes of an experiment? What is an event?

2. Explain how you would find the empirical probability of rolling an odd number on a die. What do you believe is the empirical probability of rolling an odd number? Find the empirical probability of rolling an odd number by rolling a die 40 times.

3. Explain in your own words the Law of large Numbers.

4. To determine life insurance premiums, life insurance companies must compute the probable date of death. On the basis of research done by the company, Mr. Timmons, age 53, is expected to live another 23.25 years. Does this mean that Mr. Timmons will live until he is 76.25 years old? If not, what does it mean?

Classify the following events as independent or dependent events.

5. Randomly selecting a voter who is a Democrat and randomly selecting a voter who is a Republican.

6. Winning the lottery in Virginia and winning the lottery in North Carolina.

7. In a club meeting of 20 students, 12 boys and 8 girls, the probability of a boy being selected president and then the probability of another boy being selected vice president.

8. A person lies, and a person fails a lie detector test.

Calculate each probability in percent form rounding to the nearest tenth.

9. In a sample of 60,000 births, 155 were found to have Down syndrome. Find the empirical probability that Mrs. Cosgrove's first child will be born with this syndrome.

10. A certain community recorded 4000 births last year, of which 2350 were female. What is the empirical probability that the next child born in the community will be a female? What is the empirical probability that the next child born will be a male?

11. The McCooks are participating in Project FeederWatch sponsored by the Cornell Lab of Ornithology. The last 30 birds at the McCooks' bird feeder were 20 robins, 5 cardinals, 3 blue jays, and 2 doves. Find the empirical probability that the next bird to feed from the feeder will be a cardinal. Find the empirical probability that the next bird will be a robin.

12. Some years ago, a survey was done of men aged 35 and women aged 35 who were married to each other. A total of 81,800 couples was surveyed. Thirty years later, only 64,528 men and 71,862 women were still living. Based on these limited data:

    (a) What is the probability that a male aged 35 will live another 30 years?
    (b) What is the probability that a female aged 35 will live another 30 years?

13. Of the last 70 people who went to the cash register at Wal-Mart, 12 had blond hair, 20 had

black hair, 33 had brown hair, and 5 had red hair. Determine the empirical probability that the next person to come to the cash register will have red hair.

14. Of the last 70 people who went to the cash register at Wal-Mart, 12 had blond hair, 20 had black hair, 33 had brown hair, and 5 had red hair. Determine the empirical probability that the next person to come to the cash register will have blond hair.

15. Mr. Harclerode's grade distribution over the past 3 years in Algebra I is shown in the following chart.

Grade	Number of Students
A	41
B	183
C	265
D	96
F	85
Incomplete	2

If Sharon Odell plans on taking Algebra I with Mr. Harclerode, determine the empirical probability that she will receive a grade of A.

16. If Lafawn Oliver plans on taking Algebra I with Mr. Harclerode, determine the empirical probability that she will receive a grade of C. (Use the table in the previous problem.)

17. Six friends chartered a deep-sea fishing boat for a day's fishing trip. They caught a total of 72 fish. The following chart provides information about the type and number of fish caught.

Fish	Number Caught
Shark	8
Flounder	35
Kingfish	10
Grouper	19

Determine the empirical probability that the next fish caught will be a shark.

18. Six friends chartered a deep-sea fishing boat for a day's fishing trip. They caught a total of 72 fish. Determine the empirical probability that the next fish caught will be a flounder. (Use the chart in the previous problem.)

When a new medication is being studied by a drug company, test groups are asked to report side effects and probabilities of those side effects are calculated based on the results from the test group. A new antidepressant is being tested using a group of 120 people. The number of people reporting specific side effects during the study are listed in the table below. Use this table for problems 19 to 22. Express all probabilities as percents.

Side Effects	Number of Participants Reporting a Side Effect
Headache	27
Nausea	24
Insomnia	15
Fatigue	10
Dizziness	6

19. Using the numbers reported in the table, calculate the probability of a person experiencing insomnia while taking this medication.

20. What is the probability of someone experiencing nausea while taking this medication?

21. What is the probability of someone experiencing dizziness while taking this antidepressant? Based on this probability, if 25,000 people take this medication, how many would you expect to experience dizziness as a side effect?

22. What is the probability of someone experiencing headaches while taking this antidepressant? If 500,000 people take this medication, how many would you expect to experience headaches as a side effect?

Body mass index (BMI) is now used by many people to determine if a person is overweight. Using this scale, a BMI ≥ 25.0 classifies an individual as overweight. There were 55 new members at a women's fitness club this month. Each woman had her body mass index (BMI) calculated when she joined. The results are posted in the table below. Use this table to answer problems 23 to 26.

Body Mass Index of Women	Frequency
15.0–19.9	9
20.0–24.9	12
25.0–29.9	25
30.0–34.9	6
35.0–39.9	2
40.0–44.9	1

23. If one new member is chosen at random, what is the probability of randomly selecting a woman who has a BMI between 20.0 and 29.9?

24. If a woman with a BMI ≥ 25.0 is classified as overweight, how many of the women at this fitness club would be in this category? If you choose a woman at random, what is the probability that she will be overweight?

25. If one new member is selected to receive a free coffee mug, what is the probability of randomly selecting a woman with a BMI of at least 30.0?

26. If another new member is selected to receive a free T-shirt, what is the probability of randomly selecting a woman with a BMI of at most 34.9 given that the woman chosen was classified as overweight?

The American Red Cross reports the distribution of blood types in our population based on ethnic groups because the distributions are different. (*Source*: http://www.redcrossblood.org/learn-about-blood/blood-types) **Use this information to answer problems 27 to 30.**

	Caucasians	African American	Hispanic	Asian
O+	37%	47%	53%	39%
O−	8%	4%	4%	1%
A+	33%	24%	29%	27%
A−	7%	2%	2%	0.5%
B+	9%	18%	9%	25%
B−	2%	1%	1%	0.4%
AB+	3%	4%	2%	7%
AB−	1%	0.3%	0.2%	0.1%

27. If an Asian blood donor arrives at a Red Cross center to donate blood, what is the probability that this donor's blood type will be A+?

28. If a Caucasian blood donor arrives at a Red Cross center to donate blood, what is the probability that this donor's blood type will not be O+?

29. According to the Red Cross, the universal red cell donor has Type O− blood type. If 25 African American donors participate in a blood drive, how many of those donors would you expect to have Type O− blood?

30. According to the Red Cross, the universal plasma donor has Type AB+ blood type. If 50 Hispanic donors participate in a blood drive, how many of those donors would you expect to have Type AB+ blood?

31. The results of a survey of the service at a local restaurant are summarized in the following table.

Meal	Service Good	Service Poor	Total
Lunch	50	15	65
Dinner	45	35	80
Total	95	50	145

Find the probability that the service was rated

(a) good
(b) good, given that the meal was lunch.
(c) poor, given that the meal was dinner.
(c) poor, given that the meal was lunch.

32. Each person in a sample of 300 residents in Buncombe County was asked whether he or she favored having a single countywide police department. The county consists of one large city and a number of small townships. The response of those sampled, with their place of residence specified, is given in the following table.

Residence	Favor	Oppose	Total
Live in city	130	50	180
Live outside city	85	35	120
Total	215	85	300

If one person from the sample is selected at random, find the probability that the person

(a) favors a single countywide police force.
(b) favors a single countywide police force if the person lives in the city.
(c) opposes a single countywide police force if the person lives outside the city.

33. A group of individuals were asked which evening news they watch most often. The results are summarized in the following table.

Viewers	ABC	NBC	CBS	Other	Total
Men	35	25	45	30	135
Women	50	10	20	15	95
Total	85	35	65	45	230

If one of these individuals is selected at random. Find the probability that the person watches

(a) ABC or NBC.
(b) ABC, given that the individual is a woman.
(c) ABC or NBC, given that the individual is a man.
(d) a station other than CBS, given that the individual is a woman.

34. A survey of 150 people asked participants if they currently smoke. Results are shown in the table. Use this information to answer the following questions.

	Yes	No	Total
Male	31	51	82
Female	24	44	68
Total	55	95	150

(a) What is the probability of randomly selecting someone who does not smoke?
(b) What is the probability of randomly selecting a male who smokes?
(c) What is the probability of selecting someone who smokes given that the person selected is female?
(d) What is the probability that the person selected is male given that he smokes?

Section 7-3 Theoretical Probability

What are the chances of winning the lottery? If you go to a carnival, which games provide the greatest chance of winning? These and similar questions can be answered using theoretical probability. Recall from the previous section that theoretical probability is determined through the study of the *possible* outcomes that can occur in a given experiment. The rest of this chapter will be spent studying theoretical probability.

In Section 7-2, we learned that the results of an experiment are called the outcomes of the experiment. If each outcome has the same chance of occurring as any other outcome, they are said to be **equally likely outcomes**. For example, when you roll a die, the possible outcomes are 1, 2, 3, 4, 5, and 6. These are said to be equally likely outcomes because one has an equal chance of getting each number on any roll of the die.

The process of predicting how a series of events might occur must always begin by studying the situation. A good starting point is to make a list of all possible outcomes of a given event. This list is called the **sample space**.

Definition Sample Space

The set of all possible outcomes of a random phenomenon is called the sample space. A sample space is listed using set notation:

$$S = \{\text{list of outcomes}\}$$

Remember that when a probability experiment is performed and the results are observed, each individual result is called an event.

Example 1 Rolling One Die One Time

Pictured in Figure 7-6 are the sides of a die "unfolded" to show the number of pips on each face. We will do a simple experiment—rolling the die one time. What is the number of possible outcomes? List these outcomes as a sample space.

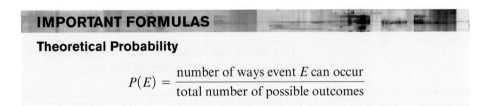

FIGURE 7-6
An "unfolded" die.

Any one of the six sides may be face up when the die is rolled. The sample space is $S = \{1, 2, 3, 4, 5, 6\}$ and the number of possible outcomes is $N = 6$.

We are now ready to state a formal definition of theoretical probability.

IMPORTANT FORMULAS
Theoretical Probability

$$P(E) = \frac{\text{number of ways event } E \text{ can occur}}{\text{total number of possible outcomes}}$$

Example 2 Calculating a Theoretical Probability

A "fair" die is to be rolled one time. Calculate each of the following probabilities:

(a) $P(2)$
(b) $P(\text{even number})$
(c) $P(7)$
(d) $P(\text{number less than } 7)$

(a) There is only one favorable outcome out of six. We have one chance to roll a 2, and there are six possible outcomes $\{1, 2, 3, 4, 5, 6\}$.

$$P(2) = \frac{\text{number of ways to roll a 2}}{\text{total number of possible outcomes}} = \frac{1}{6}$$

or one chance in six.

The value of P may be given as a fraction, decimal (rounded appropriately), or percent. In this case, $\frac{1}{6}$ is a nonending, repeating decimal, so to express the value of $P(2)$ as a decimal it must be rounded. The value of $P(2)$ may thus be given as follows if decimal numbers or percentages are preferred.

$$P(2) = \frac{1}{6} \approx 0.167 = 16.7\%$$

(b) There are three even numbers on the die $S = \{2, 4, 6\}$. Any one of these three would fulfill the desired conditions for $P(\text{even number})$.

$$P(\text{even number}) = \frac{\text{number of even numbers on a die}}{\text{total number of outcomes on a die}} = \frac{3}{6} = \frac{1}{2}$$

or one chance in two. Also,
$$P(\text{even number}) = 0.5 \text{ or } 50\%$$

(c) There are no 7s on a standard die.
$$P(7) = \frac{\text{number of 7's on a die}}{\text{total number of outcomes on a die}} = \frac{0}{6} = 0$$

Also,
$$P(7) = 0\%$$

This means that it is not possible to roll a 7 at all on a single die.

(d) All numbers on the die are less than 7.
$$P(\text{number} < 7) = \frac{\text{number of outcomes} < 7 \text{ on a die}}{\text{total number of outcomes on a die}} = \frac{6}{6} = 1$$

Also,
$$P(\text{number} < 7) = 1.00 = 100\%$$

This means that no matter what number turns up, the desired result will occur. You cannot miss if you bet on this one.

The following **laws of probability** can be easily understood if you refer to the example just completed.

Rule Probability Laws

Law 1
Every probability is a number between 0 and 1 (inclusive).

Law 2
The sum of the probabilities for all outcomes for a particular sample space is exactly 1.

Law 3
The probability of an event occurring that cannot possibly happen is 0.

Law 4
The probability of an event that must occur is 1.

Example 3 Marbles in a Jar

Suppose that a jar contains 10 marbles. (See Figure 7-7.) Two are red, three are green, four are blue, and one is yellow. Each marble is the same size and weight as all of the others, so each has an equal chance of being chosen. What is the probability that you will reach into the jar without looking and draw out

(a) a red marble?
(b) a green marble?
(c) a blue marble?
(d) a yellow marble?

(e) a purple marble?
(f) a red, green, blue, or yellow marble?
(g) Referring to the laws of probability, is law 1 true here?
(h) Is law 2 true here also?

(a) There are two red marbles out of ten marbles in the jar. The probability of drawing a red marble may be abbreviated as $P(\text{red})$.

$$P(\text{red}) = \frac{2}{10} = \frac{1}{5}$$

FIGURE 7-7
Marbles in a jar.

(b) $$P(\text{green}) = \frac{3}{10}$$

(c) $$P(\text{blue}) = \frac{4}{10} = \frac{2}{5}$$

(d) $$P(\text{yellow}) = \frac{1}{10}$$

(e) $$P(\text{purple}) = \frac{0}{10} = 0$$

A probability of 0 means that the event is an *impossible event*. In this case, it means that it is not possible to draw a purple marble from the jar because there are none of that color in the jar.

(f) $$P(\text{red, green, blue, or yellow}) = \frac{10}{10} = 1$$

A probability of 1 means that the event is *certain to occur*. In this example, if a marble is drawn at all it will be one of the four colors, so we are certain to draw a red, green, blue, or yellow marble.

(g) All of the individual probabilities are between 0 and 1 (inclusive), so law 1 is true here.

(h) $$P(\text{red}) + P(\text{green}) + P(\text{blue}) + P(\text{yellow}) = \frac{2}{10} + \frac{3}{10} + \frac{4}{10} + \frac{1}{10}$$

$$= \frac{10}{10} = 1$$

Law 2 is also true here.

IMPORTANT EQUATIONS
Mathematical Relationships

$$P(A) + P(\text{not } A) = 1$$

$$P(\text{not } A) = 1 - P(A)$$

If the probability that an event will occur is $\frac{3}{8}$, then the probability that it will not occur is $1 - \frac{3}{8}$ or $\frac{5}{8}$. If the probability that some event, A, will occur is 0.25, then the probability that it will not occur is $1 - 0.25$ or 0.75. This concept is further illustrated in Example 4.

Example 4 — Using the Mathematical Relationships

(a) Referring to Example 3, what is the probability of reaching into the jar of colored marbles and *not* drawing out a red one?

In Example 3(a), $P(\text{red}) = \frac{1}{5}$. Thus, $P(\text{not red}) = 1 - P(\text{red}) = 1 - \frac{1}{5} = \frac{4}{5}$.

(b) The probability of winning a lottery game by correctly picking 5 numbers from 1 to 49 is 0.000000524. What is the probability of losing this game?

Since $P(\text{win}) = 0.000000524$, the probability of losing would be $1 - P(\text{win})$,

$$P(\text{loss}) = 1 - 0.000000524 = 0.999999476$$

Example 5 — Probability of Drawing a Particular Card

Figure 7-5 in Section 7-2 shows a standard deck of playing cards. This deck has four suits: spades, ♠; hearts, ♥; diamonds, ♦; and clubs, ♣. The spades and clubs are numbered in black, the hearts and diamonds in red. There are 13 cards in each suit and 52 cards in the entire deck. Each suit has three face cards, a king, a queen, and a jack. The other cards are numbered 1, the ace, through 10.

You are to draw one card from a well-shuffled deck of playing cards. Find each of the following.

(a) $P(\text{red card})$ (b) $P(\text{king})$ (c) $P(\text{not a face card})$

(a) All of the hearts and diamonds are red cards, so drawing any heart or diamond will satisfy this probability. There are 13 hearts and 13 diamonds in the deck, so there is a total of 26 red cards in a deck.

$$P(\text{red}) = \frac{\text{total number of red cards in a deck}}{\text{total number of cards in a deck}} = \frac{26}{52} = \frac{1}{2}$$

(b) There is one king in each of the four suits, so there are 4 kings in the deck.

$$P(\text{king}) = \frac{\text{number of kings in a deck}}{\text{total number of cards in a deck}} = \frac{4}{52} = \frac{1}{13}$$

(c) Each suit contains 3 face cards and 10 nonface cards. Thus, the deck contains 12 face cards and 40 nonface cards. The probability of not drawing a face card is the same as the probability of drawing a nonface card, and so

$$P(\text{not a face card}) = \frac{\text{number of nonface cards in a deck}}{\text{total number of cards in a deck}} = \frac{40}{52} = \frac{10}{13}$$

Practice Set 7-3

1. What are "equally likely outcomes"?
2. How does theoretical probability differ from empirical probability?
3. Can a theoretical probability ever exceed 1? Why or why not?
4. A certain baseball player has a batting average (number of hits/number of times at bat) of 0.375. What is the probability that he will not get a hit in his next at bat?

All of the letters in the word *Mississippi* are written on separate pieces of paper and put in a hat.

5. What is the probability of drawing the letter *s* from the hat?

6. What is the probability of drawing the letter *p*?

7. What is the probability of drawing the letter *a*?

8. What is the probability of drawing the letter *m* or *i*?

Each of the whole numbers 1 to 20 is written separately on 20 ping-pong balls and placed in a hat.

9. What is the probability that you will reach into the hat and draw out an even number?

10. What is the probability that you will draw a number that is evenly divisible by 5?

11. What is the probability that you will draw the number 3 on the first draw?

12. What is the probability that you will draw a number less than 5 on the first draw?

A card is drawn from a standard deck of playing cards.

13. What is the probability that the card is a 3?

14. What is the probability that the card is not a 3?

15. What is the probability that the card is a heart?

16. What is the probability that the card is a red card?

17. What is the probability that the card is a red spade?

18. What is the probability that the card is a black face card?

19. What is the probability that the card is a face card?

20. What is the probability that the card is a card with a value greater than 5 but less than 9?

21. What is the probability of drawing an ace?

22. What is the probability of drawing a red four?

23. What is the probability of drawing a queen or king of hearts?

24. What is the probability of drawing a card that is not a diamond?

A bag contains three red, four green, and five blue marbles. One marble is drawn from the bag.

25. What is the probability that the marble is red?

26. What is the probability that the marble is not blue?

27. What is the probability that the marble is red or green?

28. What is the probability that the marble is not red or green?

A fair, 6-sided die is tossed.

29. What is the probability of tossing a 5?

30. What is the probability of tossing a 4?

31. What is the probability of tossing an odd number?

32. What is the probability that the number shown is even?

33. What is the probability that the number shown is less than 3?

34. What is the probability that the number shown is greater than 6?

35. What is the probability of tossing a 1 or a 4?

36. What is the probability of tossing a number greater than 5 or an even number?

37. What is the probability of tossing an odd number or a 6?

38. What is the probability of tossing an odd number and a 6?

M&Ms® are packaged so that 24% in a given package are blue, 20% orange, 16% green, 14% yellow, 13% red, and 13% brown.

39. What is the probability that you will reach into a bag of these candies and remove a brown-colored candy?

40. What is the probability of pulling out a green M&M?

41. What is the probability of drawing a purple M&M?

42. What is the probability of drawing an M&M that is not green?

A TV remote has keys for channels 0–9. You select one key at random.

43. What is the probability that you select channel 6?

44. What is the probability that you select a channel that is an odd number?

45. What is the probability that you select a channel less than 8?

46. What is the probability that you don't select 1?

A multiple-choice test has six possible answers for each question.

47. If you guess at an answer, what is the probability that you select the correct answer for one particular question?

48. If you guess at an answer, what is the probability that you select the incorrect answer for one particular question?

49. If you correctly eliminate two of the six possible answers and guess from the remaining possibilities, what is the probability that you select the correct answer to that question?

50. If you correctly eliminate two of the six possible answers and guess from the remaining possibilities, what is the probability that you select the incorrect answer to the problem?

A traffic light is red for 35 sec, yellow for 10 sec, and green for 45 sec.

51. What is the probability that when you reach the light, the light will be red?

52. What is the probability that when you reach the light, the light will be yellow?

53. What is the probability that when you reach the light, the light will not be red?

54. What is the probability that when you reach the light, the light will not be red or yellow?

A bin contains 100 batteries (all size C). There are 40 Eveready, 24 Duracell, 20 Sony, 10 Panasonic, and 6 Rayovac batteries. One battery is selected at random from the bin.

55. Find the probability that the battery selected is a Duracell.

56. Find the probability that the battery selected is a Duracell or Eveready.

57. Find the probability that the battery selected is a Duracell, Eveready, or Sony.

58. Find the probability that the battery selected is a Kodak.

Section 7-4 Odds

The **odds** of an event occurring or not occurring are directly related to the probability of the events. *The odds in favor of* an event are expressed as the ratio of a pair of integers, which is the ratio of the probability that an event will happen to the probability that it will not happen. For example, there are six numbers on a die. The probability that you will roll the number 3 is $\frac{1}{6}$ while the probability of not rolling a 3 is $\frac{5}{6}$. The odds of rolling a 3 would be

$$\frac{p(x)}{1-p(x)} = \frac{\frac{1}{6}}{\frac{5}{6}} = \frac{1}{5} = 1:5.$$

Because the total number of outcomes is the same for rolling a 3 or not rolling a 3 (6 outcomes), the complex fraction $\frac{\frac{1}{6}}{\frac{5}{6}}$ simplifies to $\frac{1}{5}$, which we usually write as 1:5 or 1 to 5 when expressing odds. The shortcut to this complicated formula for the *odds in favor of* an event is to form the ratio of number of ways the event can occur to the number of ways it cannot occur. On a single die, there is only one 3 and there are five numbers that aren't 3, so the odds in favor of rolling a 3 are 1:5 (one way to roll a 3 compared with five ways to roll a different number). Generally, we use a colon (:) to represent the odds ratio instead of a fraction in order to distinguish it from probability.

Definition

Odds

Odds in favor of an event = event occurs : event does not occur

Odds against an event = event does not occur : event occurs

Look at the following examples that illustrate this definition.

Example 1

Odds of Drawing an Ace

What are the odds in favor of and the odds against drawing an ace from a standard deck of playing cards?

There are 52 cards in a standard deck of playing cards. Only 4 are aces and the other 48 cards are not aces (see Figure 7-5 in Section 7-2).

Therefore, the odds in favor of drawing an ace = aces : nonaces = 4 : 48 = 1 : 12. This means that the odds in favor of drawing an ace are 1 to 12.

The odds against drawing an ace = nonaces : aces = 48 : 4 = 12 : 1 (or 12 to 1 against).

Note: Do *not* write 12 : 1 as 12, because odds must be expressed as ratios even when the "fractions" can be reduced to whole numbers.

Odds are given for many games in gambling casinos and at horse races. Usually, these odds are the odds against winning. If the odds on a game are given as 14 : 1, then the game is estimated to be lost 14 times for each win. Definitely a long shot! The probability of winning can be calculated from the odds by writing the ratio of wins to total outcomes. In this case, the probability of winning would be $\frac{1}{15}$.

Example 2

Odds of Being Born with an IQ of 130 or Above

Three percent of the U.S. population is born with an IQ of 130 or above.

(a) What are the odds against a person being born with an IQ of 130 or greater?
(b) What are the odds in favor of a person being born with an IQ of 130 or greater?

(a) If 3% of the population has an IQ of 130 or greater, this means that every 3 births out of 100 will be gifted and every 97 births out of 100 will not be gifted. So the odds against = not gifted:gifted = 97 : 3.
(b) The odds in favor of being born gifted are gifted:not gifted = 3 : 97.

Odds and probabilities are related but give different information. Unfortunately, many people interchange the words when describing the chances of winning games. For example, the rules for a North Carolina's Cash 5 Lottery game, where the player chooses 5 numbers from the numbers 1 to 39, state that the odds of selecting all five numbers correctly are 1 in 575,757 (*Source:* http://www.nc-educationlottery.org/cash5.aspx). However, this is actually a probability number, since the number 575,757 represents the total number of possible selections or outcomes given the rule of selecting 5 numbers from 1 to 39. Since there is 1 winning combination for the grand prize out of 575,757 total possible outcomes, the probability of a win is $P(\text{win}) = \dfrac{1}{575{,}757}$. However, there is 1 winning combination for the grand prize and 575,756 losing combinations for the grand prize, so the odds of a win are win:loss or 1:575,756. The primary thing to remember when converting probabilities to odds or odds to probabilities is that the denominator in a probability represents all possible outcomes of the experiment while in representing odds, the comparison is between the two possible outcomes—an event occurring and an event not occurring. The total of the two numbers used to create the odds will always equal the total number of outcomes of the experiment. Look at the following examples that relate odds to probability.

Example 3 — Virginia Lottery's Pick 3

The Virginia Lottery's website http//www.valottery.com/pick3/howtoplay.asp details a game called Pick 3, where the player must pick 3 winning numbers in the same order that the numbers are drawn during the game. The odds of winning listed on the website are 1:1000. However, the total number of combinations of 3 numbers being selected in a particular order is 1000. Correctly express both the probability and odds of winning this game.

(a) Probability—The probability of an event is $P(E)$ = number of ways event E can occur/total number of possible outcomes. There is only one possible winning combination but there are 1000 possible choices. Therefore,

$$P(\text{win}) = \frac{1}{1000}$$

(b) Odds—The odds in favor of winning are the ratio of number of ways you can win to the number of ways you can lose. There is only 1 winning combination while there are 999 losing combinations. Therefore,

$$\text{Odds of winning} = 1:999$$

Example 4 — Calculating Odds from Probabilities

Today's forecast gives the probability of rain as 30%. What are the odds in favor of rain today? What are the odds against rain?

(a) Odds of rain: Since the probability of rain is 30%, this means

$$P(\text{rain}) = \frac{\text{rain}}{\text{total outcomes}} = \frac{30}{100} = \frac{3}{10}$$

Based on this probability, 3 times it will rain, and 7 times it will not rain. Therefore, the odds in favor of rain = rain:no rain = 3:7.

(b) Odds against rain: The odds against rain are the opposite of the odds in favor of rain. Therefore, the odds against rain = no rain:rain = 7:3

Example 5 — Calculating Probabilities from Odds

The National Safety Council lists the odds of dying from a lightning strike as 1 in 84,079. Use these odds to calculate the probability of dying from a lightning strike. (*Source*: http://www.nsc.org/NSC%20Picture%20Library/News/web_graphics/Injury_Facts_37.pdf)

We know that probability is defined as

$$P(E) = \frac{\text{number of ways event } E \text{ can occur}}{\text{total number of possible outcomes}}.$$

To calculate total number of outcomes of an experiment using odds, we add the two number used to create the ratio, 1 and 84,049.

$$\text{Total Outcomes} = 1 + 84{,}079 = 84{,}080$$

Therefore, $P(\text{dying from a lightning strike}) = \dfrac{1}{84{,}080}$.

Practice Set 7-4

Calculate odds in favor of an event given the following sets of probabilities.

1. $P(\text{birth of a boy}) = \dfrac{1}{2}$
2. $P(\text{rolling a 4 on a die}) = \dfrac{1}{6}$
3. $P(\text{winning a game}) = \dfrac{2}{5}$
4. $P(\text{guessing correct answer}) = \dfrac{1}{4}$
5. $P(\text{drawing an ace}) = \dfrac{1}{13}$
6. $P(\text{drawing a heart}) = \dfrac{1}{4}$

Calculate the probability given the odds in favor of each event.

7. Odds of rolling an even number on a die—$1:1$
8. Odds of rolling a 1 or 3 on a die—$1:3$
9. Odds of winning a race—$3:11$
10. Odds of guessing a correct answer out of 5 possibilities—$1:4$
11. Odds of drawing the number 10 from a deck of cards—$1:12$
12. Odds of drawing a face card from a deck—$3:10$
13. The odds in favor of winning the door prize are 3 to 17. Find the odds against winning the door prize.
14. The odds against Sparkles Pretty winning the horse race are $5:2$. Find the odds in favor of Sparkles Pretty winning.

15. A committee of 3 is to be chosen from your math class members. If there are 20 members in your class, what are the odds that you will be chosen to be on this committee?

16. One person is selected at random from a class of 19 males and 13 females. Find the odds against selecting
    (a) a female.   (b) a male.

17. You roll a fair, 6-sided die one time. Use this information to answer problems 17 to 22.
    (a) Find the probability of rolling a 5.
    (b) Find the odds of rolling a 5.

18. (a) Find the probability of not rolling a 5.
    (b) Find the odds against rolling a 5.

19. Find the odds against rolling an odd number.

20. Find the odds against rolling a number greater than 4.

21. Find the odds in favor of rolling a number less than 3.

22. Find the odds in favor of rolling a number greater than 4.

Given a standard deck of 52 playing cards, you draw one card. Use this information for problems 23 to 28.

23. (a) Find $P(\text{club})$.
    (b) Find the odds of drawing a club.

24. (a) Find $P(\text{ace})$.
    (b) Find the odds of drawing an ace.

25. Find the odds in favor of selecting a 6.

26. Find the odds in favor of selecting a heart.

27. Find the odds against selecting a face card.

28. Find the odds against drawing an ace.

29. Suppose the probability that you are asked to work overtime this week is $\frac{5}{8}$. Find the odds in favor of your being asked to work overtime.

30. Suppose that the probability that a mechanic fixes your car right the first time is 0.7. Find the odds against your car being repaired right on the first attempt.

31. One million tickets are sold for a raffle. If you purchase one ticket, find your odds against winning (odds of losing).

32. One million tickets are sold for a raffle. If you purchase 10 tickets, find your odds against winning (odds of losing).

33. You bet on a horse in a race. If the odds on the horse are $5:1$, what is the probability that the horse will win the race?

34. If the odds in a certain game are $5:9$ in your favor, what is the probability that you will win the game?

35. The odds against Julie Wilson being admitted to the college of her choice are $9:2$. Find the probability that Julie will be admitted.

36. The odds against Julie Wilson being admitted to the college of her choice are $9:2$. Find the probability that Julie will not be admitted.

37. The odds against Jason getting promoted are $5:9$. Find the probability that Jason will get promoted.

38. The odds against Jason getting promoted are $5:9$. Find the probability that Jason will not get promoted.

39. The odds in favor of Kristen winning the spelling bee are $7:5$. Find the probability that Kristen will win the spelling bee.

40. The odds in favor of Kristen winning the spelling bee are $7:5$. Find the probability that Kristen will lose the spelling bee.

41. The odds in favor of Tim winning the tennis match are $1:6$. Find the probability that Tim will win the match.

42. The odds in favor of Tim winning the tennis match are $1:6$. Find the probability that Tim will not win the match.

43. The odds against Brian winning the 100-yard dash are $5:2$. Find the probability that Brian will win.

44. The odds against Brian winning the 100-yard dash are $5:2$. Find the probability that Brian will lose.

Section 7-5 Tree Diagrams

Even when situations seem simple, the number of possible results can be quite large. It is often important to know the number of possibilities that a situation may present before beginning to calculate the probability of any individual event.

As stated earlier, the possible results of an experiment are called outcomes. To solve more difficult probability problems, you must first be able to determine all the possible outcomes of the experiment. There are two methods that may be used to help you determine the number of outcomes. These methods are the **counting principle** and tree diagrams.

Rule Counting Principle

If a first experiment can be performed in X distinct ways and a second experiment can be performed in Y distinct ways, then the two experiments in that specific order can be performed in X times Y distinct ways.

Section 7-5 **Tree Diagrams** 317

Example 1

Using the Counting Principle: A Simple Case

What is the total number of outcomes for tossing two coins?

The first coin has two possible outcomes: heads or tails. The second coin also has two possible outcomes: heads or tails. Thus, the two experiments together have 2(2) or 4 possible outcomes.

Example 2

Using the Counting Principle: A Not-So-Simple Case

A license plate is to consist of two letters followed by three digits. Use the counting principle to determine how many different license plates are possible if repetition of letters and digits is permitted.

There are 26 letters and 10 digits (0–9). We have five positions to fill, as follows:

$$L \quad L \quad D \quad D \quad D$$

Because repetition is permitted, there are 26 possible choices for the first and second positions. There are 10 possible choices for the last three positions.

$$26 \quad 26 \quad 10 \quad 10 \quad 10$$

Therefore, the counting principle says that we have 26(26)(10)(10)(10) = 676,000 different possible license plates.

Example 3

License Plates Again

Repeat the problem in Example 2, but do not allow repetition of letters and digits.

There are 26 possible choices for the first position. Because repetition of letters is not permitted, there are only 25 choices for the second position. There are 10 choices for the third position. Because repetition of digits is not permitted, there are only 9 choices for the fourth position and 8 for the fifth position.

Therefore, the counting principle says that we have 26(25)(10)(9)(8) = 468,000 different possible license plates.

A second method used to determine the sample space is to construct a **tree diagram**. As long as the number of possible events is small, tree diagrams work very well. They provide a list of all possible events as well as giving the size of the sample space. Recall from a previous section that the sample space is a list of all possible outcomes. Tree diagrams can be very helpful in helping to organize our outcomes as we determine the sample space, s. However, tree diagrams can rapidly become very large and complicated for experiments involving several possible events. Another drawback of tree diagrams is that they are very time consuming to construct.

Example 4

Constructing a Tree Diagram for Tossing Two Coins

In Example 1, we used the counting principle to help us determine the number of possible outcomes for this experiment. The tree diagram pictured in Figure 7-8 shows us that there are four outcomes and lists the individual outcomes for the experiment.

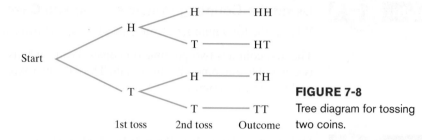

FIGURE 7-8 Tree diagram for tossing two coins.

A multipart experiment has two or more parts. Tree diagrams can become quite large when there are several parts to an experiment. Consider the experiment of having three children. There are 8 possible outcomes, which can be illustrated using a tree diagram as illustrated in Example 5.

Example 5

Another Tree Diagram

Construct a tree diagram and determine the sample space if a family has three children.

To determine the size of the sample space, we can use the Counting Principle. There are two possible outcomes for each of the three events (births)—boy or girl. Therefore, there are $2 \cdot 2 \cdot 2 = 8$ possible outcomes for this problem. Figure 7-9 illustrates these 8 outcomes. As you can see from the tree diagram, the sample space is $S = \{ggg, ggb, gbg, gbb, bgg, bgb, bbg, bbb\}$.

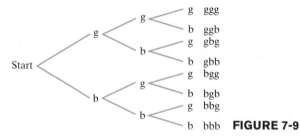

FIGURE 7-9

Example 6

Choosing Without Replacement

Two marbles are to be selected without replacement from a bag that contains one red, one purple, one green, and one blue marble.

(a) Use the counting principle to determine the number of items in the sample space.
(b) Construct a tree diagram and list the sample space.
(c) Find the probability that one purple marble is selected.
(d) Find the probability that a blue marble followed by a green marble is selected.

(a) The first selection may be any one of the four marbles. Once the first marble is selected, only three marbles remain for the second selection. Therefore, there are $4(3) = 12$ items in the sample space.
(b) See Figure 7-10. The sample space is $S = \{$RP, RG, RB, PR, PG, PB, GR, GP, GB, BR, BP, BG$\}$.

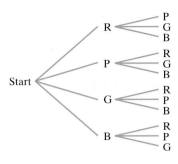

FIGURE 7-10
Tree diagram for drawing marbles.

(c) Recall that

$$P(E) = \frac{\text{number of ways event } E \text{ can occur}}{\text{total number of possible outcomes}}$$

The total number of outcomes is 12. Six outcomes have one purple marble: PR, PG, PB, RP, GP, or BP.

$$P(\text{one purple marble is selected}) = \frac{6}{12} = \frac{1}{2}$$

(d) One possible outcome meets the criteria for this problem: BG

$$P(\text{blue followed by green}) = \frac{1}{12}$$

Practice Set 7-5

Use the counting principle to determine the number of possible outcomes for each of the following.

1. How many different four-digit numbers can be formed from the digits 0–9 if the first digit must be even and cannot be 0? (You may repeat digits.)

2. How many different four-digit numbers can be formed from the digits 0–9 if the first digit must be even and cannot be 0? (You may not repeat digits.)

3. A bag contains five different lightbulbs. The bulbs are the same size, but each has a different wattage. If you select three bulbs at random, how many different outcomes will be in the sample space if the bulbs are selected with replacement?

4. A bag contains five different lightbulbs. The bulbs are the same size, but each has a different wattage. If you select three bulbs at random, how many different outcomes will be in the sample space if the bulbs are selected without replacement?

5. Two marbles are to be selected from a bag without replacement. The bag contains one red, one blue, one green, one yellow, and one purple marble. How many possible outcomes will there be for this experiment?

6. Two marbles are to be selected from a bag with replacement. The bag contains one red, one blue, one green, one yellow, and one purple marble. How many possible outcomes will there be for this experiment?

Draw a tree diagram for tossing a coin 3 times and then answer the following questions.

7. What is the size of this sample space?

8. What is the sample space for this experiment?

9. What is the probability that tossing a coin 3 times will result in one head and two tails?

10. What is the probability of tossing two or more heads?

A penny is tossed, and a fair, 6-sided die is rolled.

11. Use the counting principle to determine the number of outcomes in the sample space.

12. Construct a tree diagram illustrating all the possible outcomes and list the sample space.

13. What is the probability of tossing a head and rolling a 3?

14. What is the probability of tossing a head and rolling an even number?

15. Complete the following chart to show all of the possible outcomes for rolling two dice. Then answer the following questions.

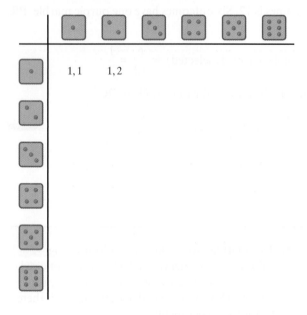

(a) What is the size of this sample space?

(b) What is the probability of rolling a total of 7 on two dice?

(c) What is the probability of rolling 13 on two dice?

(d) What is the probability of rolling a 3 on both dice?

You perform an experiment that involves flipping a coin twice and then rolling a die once.

16. What is the size of the sample space?

17. What is the sample space?

18. Find $P(HH3)$.

19. Find $P(HT$, in any order, and then roll a 5$)$.

20. Find $P(TT$ and then roll an even number$)$.

A bag contains three cards: an ace, a jack, and a queen. Two cards are to be selected at random with replacement.

21. Construct a tree diagram and determine the sample space.

22. Find the probability that two aces are selected.

23. Find the probability that a jack and then an ace are selected.

24. Find the probability that at least one queen is selected.

A couple plans to have two children.

25. Construct a tree diagram and list the sample space of the possible arrangements of boys and girls.

26. Find the probability that the family has two girls.

27. Find the probability that the family has at least one boy.

28. Find the probability that the family has at least two boys.

Suppose that you are going to make up a password for access to some files on your computer. You are going to choose three lowercase letters from the alphabet and then choose two digits (like *kuy76*).

29. How many different passwords are possible if repetition is allowed?

30. How many different passwords are possible if repetition is not allowed?

31. What is the probability that someone could guess your password on the first guess if the person knew to choose three letters and two digits?

32. What is the probability that someone could guess your password on the first guess if the person knew to choose three letters and two digits, and repetition is not allowed?

A monkey types the letters *o*, *k*, and *a* in random order. Answer the following questions.

33. How many different three-letter "words" could the monkey type using just these three letters? (You cannot repeat the letters.)

34. What is the probability that the monkey would type a meaningful word using these three letters?

Section 7-6 Or Problems

Suppose that a series of events could occur in a given situation. You might wish to calculate the probability for the occurrence of one particular series of events as opposed to other possible outcomes. For example, suppose that you wanted to find the probability of tossing a coin 3 times and having the coin come up heads every time. This is a series of three events.

There are some complications that have not yet been discussed in this chapter. First, can the possible events all occur at once or must they be entirely separate from each other? **Mutually exclusive events** are events that cannot occur together. If you are flipping coins, the events of the coin turning up heads or tails are mutually exclusive. The coin cannot turn up both sides at once.

Definition | Mutually Exclusive Events

Two events A and B are mutually exclusive if it is impossible for both events to occur at the same time.

In Section 7-2, we introduced the concept of independent and dependent events. The result of each flip of the coin in this experiment is an **independent outcome**. This means that the result of the first flip of the coin has no effect on the probability of the next flip of the coin. This concept is illustrated in the Venn Diagrams shown in Figure 7-11.

This Venn diagram illustrates the idea that the two events do not overlap, since you can't get both outcomes at the same time.

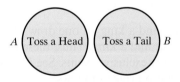

FIGURE 7-11
Mutually exclusive events—flipping a coin.

Definition | Independent Events

Two events A and B are **independent events** if the occurrence of either event in no way affects the probability of occurrence of the other.

To find the total probability of a series of events, the probability of each individual event must first be calculated. Therefore, for tossing a coin once, the probability of heads showing up on any single toss is $\frac{1}{2}$ (see Section 7-3). Thus, each time you flip the coin the probability of heads showing up will be the same, $\frac{1}{2}$.

Now what do you do with the three probabilities of $\frac{1}{2}$? Look at the original proposed experiment and the outcome desired. Are each of the events (individual coin flips) tied together in the experiment by *and* or *or*? This may seem to be a silly question, but it is not. For example, suppose someone asked these questions of you, how would you answer?

Question 1: Do you want coffee *and* cream?
Question 2: Do you want coffee *or* cream?

Answering "yes" to question 1 gets you coffee with cream mixed in it. Answering "yes" to question 2 is meaningless. Question 2 says that you can have either coffee or you can have cream, but you cannot have a mixture. There are two different rules to take care of this situation. We will examine the rule that relates to "or" problems in this section.

If the events are described as "this happens *or* that happens," then the **addition rule** tells us that we should add up the probabilities of the mutually exclusive events to determine the probability that any one of these events might occur.

This probability concept is related to the union of sets (U) in set theory that was discussed in Section 7-1. If you have two sets A and B, these two sets can be combined to form a new set called the **union** of sets A and B. This new set contains all the elements in both sets. Similarly, the probability of either of two mutually exclusive events occurring is the sum of those probabilities. Look at the problem in Example 1.

Rule — Addition Rule for Mutually Exclusive Events

$$P(A \text{ or } B) = P(A) + P(B)$$

Example 1 — Mutually Exclusive Events

If A is the event of drawing an ace from a standard deck of 52 playing cards and B is the event of drawing a king, then what is the probability of drawing an ace or a king from the deck on a single draw?

Can you draw *both* an ace and a king on one draw? No! There is no single card in the deck that is both an ace and a king. Thus, these two events, drawing an ace or drawing a king, are mutually exclusive events. See Figure 7-12. The probability can be calculated with the addition rule as follows:

$$P(A) = P(\text{ace}) = \frac{4}{52} = \frac{1}{13}$$

$$P(B) = P(\text{king}) = \frac{4}{52} = \frac{1}{13}$$

$$P(A \text{ or } B) = P(A) + P(B) = \frac{1}{13} + \frac{1}{13}$$

$$P(\text{ace or king}) = \frac{2}{13}$$

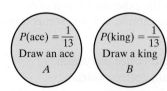

FIGURE 7-12
Mutually exclusive events.

Of course, not all events are mutually exclusive. For example, what if you were asked to choose the name of a student in your class from the class roll or choose a male student from the same class? Would it be possible to choose one name and have both events occur? Yes, you could draw the name of a class member who was also a male. Thus, the events are not mutually exclusive. This relationship is illustrated in Figure 7-13 using Venn Diagrams.

Here is a modification of the addition rule to cover this situation.

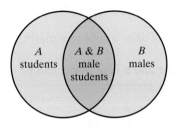

FIGURE 7-13
Events that are not mutually exclusive.

Rule | **Addition Rule for Events That Are Not Mutually Exclusive**

$$P(A \text{ or } B) = P(A) + P(B) - P(A \text{ and } B)$$

Example 2 | **Drawing Objects from a Container**

A container holds 15 marbles and 15 cubes. Seven marbles and 7 cubes are red, 4 marbles and 4 cubes are blue, and 4 marbles and 4 cubes are green. You are to reach into the container and draw out one object at random. Determine whether each of the following pairs of events are mutually exclusive and then calculate $P(A \text{ or } B)$ for each pair.

(a) P(a cube or a red object)
(b) P(a cube or a marble)
(c) P(a blue marble or a blue object)

(a) Because it is possible to draw both a cube and a red object in one draw, these events are *not mutually exclusive*.

$$P(\text{a cube}) = \frac{15}{30} = \frac{1}{2}$$

$$P(\text{a red object}) = \frac{14}{30} = \frac{7}{15}$$

$$P(\text{a cube and a red object}) = \frac{7}{30}$$

$$P(\text{a cube or a red object}) = P(\text{a cube}) + P(\text{a red object}) - P(\text{a cube and a red object})$$

$$P(\text{a cube or a red object}) = \frac{15}{30} + \frac{14}{30} - \frac{7}{30} = \frac{22}{30} = \frac{11}{15}$$

(b) Because you cannot draw one object and have it be both a cube and a marble, these are *mutually exclusive* events.

$$P(\text{a cube}) = \frac{15}{30} = \frac{1}{2}$$

$$P(\text{a marble}) = \frac{15}{30} = \frac{1}{2}$$

$$P(\text{a cube or a marble}) = P(\text{a cube}) + P(\text{a marble}) = \frac{1}{2} + \frac{1}{2} = 1$$

It is certain that you will draw a cube or a marble if there are no other objects in the container.

(c) Because a blue marble and a blue object can be drawn at the same time, these events are *not mutually exclusive*.

$$P(\text{a blue marble}) = \frac{4}{30} = \frac{2}{15}$$

$$P(\text{a blue object}) = \frac{8}{30} = \frac{4}{15}$$

$$P(\text{a blue marble and a blue object}) = \frac{4}{30} = \frac{2}{15}$$

$$P(\text{a blue marble or a blue object}) = \frac{2}{15} + \frac{4}{15} - \frac{2}{15} = \frac{4}{15}$$

The primary thing that you need to remember when finding the probability of one event or another event is to avoid double counting the probability of both occurring together. Another way to think of the $P(A \text{ or } B)$ is to find the $P(A)$ and add it to the $P(B)$ exclusive of any of event A included in B. For example, if you are asked to find the probability of drawing a heart or an ace, you can add the probability of drawing a heart to the probability of drawing an ace *that is not a heart*, thus avoiding double counting the ace of hearts.

$$P(\text{heart or ace}) = \frac{13 \text{ hearts}}{52} + \frac{3 \text{ aces that are not hearts}}{52} = \frac{16}{52} = \frac{4}{13}$$

Look at the following example illustrating the use of the formula method, as well as the use of this more intuitive method.

Example 3

Probabilities from a Table

A survey of 150 people asked participants if they currently smoke. The results are shown in the table.

	Smoker	Not a Smoker	Total
Male	31	51	82
Female	24	44	68
Total	55	95	150

If you choose one participant at random, find the probability that the participant is either a smoker or a male.

(a) Using the formula: $P(A \text{ or } B) = P(A) + P(B) - P(A \text{ and } B)$

$P(\text{smoker}) = 55$ smokers out of the 150 participants:

$$\frac{55 \text{ smokers (male and female)}}{150 \text{ participants}}$$

$P(\text{male}) = 82$ males out of the 150 participants:

$$\frac{82 \text{ males (smokers and nonsmokers)}}{150 \text{ participants}}$$

$P(\text{smoker and male}) = 31$ participants out of 150 are male smokers:

$$\frac{31 \text{ male smokers}}{150 \text{ participants}}$$

$P(\text{smoker or male}) = P(\text{smoker}) + P(\text{male}) - P(\text{smoker and male})$

$$= \frac{55}{150} + \frac{82}{150} - \frac{31}{150} = \frac{106}{150} = \frac{53}{75}$$

(b) Using an intuitive method: $P(A \text{ or } B) = P(A) + P(B)$ exclusive of any of event A included in B

$P(\text{smoker}) = 55$ smokers out of the 150 participants:

$$\frac{55 \text{ smokers (male and female)}}{150 \text{ participants}}$$

$P(\text{males who are not smokers}) = 51$ males are included in the survey but are not smokers:

$$\frac{51 \text{ nonsmoking males}}{150 \text{ participants}}$$

$P(\text{smoker or male}) = P(\text{smoker}) + P(\text{males who are not smokers})$

$$= \frac{55}{150} + \frac{51}{150} = \frac{106}{150} = \frac{53}{75}$$

Practice Set 7-6

One card is selected at random from a deck of cards. Use this information for problems 1 to 6.

1. Find the probability of selecting a king or a queen. Are these events mutually exclusive? Why or why not?

2. Find the probability of selecting an ace or a club. Are these events mutually exclusive? Why or why not?

3. Find the probability of selecting a black card or a spade. Are these events mutually exclusive? Why or why not?

4. Find the probability of selecting a red card or a black card. Are these events mutually exclusive? Why or why not?

5. Find the probability of selecting a face card or a black card. Are these events mutually exclusive? Why or why not?

6. Find the probability of selecting a 2 or a 5. Are these events mutually exclusive? Why or why not?

7. Each of the numbers 0, 2, 4, 6, 8, 10, 12, and 14 is written on a separate sheet of paper and placed into a box. One piece of paper is selected at random. Find the probability that the piece of paper selected contains the number 2 or a number greater than 2.

8. Each of the numbers 0, 2, 4, 6, 8, 10, 12, and 14 is written on a separate sheet of paper and placed into a box. One piece of paper is selected at random. Find the probability that the piece of paper selected contains the number 8 or a number less than 12.

A single die is rolled one time. Use this information for problems 9 to 14.

9. Find the probability of rolling an odd number or a number greater than 2.

10. Find the probability of rolling a number greater than 5 or less than 3.

11. Find the probability of rolling an odd number or an even number.

12. Find the probability of rolling a number greater than 6 or less than 1.

13. Find the probability of rolling a 1 or a 6.

14. Find the probability of rolling an even number or a 1.

The table below displays enrollment figures for several programs at a local community college. One student is selected at random from one of these programs and receives a gift card to the student bookstore. Use the information in this table for problems 15 to 20.

Major	Male	Female	Total
Criminal Justice	48	25	73
Early Childhood	3	47	50
Automotive	32	3	35
Culinary	27	20	47
Total	110	95	205

15. Find the probability that the student selected is male.

16. Find the probability that the student selected is a culinary student.

17. Find the probability that the student selected is an automotive student or an early childhood major.

18. Find the probability that the student selected is a culinary student or an automotive student.

19. Find the probability that the student selected is male or a criminal justice student.

20. Find the probability that the student selected is female or an early childhood major.

A recent poll of Americans was used to present the public's opinions concerning the death penalty. The question asked was, "Are you in favor of the death penalty for someone convicted of murder?" The table below lists the results of the poll based on where each person lives. Use the information in this table for problems 21 to 26.

	Approve	Disapprove	No Opinion	Total
East	171	120	8	299
South	165	70	15	250
West	168	95	11	274
Midwest	211	90	16	317
Total	715	375	50	1140

21. If a survey participant is selected at random, what is the probability that the respondent approves of the death penalty?

22. If a survey participant is selected at random, what is the probability that the respondent has no opinion about the death penalty?

23. If a survey participant is selected at random, what is the probability that the respondent is from the East or the South?

24. If a survey participant is selected at random, what is the probability that the respondent is from the Midwest or the West?

25. If a survey participant is selected at random, what is the probability that the respondent is from the South or disapproves of the death penalty?

26. If a survey participant is selected at random, what is the probability that the respondent is from the West or approves of the death penalty?

27. What is the probability that the respondent approves of the death penalty given that the respondent lives in the South?

28. What is the probability that the respondent disapproves of the death penalty given that the respondent lives in the West?

29. The Music Club consists of 10 male freshmen, 15 female freshmen, 23 male sophomores, and 12 female sophomores. If one person is randomly selected from the group, find the probability of selecting a freshman or a female.

30. The History Club consists of 5 male freshmen, 12 female freshmen, 13 male sophomores, and 8 female sophomores. If one person is randomly selected from the group, find the probability of selecting a sophomore or a male.

Section 7-7 *And* Problems

In the last section, we looked at the probability that event *A or* event *B* occurs in an experiment with one event. In this section, we will look at the probability of event *A and* event *B* both occurring in a series of events. If the events are described as "this happens *and* that event happens," then the **multiplication rule** tells us to multiply the probabilities of the outcomes for each event in the series to find the probability that these events will all occur together. *Note:* We must always assume that event *A* has occurred before calculating the probability of event *B*.

Rule Multiplication Rule

$$P(A \text{ and } B) = P(A) \cdot P(B)$$

Probability problems requiring the multiplication rule are related to the concept of intersection (∩) in set theory. In each case, specific requirements must be fulfilled in both sets or in both events.

To use the multiplication rule correctly, we must know if the events **are independent events** or **dependent events**. If the events are independent, multiply the individual probabilities together. If the probability of the second event is dependent on the outcome of the first event, multiply the individual probabilities taking into account the conditional probability that event *A* occurred prior to event *B*.

Example 1 Selecting Cards from a Deck with Replacement

Two cards are selected at random from a standard deck of 52 playing cards. Find the probability that two aces will be selected in a row if the first card is replaced into the deck before drawing the second card.

We are asked to calculate the following:

$$P(\text{ace and ace, with replacement})$$

Since the card is being replaced, the probability of drawing an ace the second time is not affected by the ace that is drawn the first time.

$$P(\text{ace}) = \frac{4}{52} = \frac{1}{13}, \text{ each time}$$

$$P(\text{ace and ace, with replacement}) = \left(\frac{1}{13}\right)\left(\frac{1}{13}\right) = \frac{1}{169}$$

Example 2 — Selecting Cards Without Replacement

Two cards are selected at random from a standard deck of 52 playing cards. Find the probability that two aces will be selected in a row if the first card is *not* replaced into the deck before drawing the second card.

We are asked to calculate the following:

$$P(\text{ace and ace, without replacement})$$

Because these are dependent events, the probability of drawing a second ace will change. We are to assume that an ace is selected on the first draw before calculating the probability of the second event. Because there are only 3 aces remaining in the deck of 51 cards, the probability of drawing a second ace is changed.

$$P(\text{first ace}) = \frac{\text{number of aces in a deck}}{\text{total number of cards in a deck}}$$
$$= \frac{4}{52} = \frac{1}{13}$$

$$P(\text{second ace, without replacement}) = \frac{\text{number of aces remaining in the deck}}{\text{number of cards remaining in the deck}}$$
$$= \frac{3}{51} = \frac{1}{17}$$

$$P(\text{ace and ace, without replacement}) = \left(\frac{1}{13}\right)\left(\frac{1}{17}\right) = \frac{1}{221}$$

Compare the result for this example with the previous one and note the differences.

Example 3 — Calculating the Probability of a Series of Events

What is the probability that you would flip a coin twice, getting tails both times, and then roll a 5 on a die?

Is this an "and" or an "or" type of problem? It can be restated this way: The result desired is tails and another tails and a 5. Therefore, use the multiplication rule. These are independent events, so determine all of the individual probabilities and multiply them together.

$$P(T) = \frac{1}{2} \text{ each time}$$
$$P(5) = \frac{1}{6}$$
$$P(T \text{ and } T \text{ and } 5) = \left(\frac{1}{2}\right)\left(\frac{1}{2}\right)\left(\frac{1}{6}\right) = \frac{1}{24}$$

Practice Set 7-7

Identify the following events as independent events or dependent events.

1. A card is drawn from a deck of cards, put back into the deck, and a second card is drawn.

2. A single die is rolled, the answer recorded, and the die is rolled a second time.

3. A bag contains 5 red marbles and 4 blue marbles. A marble is drawn and put on the table. A second marble is drawn.

4. A baby is born into a family followed by a second birth two years later.

5. A die is rolled and a coin is flipped.

6. A card is drawn from a deck, placed on a table, and a second card is drawn.

Calculate the following probabilities.

7. Calculate the probability of drawing a king from the deck followed by another king, if the first card is replaced after it is drawn.

8. Calculate the probability of rolling a 3 on a die followed by another 3 on the same die.

9. A bag contains 5 red marbles and 4 blue marbles. Calculate the probability of drawing a red marble followed by a blue marble if the first marble is not replaced after it is drawn.

10. Find the probability of a family having a girl followed by another girl two years later.

11. Find the probability of rolling a 4 on a die and getting a heads when a coin is flipped.

12. Calculate the probability of drawing a three from a deck and then drawing another three from the same deck if the first card is not replaced.

A package of 24 zinnia seeds contains 7 seeds for red flowers, 12 seeds for white flowers, and 5 seeds for yellow flowers. Three seeds are randomly selected and planted.

13. Find the probability that all three seeds will produce white flowers.

14. Find the probability that the first seed selected will produce a yellow flower, the second seed will produce a red flower, and the third seed will produce a yellow flower.

15. Find the probability that none of the seeds will produce red flowers.

16. Find the probability that the first seed selected will produce a red flower, the second seed will produce a yellow flower, and the third seed will produce a white flower.

Dan is a good, but not great, chess player. Based on past tournament experience, his chances of winning any single game are $\frac{2}{5}$.

17. What is the probability that Dan will win the second and third games of a five-game match and lose all the others?

18. What is the probability that Dan will win the first game and lose the last two games of a three-game match?

19. What is the probability that he will win all five games?

20. What is the probability that he will lose all five games?

21. Each letter of the word *Mississippi* is placed on a piece of paper, and all 11 pieces of paper are placed in a hat. Three letters are selected at random from the hat and are not replaced.
    (a) Find the probability that three consonants are selected.
    (b) Find the probability that the first letter selected is a vowel, the second letter is a consonant, and the third letter is a vowel.

22. Repeat problem 21 with replacement.

A family has three children. Assuming independence, find the following probabilities in problems 23 to 26.

23. All three are boys.

24. The youngest child is a boy, and the older children are girls.

25. The youngest child is a girl, the middle child is a girl, and the oldest child is a boy.

26. All three are girls.

27. A club has 15 female members and 12 male members. If three members are selected at random for a subcommittee, what is the probability that they are all male?

28. A club has 15 female members and 12 male members. If three members are selected at random for a subcommittee, what is the probability that they are all female?

29. What is the probability that two people were both born on April 1? Ignore leap years.

30. What is the probability that two people were both born on April 1, given that they were both born in April?

Section 7-8 The Counting Principle, Permutations, and Combinations

An alternative to creating tree diagrams is to mathematically calculate the number of possible outcomes for a single experiment (tossing a coin once) or for each experiment in a series (tossing a coin twice) and then use these calculated values to count the number of items in the sample space. Using the appropriate formula will require that you first understand some new terms.

The counting principle was introduced in Section 7-5. It is repeated here for your convenience.

Rule Counting Principle

If a first experiment can be performed in X distinct ways and a second experiment can be performed in Y distinct ways, then the two experiments in that specific order can be performed in X times Y distinct ways.

Example 1 Arranging Books on a Shelf: The Counting Principle

In how many ways can nine different books be arranged on a shelf?

Because there are nine different books, by the counting principle the number of possible arrangements will be $9(8)(7)(6)(5)(4)(3)(2)(1) = 362{,}880$.

Another way to calculate $9(8)(7)(6)(5)(4)(3)(2)(1)$ is to use what is called a **factorial**. A factorial is an algebraic shorthand. Nine factorial is written as $9!$.

IMPORTANT FORMULAS
Factorial

If n is a positive integer, then n factorial ($n!$) is given by

$$n! = n(n-1)(n-2)(n-3) \ldots (2)(1)$$

This definition is usually extended to include zero factorial, although it is not usually needed. By definition, $0! = 1$.

CALCULATOR MINI-LESSON
Factorial

Any *standard scientific calculator* has a key that will automatically calculate the value of a factorial. Look for a key or the second function of a key labeled $\boxed{x!}$. To test this key, simply type in a number like 9 and then press the $\boxed{x!}$ key. The calculator should show you the same value as was obtained in Example 1.

A *graphing calculator (TI-84 Plus)* also has a factorial function, but it is harder to find. Type in a number like 9. Press the $\boxed{\text{MATH}}$ key and use the right arrow key to highlight $\boxed{\text{PRB}}$ at the top of the screen. Use the down arrow key to highlight $\boxed{4{:}!}$. Press $\boxed{\text{ENTER}}$ and then press $\boxed{\text{ENTER}}$ again to see the result.

Section 7-8 **The Counting Principle, Permutations, and Combinations** 331

Example 2 **Calculating Factorials**

Find the value of each of the following.

(a) $3! = ?$
$3! = 3(2)(1) = 6$

(b) $7! = ?$
$7! = 7(6)(5)(4)(3)(2)(1) = 5040$

Example 3 **More on Factorials and the Calculator**

Using your calculator, find the value of each of the following.

(a) $5! = ?$ (b) $9! = ?$ (c) $12! = ?$
(d) $3.8! = ?$ (e) $0! = ?$

Here are the answers that the calculator should give you:

(a) $5! = 120$
(b) $9! = 362,880$
(c) $12! = 479,001,600$
(d) $3.8! = $ E or Error (remember, nonnegative integers only)
(e) $0! = 1$

We must also consider the idea of the *order* in which events occur. Consider the experiment of tossing a coin twice, as shown in Example 4 in Section 7-5. The sample space is $S = \{HH, HT, TH, TT\}$. If you determine the probability of tossing heads once and tails once (and it does not matter which is tossed first), then $P(HT \text{ or } TH) = \frac{2}{4} = \frac{1}{2}$. However, if you wish to determine the probability of tossing heads on the first of two tosses and tails second (i.e., order does matter), then $P(HT) = \frac{1}{4}$. It is clear that whether or not order is important does affect the probability of individual events.

IMPORTANT EQUATIONS

Permutations

If there are n distinct objects or events possible and you are going to choose r objects or events ($r \leq n$), then the number of permutations is given by

$$_nP_r = \frac{n!}{(n-r)!}$$

Note: Order is important here.

Combinations

If there are n distinct objects or events possible and you are going to choose r objects or events ($r \leq n$), then the number of combinations is given by

$$_nC_r = \frac{n!}{r!(n-r)!}$$

Note: Order is not important here.

The number of different ways that a set of objects or events can be placed in a specific order is called a **permutation**, and the number of ways that a set of objects or events can be selected without regard to order is called a **combination**. Referring to the example mentioned previously, HT and TH are two different permutations of the same combination. If order matters, then the number of permutations will be calculated, and if order does not matter, then the number of combinations will be calculated.

Example 4 — Combinations and Permutations Using the Formula

(a) A local pizza parlor is running a weekend special on large pizzas. There are 10 possible toppings on the menu and each customer can choose any 3 different toppings to get the special price. How many possible combinations of toppings are possible?

Because the order of toppings on a pizza is not important and the toppings are all different, we will use the formula for a combination to answer the question. There are 10 possible toppings and the customer can choose any 3, so $n = 10$ and $r = 3$.

$$_nC_r = \frac{n!}{r!(n-r)!} = {_{10}C_3} = \frac{10!}{3!(10-3)!} = \frac{10 \cdot 9 \cdot 8 \cdot 7!}{3!7!} = \frac{10 \cdot 9 \cdot 8}{3 \cdot 2} = \frac{720}{6}$$

$$= 120 \text{ possible pizzas}$$

(b) A computer password is to consist of 4 different letters. How many different passwords are possible?

Because the order of the letters is important in a password and repetition is not allowed, we will use the formula for a permutation to answer the question. There are 26 letters in the alphabet, and 4 letters are to be chosen. So, $n = 26$ and $r = 4$.

$$_nP_r = \frac{n!}{(n-r)!} = {_{26}P_4} = \frac{26!}{(26-4)!} = \frac{26 \cdot 25 \cdot 24 \cdot 23 \cdot 22!}{22!}$$

$$= 358{,}800 \text{ possible passwords}$$

The formulas for combinations and permutations are relatively complicated. However, your calculator has function keys that will allow you to enter the information from a problem and calculate the answer easily. Study the directions in the Calculator Mini-Lesson and Examples 5 and 6.

CALCULATOR MINI-LESSON

Combinations and Permutations

There are two keys on a *standard scientific calculator* that are used to determine the number of combinations or permutations. These keys have the same symbols on them as shown in the definitions given previously, $\boxed{_nC_r}$ and $\boxed{_nP_r}$. Using these keys is simple. First, enter the value of n. Next, press the $\boxed{_nC_r}$ or $\boxed{_nP_r}$ and then enter the value of r. The calculator should then display the value.

On a *graphing calculator (TI-84 Plus)*, enter the value of n. Press $\boxed{\text{MATH}}$ and use the right arrow key to highlight $\boxed{\text{PRB}}$. Move down to $\boxed{3:_nC_r}$ or $\boxed{2:_nP_r}$ and press $\boxed{\text{ENTER}}$. Type in the value of r and then press enter to display the answer.

Example 5 — Choosing a Committee (Combinations)

Suppose that a committee is to be chosen by drawing names from members of your math class. We will say that there are 20 persons in your class and that a committee will consist of three names drawn at random. If the order in which names are to be drawn has no effect on the final makeup of the committee, then how many different committees are possible?

Because order is unimportant, the number of possible combinations should be calculated.

$$n = 20 \quad r = 3$$

$$_nC_r = \frac{n!}{r!(n-r)!} = \frac{20!}{3!(20-3)!} = 1140$$

As complicated as this looks, wouldn't a shortcut or some mechanical assistance be nice?

Using the calculator as described in the last Calculator Mini-Lesson,

$$_nC_r = {}_{20}C_3 = 1140$$

This number, 1140, seems large, but it is the number of different possible committees of three persons that can be chosen from a class of 20.

Example 6 — Choosing Officers for a Committee (Permutations)

Suppose that officers are to be chosen by drawing names from members of your math class. We will say that there are 20 persons in your class and that the officers will consist of three names drawn at random. If the order in which names are to be drawn has some effect on the final makeup (e.g., the first name drawn will be chair, the second the secretary, and the third person will do all the work), then how many different sets of officers are possible?

Because order does make a difference, we will calculate the number of permutations possible.

$$_nP_r = {}_{20}P_3 = 6840$$

This number is a much larger number than in the previous example. Why is that? If order does not matter (see Example 5), then if James, Jane, and Joe are chosen, it will be the same committee no matter whose name is chosen first, second, or third. However, if order matters because of the office and duties assigned, then the list of possibilities for James, Jane, and Joe actually forms six *different* sets of officers.

Example 7 — Winning the Lottery: Fat Chance!

What are your chances of winning the Texas Lottery? The Texas Lottery Cash Five game requires that you choose five numbers from 1 to 37. You may not choose the same number twice.

(a) How many different combinations of the 37 numbers are possible?
(b) What is the probability that your five numbers would win?

(a) $_nC_r = {}_{37}C_5 = 435{,}897$ different lottery tickets are possible.
(b) If you buy one ticket, your probability of picking the correct five-number combination is 1 in 435,897. (In other words, you are wasting your money to enter because you are probably not going to win.)

$$P(\text{win}) = \frac{1}{435{,}897}$$

Example 8 — Letter Arrangements

Consider the five letters a, b, c, d, and e. In how many distinct ways can three letters be selected and arranged if repetition is not allowed?

This is a permutation of five things taken three at a time.

$$_5P_3 = 60$$

We could also solve this problem using the counting principle. There are five possible letters for the first choice, four for the second choice, and three for the third choice:

$$5(4)(3) = 60$$

Therefore, there are 60 possible ordered arrangements or permutations.

There are many applications of probability in our daily lives. The weather forecast may give the probability of rain for today. Prescription information may give the probability of a particular side effect occurring if you use that drug. Calculating probabilities can be quite complicated. In this section, we will demonstrate some simple applications of the probability techniques that we have studied.

Example 9 — Delegations

The National Honor Society at Williams High School has 18 members. A delegation of four members must be chosen to attend an upcoming national meeting. The members decide to draw names at random.

(a) How many different delegations are possible?
(b) How many different delegations that include José are possible?
(c) How many possible delegations do not include José?

(a) Order is not a factor, so the number of possible combinations should be calculated as follows:

$$_nC_r = {}_{18}C_4 = 3060$$

(b) If José is sure to be in the delegation, then the other three members must be chosen at random from the 17 remaining members as follows:

$$_nC_r = {}_{17}C_3 = 680$$

(c) If José is sure not to be in the delegation, then it is as if José's name was not even in the hat to be drawn. The delegation of 4 will be drawn from the other 17 members who are not José as follows:

$$_nC_r = {}_{17}C_4 = 2380$$

Note: The total of all delegations with José plus the total of all delegations without José equals the total number of possible delegations that could be chosen:

$$680 + 2380 = 3060$$

Example 10 A Quality Control Problem

A manufacturer of computer chips requires that four chips from each lot of 25 be tested for proper function by the company's quality control department. If one or more of the chips tested are defective, then the entire lot of 25 will be rejected. If a certain lot were to contain 3 defective chips and 22 good chips, what is the probability that this lot would pass inspection?

First, what is the size of the sample space? In other words, how many different combinations of four computer chips can be chosen?

$$_nC_r = {_{25}C_4} = 12{,}650$$

Second, how many different ways can four chips be chosen involving only the 22 good ones?

$$_nC_r = {_{22}C_4} = 7315$$

Thus, the probability that the lot will pass inspection is

$$P(\text{lot passes}) = \frac{7315}{12{,}650} \approx 0.578 \text{ (or about a 58\% chance)}$$

Example 11 Choosing a Cereal

The Kellogg Company is testing 14 new cereals for possible production. It is testing 4 oat cereals, 5 wheat cereals, and 5 rice cereals. If we assume that each of the 14 cereals has the same chance of being selected and 4 new cereals will be produced, find the probability that no wheat cereals will be selected for production.

If no wheat cereals are to be selected, then only oat or rice cereals can be selected. A total of 9 cereals are oat or rice. So the number of ways that 4 oat or rice cereals may be selected from the 9 possible oat or rice cereals is $_9C_4$. The total number of possible selections is $_{14}C_4$.

$$P(\text{no wheat cereals}) = \frac{_9C_4}{_{14}C_4} = \frac{126}{1001} = \frac{18}{143} \approx 0.126 \text{ (or about a 13\% chance)}$$

Practice Set 7-8

Determine each of the following combinations, permutations, and factorials.

1. $_7C_2$
2. $_{15}C_5$
3. $_7P_2$
4. $_{13}P_8$
5. $9!$
6. $18!$
7. $(-8)!$
8. $5.65!$

9. Describe the differences between combinations and permutations.

10. A student must select and answer any four of five essay questions on a test. In how many ways can she do so?

11. There are 10 girls on the golf team at East High. How many different four-girl teams may be chosen by the coach if Jane must be on the team?

12. A standard deck of playing cards has 52 different cards. A poker hand contains five cards. How many different poker hands are possible?

13. You are the coach of a nine-person baseball team. How many different batting orders are possible?

14. You have seven different books. How many different ways may they be arranged on a bookshelf?

15. Angelo's Pizza offers its customers a choice of pizza toppings. You may choose from onions, green peppers, hot peppers, mushrooms, and pepperoni. The base for all of the pizzas is

cheese. A pizza may have any number of toppings that the customer desires. How many different pizzas are possible?

16. To win the jackpot of a state lottery, the six numbers chosen by a person, from 1 through 54, must be the same as the six numbers selected at random by the state lottery system. On each lottery ticket purchased, there are two separate selections for the six numbers. (Repetition of numbers is not permitted.) What is the probability of winning the jackpot with one ticket?

17. In how many different ways may four persons be seated in a row of four rocking chairs?

18. In how many different ways may six students be seated in a row with eight seats?

19. Using the letters of the word *algorithm* and calling any arrangement of letters a word:
    (a) How many different nine-letter words may be formed?
    (b) How many different four-letter words may be formed?

20. In how many different ways may three boys and three girls be seated in a row if they must alternate boy, girl?

21. In boating, colored flags, arranged in various orders and combinations, have been used for centuries to convey messages to other boats and ships or the persons on shore. If you have three red, three green, and two blue flags, how many different messages (arrangements) could you put up on the mast of the ship if you used
    (a) all eight flags at one time?
    (b) only three flags at one time?
    (c) only four flags at one time?

22. Alvin Community College's identification number consists of a letter followed by four digits. How many different identification numbers are possible if repetition is not permitted?

23. The daily double at most racetracks consists of selecting the winning horse in both the first and the second races. If the first race has eight entries and the second race has nine entries, how many daily-double tickets must you purchase to guarantee a win?

24. The Stereo Shop's warehouse has 12 different receivers in stock. The owner of the chain calls the warehouse requesting that a different receiver be sent to each of five stores. How many ways can the receivers be distributed?

25. A social security number consists of nine digits. How many different social security numbers are possible if repetition of digits is permitted?

26. An ice cream parlor has 31 different flavors. Maribeth orders a banana split and has to select three different flavors. How many different selections are possible?

27. At a car rental agency, the agent has 10 midsize cars on his lot and six people have reserved midsize cars. In how many different ways can the six cars to be used be selected?

28. If your class has 10 males and 14 females in it, how many different committees consisting of 4 males and 4 females may be chosen?

29. A committee of three judges is considering nine people for a reality show. The committee has decided to select three of the nine for further consideration. In how many ways can it do so?

30. At the beginning of the semester, your teacher informs the class that she marks on a curve and exactly 5 of the 26 students in the class will receive a final grade of A. In how many ways can this result occur?

31. A bag contains five red marbles and four blue marbles. You plan to draw three marbles at random without replacement. Find the probability of selecting three red marbles.

32. Repeat problem 31 with replacement.

33. Each of the numbers 1–6 is written on a piece of paper, and the six pieces of paper are placed in a hat. If two numbers are selected at random without replacement, find the probability that both numbers selected will be even.

34. Repeat problem 33 with replacement.

35. A box contains four good and four defective batteries. If you select three batteries at random without replacement, find the probability that you will select three good batteries.

36. Repeat problem 35 with replacement.

37. Dan's wallet contains eight bills of the following denominations: four $5 bills, two $10 bills, one $20 bill, and one $50 bill. If Dan selects two bills

at random without replacement, determine the probability that he will select two $5 bills.

38. Repeat problem 37 with replacement.

39. How many different "words" (the "words" do not have to be meaningful) can be made using
    (a) all the letters in the word *modern*?
    (b) only four letters at a time?

40. You have a penny, a nickel, a dime, and a quarter in your pocket and no other coins. You reach into your pocket and remove two coins at random.
    (a) What is the sample space?
    (b) What is the probability that exactly one of the coins is a nickel?
    (c) What is the probability that the total value of the two coins is greater than 10 cents?

41. An airplane has 10 empty seats left: six aisle seats and four window seats. If four people about to board the plane are given seat numbers at random, what is the probability that they are all given aisle seats?

42. Some slot machines in Las Vegas, Nevada, have three wheels that spin when you pull the "arm" of these one-armed bandits. To win, each wheel must show a picture of an apple when they all stop turning. Each wheel has 20 pictures on it. Wheel 1 has five apples on it, wheel 2 has three apples on it, and wheel 3 has two apples on it. What is the probability that you will win on the first pull?

43. Lowest Bidder Computer Chip Company ships out its memory chips in lots of 100. Three memory chips chosen at random from each lot are inspected by its quality control department. If any chip inspected is found to be defective, then the entire lot is rejected.
    (a) What is the probability that a lot containing two defective chips will pass this inspection?
    (b) What do you think of this quality control procedure?

44. Four friends join the army at the same time. They will be assigned duties at random in either the infantry or the artillery. What is the probability that two of these friends will be assigned to each service area?

45. Suppose that you have invited 10 relatives to a party at your house. You invite your mom, two uncles, three brothers, and four cousins. What is the probability that the first guest to arrive will be one of your uncles?

46. You are dealt five cards from a deck of 52 cards. Find the probability that you are dealt five red cards.

47. A committee of four is to be randomly selected from a group of seven administrators and eight instructors. Find the probability that the committee will consist of four instructors.

48. A three-person committee will be selected at random from five Democrats and three Republicans. Find the probability that all three selected will be Democrats.

49. Each of the digits 0, 1, 2, 3, 4, 5, 6, 7, 8, and 9 is written on a slip of paper, and the slips are placed in a hat. If three slips of paper are selected at random without replacement, find the probability that the three numbers selected will be greater than 4.

50. Repeat problem 49 with replacement.

Investigate each of the following situations involving telephone numbers. In each case, determine the total number of phone numbers that could be made available for use.

51. At Laredo Community College (LCC), each office extension begins with a 5 and has four digits. How many different telephone extensions can LCC have?

52. Local phone numbers are seven digits long. No local phone number may begin with a 0 or 1. How many different local phone numbers are possible within a given area code?

53. Recently the central area of North Carolina was split into two area codes by BellSouth. The old area code was 910 and the new area code is 336. Some people had to change their area code to 336 and some kept the 910 area code. Speculate about why this was done.

54. If, in the not-too-distant future, a worldwide government wanted each individual to have his or her own personal phone number, would the current 10-digit system (counting the three area code digits) work? Explain your answer.

Chapter 7 Summary

Key Terms, Properties, and Formulas

- addition rule
- combination
- conditional probability
- counting principle
- dependent events
- disjoint sets
- elements
- empirical probability
- empty set
- equal
- equally likely outcomes
- event
- experiment
- factorial
- finite
- independent events
- independent outcome
- infinite
- intersection
- law of large numbers
- laws of probability
- multiplication rule
- mutually exclusive events
- null set
- odds
- outcomes
- permutation
- probability
- random
- sample space
- set-builder notation
- sets
- set notation
- subset
- theoretical probability
- tree diagram
- union
- universal set
- Venn diagram

Formulas to Remember

Empirical probability:

$$P(E) = \frac{\text{number of times event } E \text{ has occurred}}{\text{total number of times the experiment has been performed}}$$

Theoretical probability:

$$P(E) = \frac{\text{number of ways event } E \text{ can occur}}{\text{total number of possible outcomes}}$$

Calculating a factorial:

$$n! = n(n-1)(n-2)\ldots(2)(1)$$

Zero factorial—definition:

$$0! = 1$$

Permutations:

$$_nP_r = \frac{n!}{(n-r)!}$$

Combinations:

$$_nC_r = \frac{n!}{r!(n-r)!}$$

Odds in favor of an event = event occurs : event does not occur

Odds against an event = event does not occur : event occurs

Probability laws

Law 1:
Every probability is a number between 0 and 1 (inclusive).

Law 2:
The sum of the probabilities for all outcomes for a particular sample space is exactly 1.

Law 3:
The probability of an event occurring that cannot possibly happen is 0.

Law 4:
The probability of an event that must occur is 1.
Mathematical relationships:

$$P(A) + P(\text{not } A) = 1$$

$$P(\text{not } A) = 1 - P(A)$$

Chapter 7 Review Problems

Let $U = \{a, b, c, d, e, f, g, h\}$; $A = \{a, b, c, d, e, f\}$; $B = \{a, b, c\}$; $C = \{a, b, e, g\}$; and $D = \{f, h\}$

1. Draw a Venn diagram illustrating the relationship between sets A and B.
2. Find $B \cup C$
3. Find $A \cap D$
4. Find $B \cap D$
5. Find $C \cup D$
6. Is $A \subseteq B$? Why or why not?
7. Is $C \subseteq A$? Why or why not?
8. Is A a finite set? Why or why not?
9. Is D an infinite set? Why or why not?
10. What is probability?
11. Two of the basic rules used in computing probabilities are the addition rule and the multiplication rule. Explain when each of these rules should be used.
12. What is the difference between the "probability" of an event and the "odds in favor" of that same event?
13. If you toss a pair of dice, what is the probability that the sum of the two numbers shown will be 8?
14. If you toss a pair of dice, what is the probability that the sum of the two numbers shown will be 13?
15. If you toss a pair of dice, what is the probability that the sum of the two numbers shown will be less than 7?

A small box contains 20 marbles. Eight of them are green, seven are red, and five are blue.

16. What is the probability that you will reach into the box and draw out a red marble?
17. What is the probability that you will reach into the box and draw out a green or a blue marble?
18. If you are going to draw out three marbles, one at a time, and you are not going to replace the marbles in the box once they are drawn, what is the probability that all three will be blue?

A group of students chosen at random were asked if they drank root beer or not. The following table summarizes the results of this brief survey. Use this table for problems 19 to 22.

	Drinkers	Nondrinkers	Total
Men	63	50	113
Women	39	48	87
Total	102	98	200

19. If you chose one of these students at random, what is the probability that the person chosen will be a male who drinks root beer?
20. What is the probability that you will choose a female who does not drink root beer?
21. What is the probability of choosing a student who drinks root beer, given that the person chosen is a male?
22. What is the probability of choosing a woman, given that the person chosen is a nondrinker?

23. What are the odds in favor of drawing a face card from an ordinary deck of 52 cards?

A social security number has nine digits, such as 000-00-0000. Use this information for problems 24 and 25.

24. How many different social security numbers are possible?

25. If the population of the United States in 2000 was about 300 million people, and we assume that the population will double each 25 years, will this numbering system need to be changed before 2050?

26. A certain manufacturing operation is known to produce two bad parts out of 75, on the average. If you walk up to the production line and choose one part, what is the probability that it is a good part?

27. Last year it rained on 92 days in Midville, Arkansas. If you pick a calendar date from last year at random, what is the probability that it rained on that day?

The following table gives the results of a poll concerning the evening news watched most often. Use this table for problems 28 to 30. Round to the nearest tenth of a percent.

28. If one of the individuals involved in this poll is selected at random, find the probability that this person watches ABC or NBC.

29. If one of the individuals involved in this poll is selected at random, find the probability that this person watches ABC or the individual is a woman.

30. If one of the individuals involved in this poll is selected at random, find the probability that this person watches Fox, given that the person is a male.

31. A $100, a $50, and a $20 prize are to be awarded to three different people. If seven people are being considered for the prizes, how many different arrangements are possible?

32. How many different committees can be formed from six teachers and 50 students if the committee is to consist of two teachers and three students?

33. An ice cream parlor has 15 different flavors. Cynthia orders a banana split and has to select three different flavors. How many different combinations are possible?

34. Alan bought eight raffle tickets to try to win a TV. What are his odds of winning if 1500 tickets were sold altogether?

	ABC	NBC	CBS	Fox	Total
Men	30	20	40	25	115
Women	50	10	20	20	100
Total	80	30	60	45	215

Find the numerical value of each of the following.

35. $4!$
36. $(-2)!$
37. $5.2!$
38. $_7C_3$
39. $_7P_5$
40. $_{23}C_3$

Chapter 7 Test

Let set $U = \{1, 2, 3, 4, 5, 6, 7, 8, 9, 10, 11, 12, 13, 14, 15\}$; $A = \{5, 6, 7, 8, 9, 10, 11\}$; $B = \{1, 3, 5, 7, 9\}$; $C = \{10, 11, 12, 13\}$; and $D = \{15\}$

1. Draw a Venn diagram to illustrate the relationship between sets A and B.
2. Find $A \cap B$
3. Find $B \cap C$
4. Find $C \cup D$

You are going to flip a coin once and roll a fair, 6-sided die once.

5. What is the sample space for this experiment?
6. What is the probability of flipping the coin and getting heads and rolling a 5 on the die?
7. What is $P\{\text{tail, and an even number}\}$?

One hundred and fifty residents of Lamb County were surveyed. They were asked first whether or not they

lived within the city limits of the city of Littlefield. Second, they were asked if they thought that they had adequate fire protection services available in the event of a fire at their home. The city of Littlefield has a full-time fire department with several fire stations, whereas the county has only volunteer fire protection services available. The following table gives the results of this survey. Use this table for problems 8 to 10.

Residence	Adequate	Inadequate	Totals
In city	56	34	90
In county	29	31	60
Total	85	65	150

8. Find P(adequate fire protection is available).

9. Find P(adequate fire protection is available | the person lives in the county).

10. Find P(inadequate protection | the person lives in the city).

11. A code is to consist of three digits followed by three letters. Find the number of possible codes if the first digit cannot be a 0 or 1 and replacement is not permitted.

A friend bets you that he can roll a fair, 6-sided die one time and have an even number show up. Use this information for problems 12 to 13.

12. What are the odds that you will win if you take the bet?

13. What is the probability of rolling an even number?

14. One of three accountants, Bill, Jane, or Doris, will be promoted in the CPA firm where they work. If the probability that Bill will be promoted is 0.25 and the probability that Jane will be promoted is 0.34, then what is the probability that Doris will be promoted?

Bob has three Scrabble tiles with the letters *t*, *o*, and *a* on them. Use this information for problems 15 to 17.

15. How many three-letter "words" can Bob make using just these letters?

16. How many of these "words" are meaningful in English?

17. If Bob closes his eyes and places these three tiles on the game board, what is the probability that he will create a meaningful word?

18. The state of "Confusion" issues license plates with four digits on them. The digits used are 1–9 (no 0s) and digits may be repeated. You are going to get a new license plate for your new car. What is the probability that the new plate will have all four numbers the same (like 8888)?

A standard deck of playing cards has four suits and two colors as follows: spades (♠) are black, hearts (♥) are red, diamonds (♦) are red, and clubs (♣) are black. Each suit has 13 cards: an ace (A); cards numbered 2–10; and three face cards, a king (K), a queen (Q), and a jack (J). Use this information for problems 19 to 22.

19. What is the probability of drawing a red card?

20. What is the probability of drawing a queen?

21. What is the probability of drawing the 3 of diamonds, replacing it in the deck, shuffling, and then drawing the 3 of diamonds again?

22. What is the probability of drawing a face card or a diamond?

23. At a horse race, such as the Kentucky Derby, the tote board gives those betting the odds assigned to each horse. These are the odds against winning. Using the following tote board, calculate the probability of winning for a bet placed on each horse.

Horse Name	Odds
Ringo	3 to 1
Bonnie	15 to 1
Clyde	8 to 5
Zero	3 to 2

24. A new family has just moved into the house next door. You have heard that they have four children. What is the probability that they have two boys and two girls?

25. How many different ways can six photos be arranged from left to right on a hallway wall?

26. If the probability of an event is 1, what is the probability that it will not occur?

27. There are 10 true/false questions on a test. If you guess the answers to all 10, what is the probability that you will miss all of the questions, resulting in a grade of 0?

A group of 60 married couples were surveyed and then went to marriage counseling classes. Thirty-six of the 60 couples reported that their marital relations were improved by the counseling. Of these 36 couples, 27 had children. Use this information for problems 28 to 30.

28. What is the probability of selecting a couple at random that reported no improvement in their relations due to counseling?

29. What is the probability of selecting a couple with children if your choices are limited to those who reported an improvement?

30. What is the probability of selecting, from among all 60 couples, a couple who reported improvement but has no children?

Suggested Laboratory Exercises

Lab Exercise 1 A Simple Survey

Certain physical traits are not evenly distributed in the general population. For example, there are more right-handed than left-handed people. Conduct a survey of persons chosen at random at your high school. Do *not* ask any members of your class itself. To do this survey in a short time, the questions will be limited to three: (1) Are you right- or left-handed? (2) What color is your hair? (3) What color are your eyes? Complete the following table by filling in all of the data collected. Ask a total of at least 100 students.

Handedness	Number of Persons
Right-handed	
Left-handed	
Eye Color	**Number of Persons**
Brown	
Blue	
Green	
Other	
Hair Color	**Number of Persons**
Brown	
Blond	
Black	
Red	

Now compare your class members to the survey results. What proportion of your class should be right-handed? Does this match your class? What is the probability that a student will have brown hair? How many students in your class should have brown hair? How many actually do? Answer the same questions about eye color (you pick the color).

Lab Exercise 2

Devising a Game

Make up a game using dice like a game of 8s. If you roll a total of 8 on two dice, you win. Find the probability of rolling an 8 and see if this probability seems to be true by rolling the dice a large number of times. Will you win a lot of money at this game?

Lab Exercise 3

Pair 'em Up

It's early in the morning. You aren't really awake yet and it's dark in your bedroom. You are groping in your sock drawer for a pair of socks to wear. Without the lights on, they all may look black, but only a few are. Let's assume that you have four black socks, three white, and two blue socks in the drawer. Work as a group and perform an experiment using black, white, and blue socks in a paper bag to simulate this problem and answer these questions. How many socks must you pull out of the drawer to be sure that you have at least one matching pair when the lights come on? How did your group arrive at this answer? Write a paragraph explaining your answer.

Lab Exercise 4

Don't Lose Your Marbles

Get a jar and a set of colored marbles, such as five red, two blue, and three yellow. Do an experiment to show that the probability of drawing three red marbles in a row is different if you keep each marble drawn rather than replacing each one before drawing the next.

Lab Exercise 5

Walk On By

Pick a well-traveled hallway or sidewalk at your school. Choose a short time period (5 or 10 minutes) and count the number of persons walking past during each time interval. It may take several days, counting for an hour or so at the same time each day, to get sufficient data. The data may look something like the following data set.

Number of Students Walking Past	Number of 10-Minute Periods
0–5	6
6–10	15
11–15	17
16–20	15
21–25	5
26 or more	2
Total	60

(a) Determine the probability for each count. For example, 0–5 persons walked by during 6 of the 60 time periods $= \frac{6}{60} = \frac{1}{10}$.

(b) Determine the probability that 16 students will walk by during the next 10-minute counting period.

(c) Determine the probability that more than five persons will walk by during the next 10-minute counting period.

(d) Determine the probability that more than 5 but fewer than 16 persons will walk by during the next 10-minute counting period.

(e) What are the important considerations in setting up an experiment such as this?

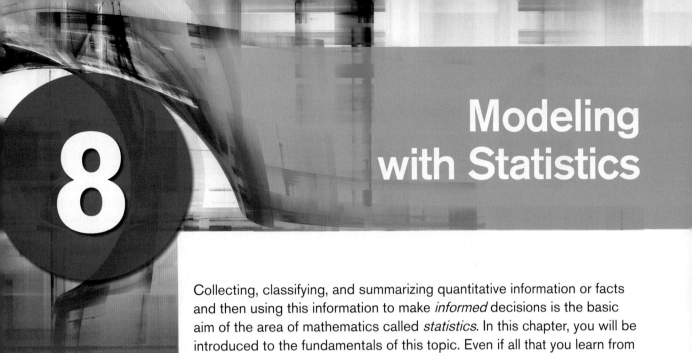

8 Modeling with Statistics

Collecting, classifying, and summarizing quantitative information or facts and then using this information to make *informed* decisions is the basic aim of the area of mathematics called *statistics*. In this chapter, you will be introduced to the fundamentals of this topic. Even if all that you learn from this chapter is how to recognize when statistical procedures have been misused by s ome group, business, organization, or government agency, then this chapter will have been worth your time.

> *"He uses statistics as a drunken man uses lampposts—for support rather than illumination."*
> – Andrew Lang

In this chapter

- 8-1 Introduction to Statistics
- 8-2 Frequency Tables and Histograms
- 8-3 Reading and Interpreting Graphical Information
- 8-4 Descriptive Statistics
- 8-5 Variation
- 8-6 Normal Curve
- 8-7 Scatter Diagrams and Linear Regression

  Chapter Summary

  Chapter Review Problems

  Chapter Test

  Suggested Laboratory Exercises

Section 8-1 Introduction to Statistics

Statistics is the area of mathematics that is involved with collecting, classifying, summarizing, and presenting data that have been collected. Businesses of all kinds collect data about sales, customer preferences, national trends, and their own costs. Manufacturers collect data about the quality and reliability of their products and compare theirs with those of the competition. Researchers in medicine, environmental areas, computer science, physics, astronomy, and so forth all collect volumes of data. Advertisers and politicians use polls to collect data that are then used to fine-tune their messages. Sports fans collect reams of data on their favorite sports or sports figures.

We are surrounded by data. A fundamental knowledge of statistics will allow us to deal with data in an honest and skillful manner. It will also help us to look objectively and critically at someone else's presentation of data and form our own conclusions. We must become good consumers of data.

This chapter is not intended to be a complete course in statistics. It will just introduce the topic and explore some of the fundamentals of statistics. The basics of data collection and analysis will be studied. A lot of new terminology will be used, but even though the words may be strange, the meanings are usually fairly simple.

There are two broad areas of statistics, **descriptive statistics** and **inferential statistics**. The area of descriptive statistics classifies, sorts, and summarizes data. Generally speaking, **data** are numerical facts or information. Once the data have been presented in a uniform way, they may be used to calculate averages and other numerical values as well as to construct tables or graphs. As long as only the quantitative *facts* are presented, you are working in the area of descriptive statistics. Most of this chapter deals with the area of descriptive statistics.

Although the area of descriptive statistics has been around for centuries, the area of inferential statistics is a 20th-century development. An inference is a conclusion or presumption that is made based on reasoning from the evidence that is available. Inferential statistics involves drawing a conclusion or generalizing based on sample data. It involves estimation and "guesstimation." It is from this area of statistics that most of the confusion arises.

There are also two categories into which data fall. **Quantitative data** are numbers that indicate amounts, differences, or counts. This is the easiest type of data to deal with. The second type of data is called **qualitative data**. These types of data are generally words (though some *numerical codes* are used) that indicate observations and are indicative of differences in kind. Before beginning to analyze a set of data, it is important to know which type of data you have. We cannot deal with both types of data in the same way.

Example 1

Quantitative versus Qualitative Data

Suppose that you are preparing questions for a new admissions form for your school. If you ask questions to gain information about students in the following areas, would you expect to get quantitative or qualitative data as a response?

(a) What is your age? The data provided in this instance should be numbers of years and thus would be quantitative.

(b) What is your intended major? The data in this instance should be the name of an area of study such as horticulture or business administration. This type of data is qualitative.

(c) What is your SAT score? The word *score* (and similar words like *scale*) typifies quantitative data.

(d) What is your sex? We would expect the answer to be male or female. This is qualitative data.

(e) What is your telephone number? Although the answer to this question involves numbers, they do not represent a count or amount—only a particular phone line or location. This is qualitative data.

Two more commonly used terms in statistics are *sample* and *population*. When you are gathering information for a statistical analysis, you must be aware of which of these two categories your data will fall into.

Suppose that you were doing a survey to find the average age of the students in your math class. It would be easy to pick a day when 100% of the class was in attendance, ask everyone their ages, and then find the average. This set of data would be called a population. If you are sure that you can get the required information from absolutely all possible sources, missing none, then you have a set of data called a **population**.

However, if you wanted to calculate the average age of persons living in your county, it would be virtually impossible to get the age of every individual person living in the county at any given time. You might be able to get a large number of ages but certainly not all of them. This set of data would be called a **sample**.

Definition **Population versus Sample**

Population: a complete set of observations
Sample: a subset of observations taken from a population

In most situations, we will be dealing with samples of data and will be calculating sample statistical values. The formulas and rules used are very similar for both sample and population statistics, but there are some differences to allow for the fact that sample values are based on incomplete information. This means that a "fudge factor" is built into many of the formulas for sample statistics.

To come up with a sample that is representative of the population, we will look at several different sampling techniques. Because samples are used to draw conclusions about the whole population, you must be careful to avoid **bias**. Bias is the difference between the results obtained by sampling and the truth about the whole population. Many statisticians use **random sampling** to help eliminate bias. The most common way to satisfy this requirement is to select the sample in such a way that every different member of the population has an equal probability (or chance) of being selected. To produce a random sample, statisticians often use a random number table or random number generator such as the one on your graphing calculator. Random sampling is one of the most widely used sampling techniques, but there are other techniques that may be used. We will look more closely at the following types of sampling: stratified, systematic, cluster, and convenience.

Groups or classes inside a population that share a common characteristic are called **strata**. In the method of **stratified sampling**, the population is divided into distinct strata. For example, in the population of all undergraduate college students, some strata might be freshmen, sophomores, juniors, or seniors. Other strata might be men or women, in-state students or out-of-state students, and so forth. Then a simple random sample of a certain size is drawn from each stratum and the information obtained is carefully adjusted or weighted in all resulting calculations.

Another type of sampling is **systematic sampling**. In this method, it is assumed that the elements of the population are arranged in some natural sequential order.

Then a random starting point is selected and we select every "nth" element for our sample. For example, to obtain a systematic sample of 20 students going through the cafeteria line at lunch, we could pick a random number between 0 and 10. Then between 12:00 noon and 12:10 (the actual time determined by our random number), we could begin the sampling process of taking every fifth student in line. The advantage in systematic sampling is that it is easy to get. However, there are dangers in using systematic sampling. When the population is repetitive or cyclical in nature, systematic sampling should not be used. For example, consider a fabric mill that produces dress material. Suppose the loom that produces the material makes a mistake every 17th yard, but we only check every 16th yard with an automated scanner. In this case, a random starting point may or may not result in detection of the fabric flaws before a large amount of fabric is produced.

Cluster sampling is used extensively by government agencies and certain private research organizations. In cluster sampling, we begin by dividing the demographic area into sections. Then we randomly select sections and survey all individuals in those sections. The resulting calculations are weighted to correct for disproportionate representation of groups. For example, in conducting a survey of schoolchildren in a large city, we could first randomly select 30 schools and then poll all of the children at each school.

Finally, **convenience sampling** uses results or data that are conveniently and readily obtained. In some cases, this may be all that is available, and in many cases it is better than no information at all. However, convenience sampling does run the risk of being severely biased. For instance, consider a newsperson who wishes to get the "opinions of the people" about a proposed seat tax to be imposed on tickets to all sporting events. The revenues from the seat tax will then be used to support the local symphony. The newsperson stands in front of a classical music store at noon and surveys the first five people coming out of the store who will cooperate. This method of choosing a sample will produce some opinions, and perhaps some human interest stories, but it certainly has bias. It is good advice to be very cautious when the data come from the method of convenience sampling.

Example 2

Which Type of Sampling Method?

Memorial Mission Hospitals (MMH) is a national for-profit chain of hospitals. Management wants to survey patients discharged this past year to obtain patient satisfaction profiles. It wishes to use a sample of such patients. Several sampling techniques are described next. Categorize each technique as simple random sampling, stratified sampling, systematic sampling, cluster sampling, or convenience sampling.

(a) Obtain a list of patients discharged from all MMH facilities. Divide the patients according to length of hospital stay (3 days or less, 4–7 days, 8–14 days, more than 14 days). Draw random samples from each group.

—**Stratified sampling (discharged patients divided into groups depending on number of days in hospital)**

(b) Obtain lists of patients discharged from all MMH facilities. Number these patients and then use a random number table to obtain a sample.

—**Simple random sampling (random number table)**

(c) Randomly select some MMH facilities from each of five geographic regions and then survey all patients discharged from each of these hospitals.

—**Cluster sampling (geographic regions)**

(d) At the beginning of the year, instruct each MMH facility to survey every 500th patient discharged.

—**Systematic sampling (every 500th patient)**

(e) Instruct each MMH facility to survey 10 discharged patients this week and send in the results.

—**Convenience sampling (any 10 discharged patients in one week)**

Misuses of Statistics

As you learn more about statistics, you will begin to see that statistics can be used to mislead as well as inform the public. Mark Twain said, "Get your facts first and then you can distort them as much as you please." For example, the way that questions are asked in a survey, or the order in which they are asked, can dramatically change the results. Respondents can be primed to answer questions on a survey in a way that favors the view of the survey's sponsor. These responses can then lead to false conclusions that support the sponsor's agenda. Small sample sizes can also give results that may not be representative of a larger population. If you are suspicious of the results from a survey, try to determine the questions that were asked, the order in which they were presented, how many people participated in the survey, and who sponsored it. If the sponsoring agent has a vested interest in the outcome of the survey, the potential for bias in the survey itself increases. For example, it is possible that the results of a survey on nuclear energy conducted by the government might be different from the results of a similar survey conducted by the Sierra Club, a group who claims on their website to be "America's oldest, largest and most influential grassroots environmental organization."

When choosing a sample for a survey or experiment, it is very important to choose a sample of sufficient size that is representative of your population of interest. The results from a group that is selected for a poll or for an experiment can be inaccurate if the group is biased in some way or different from the population that you are targeting. For example, if a new medication is tested on a sample of male heart patients and found to be beneficial, then using inferential statistics, the results can be generalized to the population of all male heart patients. However, it would not be statistically correct to generalize these results to all heart patients, male and female, since the original sample did not include female participants. There are formulas to help you determine how large a sample should be and, in fact, samples of 50 to 100 participants can be adequate if the sample is chosen carefully. In this section, we have already discussed some common methods for selecting representative samples: random sampling, cluster sampling, systematic sampling, and stratified sampling. Generally, convenience sampling will not provide a representative sample, since it is usually voluntary.

There are many graphs used today in media publications to display the results of polls. You should look at these figures closely if the results seem sensational or widely different. Misleading graphs can be created from the results of a survey by simply manipulating the scale on the graph. Look at the two graphs pictured in Figure 8-1. Random samples from the Democratic, Republican, and Independent parties were asked if they agreed with a certain proposal. The results showed that there was little difference in the outcomes among the three parties since the margin of error for the survey was ±7 points. That means that 62% ± 7% or somewhere between 55% and 69% of Democrats agreed with the proposal. At the same time, 54% ± 7% or 47% to 61% of the Republicans and Independents agreed with the

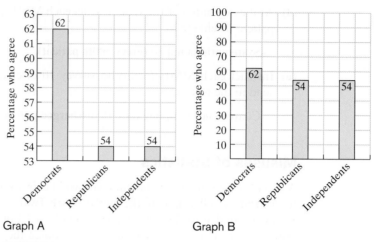

FIGURE 8-1
Misleading graphs.

proposal. Because these ranges overlap when you include the margin of error, you cannot say that there is a statistical difference among the three parties. However, in Graph A, the scale on the vertical axis increases in increments of 1% with a range of only 10 percentage points and the result seems to indicate a large difference in the results among the parties. When the scale is revised, as shown in Graph B, to a range of 100 percentage points labeled in increments of 10%, the differences in the results don't seem so extreme.

Percentages can also be used to sensationalize results. When an instructor announced proudly that 100% of her students passed her networking class with an A last semester, she was asked how many students she had. She responded that there were 3 students in the class. While 3 out of 3 is 100%, it doesn't seem as amazing that all three would receive A's in her class because she should have been able to give each of them a considerable amount of assistance and instruction during the semester. If the class size had been 30 with 100% of the class receiving A's, this would have been more unusual.

On the other hand, people can be fooled by numbers that sound precise but are really not valid. For example, I sleep 6.45 hours per night. That figure gives the impression that I measure my sleeping time in very precise units. It is more likely that I estimate my sleeping time in less precise units by stating that I sleep about $6\frac{1}{2}$ hours each night. When numbers sound precise, we can be fooled into thinking that the presenter has accurate information but that information may or may not be correct. You should usually investigate these numbers a little more closely.

Another Mark Twain quotation states, "Facts are stubborn things, but statistics are more pliable." Do numbers lie? The numbers themselves may not lie but because the method of collection of numbers may be flawed, the statistics can become pliable. The careless collection and use of data can lead us to false conclusions, whether deliberately or inadvertently. When evaluating a statistical study, remember to review the method used to select the sample, the size of the sample, and the questions that were asked. It is also a good idea to determine who sponsored the survey to see if the sponsor has a vested interest in the outcome. While the majority of polls and surveys are probably conducted using proper statistical procedures, statistics can be manipulated in such a way as to mislead the public.

> **"In earlier times they had no statistics, and so they had to fall back on lies. They did it with lies, we do it with statistics; but it's all the same."**
>
> – Stephen Leacock

Practice Set 8-1

Decide whether each of the following is in the area of descriptive or inferential statistics.

1. A survey recently revealed that 63% of Americans feel the voting process in America is fair.
2. The age of students at a particular community college ranges from 18 to 78 years.
3. My GPA has been near 3.5 for the past 3 years.
4. The state Highway Safety Council predicts that 15 persons will die in traffic accidents over the Labor Day weekend.
5. Sammy Sosa's batting average is 0.455 this week.
6. In a recent study of a sample of 100 volunteers, eating garlic was shown to lower blood pressure.
7. The average number of students in a class at Alvin Community College is 22.6.
8. The New Mexico Department of Transportation has predicted that the average number of automobiles per household will increase by 0.2% next year.
9. The total attendance for all of the football games held at Southern High School this season was 8653.
10. If you live your entire life in New York City, the chances that you will be robbed at least once are about 15%.

Each of the following words or terms describes some data that are to be collected. Will the data be quantitative or qualitative in each case?

11. Years of education that you have had
12. Your favorite TV show
13. Your daily caloric intake
14. Your place of birth
15. Your blood type
16. Your favorite brand of candy bar
17. Your student number
18. The high temperature at Salt Lake City, Utah, each day
19. The average time you spend standing in the checkout line of a grocery store
20. Your telephone number, including area code

Categorize each of the following as simple random sampling, stratified sampling, systematic sampling, cluster sampling, or convenience sampling.

21. Randomly select five high schools from different geographic regions and survey all students at each of the chosen schools.
22. Survey each 10th student who walks in the front entrance of the school.
23. Obtain a list of students enrolled at Western High School. Number these students and then use a random number table to obtain the sample.
24. Obtain a list of students enrolled at a community college. Draw a random sample from the horticulture students, transfer students, and nursing students.
25. Survey the first five students who walk in the classroom.
26. Once every year, a popular magazine surveys its readers using a mail-in response card, and, based on the results that are obtained from the responses, the magazine names the best- and worst-dressed celebrities of the year.
27. An instructor at Texas Tech teaches five classes in a semester. Two of the classes are chosen by the dean, and all students in those classes fill out an instructor evaluation form about that instructor.
28. Students at Hampton University obtain the voting records for Norfolk, Virginia. They randomly select 60 Democrats, 50 Republicans, and 10 Independents (the numbers chosen are based on the percentage of registered voters in each category in the county) and survey them concerning their favorite candidate for mayor in the upcoming fall election.

29. At a club meeting, the names of all members are written on identical slips of paper and placed in a bowl. The names of five members are drawn and these will participate in face-to-face interviews with a pollster.

30. You are approached by a lady at the shopping mall who asks you to answer a few questions about the facilities at the mall.

Identify the possible misuses of statistics in the following problems. What questions should you ask if presented with these statements?

31. The president of Acme Toy Company used this graph to illustrate to the Board of Directors the significant increase in monthly salaries over the past 10 years.

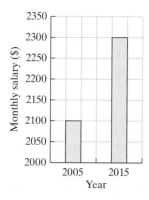

32. The Carolina Savings Bank sent this graph in some advertising material to inform investors that their interest rate was significantly higher than their competitor's rate.

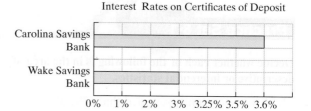

33. A report published in a recent magazine states that 4 out of 5 dentists prefer Brite-White Toothpaste.

34. A recent newspaper article stated that 1 in 4 college age women has been the victim of rape.

35. People spend an average of $543.16 on Christmas presents each year.

36. Freshmen in college gain an average of 9.2 lbs during their first year in college.

37. A newspaper headline stated recently that a special school for troubled students had a 126% drop-out rate last year.

38. Reporting on population declines in rural counties, a recent report stated that the town of Rose City had a 20% decline in its population each year for the last 10 years.

39. Based on a survey of senior citizens, Jerry Diaz was predicted to win the mayor's race in his home town in California.

40. R2 cough medication is shown to be effective in stopping the cough of adult patients so a mom buys it to give to her 12-year-old son when he catches a cold and begins to cough.

Section 8-2 Frequency Tables and Histograms

In statistics, it is important to be able to organize and present your data in an effective manner. A table or graph is a much more effective way of presenting information than a long list of numbers. One type of table that is used to present quantitative data is a **frequency table**, also called a **frequency distribution**. A frequency distribution is used when the amount of data is very large. Individual data items are grouped in "classes" to help in the analysis of the distribution of the sample.

Section 8-2 Frequency Tables and Histograms

> **Definition** **Frequency Distribution**
>
> A frequency distribution is a listing in chart form of the observed values in a set of data and the corresponding frequency for each value.

When constructing a frequency distribution, there are several guidelines that should be followed:

1. Each class should be the same "width"—that is, span the same number of possible data items.
2. Classes should not overlap. Each data item should belong to only one class.
3. Frequency distributions should contain approximately 5 to 12 classes.
4. Use an even or "easy to count" class width whenever possible.

Look at the following example that constructs a frequency distribution for a set of 50 final exam grades in a biology class.

Example 1

Constructing a Frequency Distribution

Construct a grouped frequency distribution of the following final exam grades of 50 biology students.

45	85	63	72	60	49	51	68	76	65
48	73	65	66	72	71	78	65	84	80
73	61	94	40	59	75	73	76	85	80
91	72	60	80	58	51	78	72	68	71
85	79	83	64	69	73	50	60	75	82

1. Identify the high and low scores and find the range.

$$\text{High} = 94 \quad \text{Low} = 40$$
$$\text{Range} = 94 - 40 = 54 \text{ points}$$

Counting by 5s will be a convenient unit for a class width. If we divide 54 by 5, the result is 10.8 or 11 classes. This falls within the recommended 5 to 12 classes for a frequency distribution.

2. Pick a starting point that is consistent with the data. The lowest class must include the smallest data item (40), and the largest class must include the highest data item (94). The numbers used to define each class should not overlap.

Classes

| 90–94 |
| 85–89 |
| ↑ |
| 55–59 |
| 50–54 |
| 45–49 |
| 40–44 |

3. Now set up a table using these classes and sort the data into those classes. Use a tally column to count items and total these in a frequency column.

Final Exam Grades

Classes	Tally	Frequency
90–94	11	2
85–89	111	3
80–84	⊞ 1	6
75–79	⊞ 11	7
70–74	⊞ ⊞	10
65–69	⊞ 11	7
60–64	⊞ 1	6
55–59	11	2
50–54	111	3
45–49	111	3
40–44	1	1
	Total	50

Once the data have been classified in a frequency distribution, we lose some information. Individual data items are no longer identifiable. However, we can look at the range of numbers in each class to answer specific questions about the data. In the frequency distribution given previously, the class 70–74 is the **modal class** because it is the class with the greatest frequency. We can also see that a majority of the data fell in the classes that include grades ranging from 60 to 84.

A histogram is a bar graph used to picture a frequency distribution. Each bar represents a class of numbers in the frequency distribution. The height of each bar indicates the frequency in that particular class. The bars should always touch because the width of a bar represents a range of numbers (a quantitative value). Because each bar represents a range of values, it is usually labeled with one number called the **midpoint of the class**. This number is calculated by adding the lower and upper class limits of one class and dividing by 2.

Example 2

Drawing a Histogram

Use the frequency distribution in Example 1 to draw a histogram.

1. Calculate the class midpoints for each class

$$\frac{40 + 44}{2} = \frac{84}{2} = 42$$

$$\frac{45 + 49}{2} = \frac{94}{2} = 47$$

and use these to label each bar.

2. A histogram of final exam grades for biology students is presented in Figure 8-2.

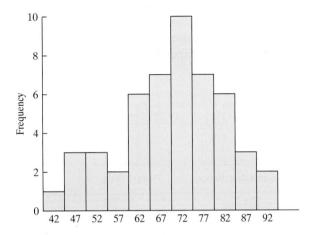

FIGURE 8-2
Histogram of exam grades.

Notice that a histogram is different from a bar graph. The bars in a histogram represent classes of data on a number line. Therefore, there are no gaps between the bars unless an intervening class has a frequency of 0. The width of the bars on a histogram is determined by the class width and the scale of the horizontal number line for the graph.

CALCULATOR MINI-LESSON

TI-84 Plus Histograms

Your TI-84 Plus calculator will fairly easily give you a histogram of data that you enter in a list. Just for practice, let's try one.

Start by entering your data in list 1 (L_1): Press STAT, then ENTER (Edit) to get to L_1. Enter the following data set (20 numbers):

$$2, 2, 3, 4, 4, 5, 5, 5, 5, 6, 6, 7, 7, 7, 7, 8, 8, 9, 10, 10$$

Now go to STAT PLOT by pressing 2nd Y=. First, be sure that only Plot 1 is on. If Plot 2, 3, or 4 is on, arrow down to that plot, ENTER, and arrow over to **Off**. Arrow up to Plot 1 and ENTER, highlight **On**, arrow down to type and over to the **histogram symbol**.

Now press ZOOM 9 (zoom stat) and your histogram will appear, automatically scaled to fit the screen.

Press TRACE and the calculator will tell you how many pieces of your data are represented by the first bar and then press the ▶ to shift from bar to bar.

If the scale of the graph doesn't suit you, you can change it manually by pressing WINDOW and changing the values shown.

This is just a brief look at the process. Your instructor may want you to try to reproduce some of the graphs in this chapter on your calculator.

Histograms are valuable tools. For example, the histogram of a sample should have a distribution shape that is similar to that of the population from which the sample was taken. Many distributions are normal or bell shaped with a symmetrical distribution. Others are skewed where a few very high or very low values cause one tail of the distribution to be longer than the other. A bimodal distribution has two high-frequency classes separated by classes with lower frequencies. Look at the typical distribution shapes illustrated in Figure 8-3.

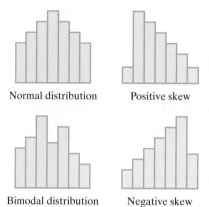

FIGURE 8-3
Various histogram shapes.

Test scores, IQ scores, measurements of height and weight, and thousands of other things are normally distributed in the general population. Therefore, we would expect a random sample of IQ scores to also approximate a normal distribution. In a classroom, students usually hope that the distribution of grades will be negatively skewed—lots of higher grades and only a few low ones!

Some statistical analysis involves the comparison of a sample with the distribution shape of the underlying population. For example, test scores on standardized tests are normally distributed or bell shaped when graphed. If the scores from a sample of students in a particular school are positively skewed (lots of low scores and a few high ones), the administration may ask questions about the difference in actual outcome from the expected bell-shaped outcome. Underlying causes of the difference will be closely examined and remedied if possible.

Practice Set 8-2

1. The speeds of 50 cars were measured by a radar device on a city street with a posted limit of 25 mph as follows:

27	23	22	38	43	24	35	26	28	25
23	22	52	31	30	41	45	29	27	43
29	28	27	25	29	36	28	37	28	29
28	26	33	25	27	25	34	32	22	32
21	23	24	18	48	23	16	26	21	23

   (a) Classify these data into a grouped frequency distribution (lowest class 15–19).

   (b) Construct a histogram of these data with the same number of classes as in the frequency distribution.

   (c) Does this histogram approximate any of the theoretical curves we have studied? If so, which one (skewed, normal, bimodal)? Is this the shape you would expect from this problem? Why or why not?

2. Trade winds are one of the beautiful features of island life in Hawaii. The following data represent

total air movement in miles each day over a weather station in Hawaii, as determined by a continuous anemometer recorder. The period of observation is January 1 to February 14.

26	28	18	14	26	40	13	22	27
28	50	42	15	13	16	33	18	16
26	11	16	17	37	10	35	20	21
18	28	18	28	21	13	19	25	19
19	15	29	14	25	54	32	34	22

(a) Classify these data into a grouped frequency distribution with the first class of 10–14.
(b) Construct a histogram of these data with the same number of classes as in the frequency distribution. Does this approximate any of the curves we have studied?

3. The following list of the weights of bears was compiled by wildlife officers in eastern North Carolina over the course of several years.

80	344	416	348	166	220	262	360	204
140	180	105	166	204	26	120	436	125
220	46	154	116	182	150	65	356	316
270	202	202	365	79	148	446	62	236
114	76	48	29	514	140	94	86	150
144	332	34	132	90	40	212	60	64

(a) Make a frequency distribution of this information. Begin with the lowest class (0–49).
(b) Use the frequency distribution to construct a histogram.

4. A task force carried out a study of one-way commuting distances in miles for workers in the downtown Dallas area. A random sample of 60 of these workers was taken. The commuting distances of the workers in the sample are as follows:

13	47	10	3	16	20	17	40	4	2
8	21	19	15	3	17	14	6	12	45
4	16	11	18	23	12	6	2	14	13
46	12	9	18	34	13	41	28	36	17
29	9	14	26	10	24	37	31	8	16
7	26	1	8	7	15	24	2	12	16

(a) Make a frequency distribution to represent this information using 10 classes.
(b) Use this distribution to construct a histogram. Which of the shapes we have studied is most like the shape of this distribution?

5. Northeast Texas Community College holds evening classes on United States history. The director of studies, who wants to know the age distribution of students in the class, obtained the following information (ages rounded to nearest years):

41	28	22	45	30	43	46	35	56
20	31	19	22	32	55	54	31	63
19	27	53	51	33	58	29	65	53
27	18	29	51	39	29	56	19	63

(a) Construct a frequency table of these data using seven classes.
(b) Construct a histogram to display this frequency table.

6. Pleasant temperatures are one of the beautiful features of island life in Tahiti. The following data represent the high temperature each day at the weather station in Tahiti. The period of observation is January 2013.

76	74	78	74	81	70	73	82	77
77	72	75	73	76	83	78	76	72
76	81	77	80	75	80	81	74	78
80	76	77	78					

(a) Classify these data into a grouped frequency distribution with a first class of 70–74.
(b) Construct a histogram of these data with the same number of classes as in the frequency distribution.

7. The speeds of 50 cars were measured by a radar device on a Presidio city street with a posted limit of 30 mph as follows:

27	23	22	38	43	24	35	26	28	25
52	31	30	41	45	29	27	43	29	28
29	36	28	37	28	29	28	26	33	25
34	32	22	32	21	23	24	18	48	23
21	23	23	22	27	25	27	25	16	26

(a) Classify these data into a grouped frequency distribution.
(b) Construct a histogram of these data with the same number of classes as in the frequency distribution.

8. A random sample of 30 customers at the grand opening of a new clothing store gave the following information about the age of each customer.

36	39	40	35	21	32
51	54	18	23	19	16
26	25	21	43	33	18
48	47	38	41	50	19
32	21	40	36	39	35

(a) Construct a frequency distribution of this data using 5 classes.

(b) Construct a histogram of these data with the same number of classes as in the frequency distribution.
(c) Does this histogram approximate any of the theoretical curves we have studied? If so, which one?

9. A random sample of 40 days gave the following information about the number of people treated each day at Memorial Mission Hospital emergency room.

(a) Construct a frequency distribution of these data using 8 classes.
(b) Construct a histogram of these data with the same number of classes as in the frequency distribution.
(c) Does this histogram approximate any of the theoretical curves we have studied? If so, which one?

10. Survey the members of your class and find out each person's favorite sport to watch on TV. Construct an appropriate graph for your data. Justify your choice of graph type.

Section 8-3 Reading and Interpreting Graphical Information

To conduct a reliable experiment, a statistician may need to analyze thousands of pieces of data. Usually, large amounts of data need to be condensed into more manageable forms so that a good evaluation is possible. This may be accomplished in a number of different ways, depending on the desired end use of the data and whether the data are *qualitative* or *quantitative* in nature.

First, we will look at methods of condensing and graphing large amounts of *qualitative data*. The problem with this type of data is that things such as mean averages and meaningful ordering of such data may not be possible. One way to handle such data is to sort the data into categories and then use these categories to draw a bar graph. Another option is to determine the percent of data in each category and create a **circle graph** or **pie graph**.

Example 1 Constructing a Bar Graph of Qualitative Data

Mrs. Johnson's math class had a test last week. After grading all the papers, she sorted the results for her 31 students and found that 3 students earned an A, 8 earned a B, 10 earned a C, 6 earned a D, and 4 earned an F. Make a bar graph of these data.

First, if Mrs. Johnson had not already done so, the data would need to be sorted into meaningful categories. The grades have been placed in chart form as follows.

Grade	Frequency
A	3
B	8
C	10
D	6
F	4
Total	31

Now each row of the chart will become a bar on our bar graph, as shown in Figure 8-4. The height of each bar is determined by the frequency in that one category.

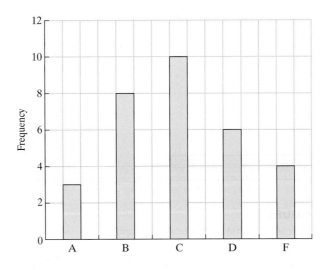

FIGURE 8-4
Grades on a math test.

Example 2 Constructing a Circle Graph

Where do you hide your "mess" when unexpected company shows up at your front door? Believe it or not, statistical data do exist related to this question. In *USA Today* some time ago, the following results were reported based on a survey of about 1000 adults. It was found that about 68% quickly throw stuff in the closet before answering the door, 23% shove things under the bed, 6% throw items into the bathtub (that is kind of odd!), and 3% put stuff in the freezer or some other strange place. Make a circle graph of these data.

To do this by hand with a calculator, a protractor for measuring angles, and a pencil and paper can be time consuming, but, in brief, this is how to do it.

First, recall that a full circle measures 360°. Find and measure on your circle 68% of 360°.

$$(0.68)(360°) \approx 245°$$

Then find and measure "pie" slices that correspond to each of the other percentages. The size of the pie slice corresponds to the percentage for each category.

This procedure is much more easily done with a computer program or spreadsheet such as Excel, with the results as shown in Figure 8-5.

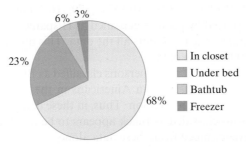

FIGURE 8-5
Typical circle graph.

These same data would also look good in the form of a horizontal or vertical bar graph, as shown in Figure 8-6. Bars on a bar graph represent categories, so each bar is separate from the others. The bars are of uniform width, but the actual width chosen is arbitrary.

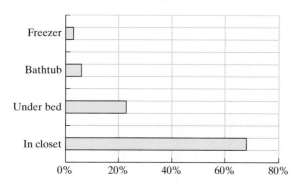
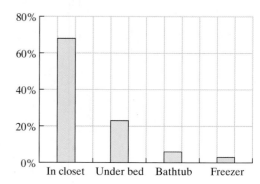

FIGURE 8-6
Typical horizontal and vertical bar graphs.

Extracting Information from a Circle Graph

In the 2008 edition of the *Statistical Abstract of the United States*, the number of AIDS cases estimated in 2005 was broken down by race. These data are represented in the form of the circle graph shown here.

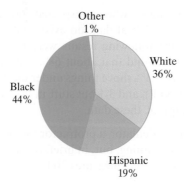

(a) If the total number of cases represented by the circle graph is 419,557, approximately how many of the cases were reported by persons classified as Hispanic?
(b) What percentage does the whole circle represent?
(c) Why would the percentage of those persons with AIDS who are classified as Black be said to be "disproportionate"?

(a) 19% of 419,557 = 0.19(419,557) ≈ 79,716. Thus, about 79,716 cases of AIDS were reported by persons classified as Hispanic.
(b) The whole circle represents all the data. This means that the whole circle represents 100% of the data.
(c) The percentage of those persons classified as Black is disproportionate because the percentage of African Americans in the United States is much less than 44% of the total population. Thus, in these data, the number of cases reported by persons classified as Black appears to be too high. The reason for this cannot be determined from these data alone.

Now let us look at organizing and graphing *quantitative data*. A **stem-and-leaf display** is one way of summarizing numerical data. This method organizes the data into groups but still allows us to see the individual data items. It is a combination of a graphing technique and a sorting technique.

Rule **Stem-and-Leaf Display**

To make a stem-and-leaf display, break the digits of each data value into two parts. The leading digit becomes the stem and the trailing digit becomes the leaf. You are free to choose the number of digits to be included in the stem.

Example 4 **Stem-and-Leaf Display**

The grades on the final exam in a statistics class are given in the following table. Use a stem-and-leaf display to summarize these grades.

92	63	75	82	73	70	61	52	84	91
83	55	61	75	60	84	88	90	78	71
70	79	68	89	93	68	75	59	73	82

The grades range from a low score of 52 to a high score of 93. Because scores are all in the 50s, 60s, 70s, 80s, and 90s, we will use the first digits (tens place) as our stems and the trailing digits (ones place) as our leaves. Draw a vertical line and place the stems in order, to the left of the line.

```
5 |
6 |
7 |
8 |
9 |
Stem  Leaf
```

Now look at each individual grade and place the ones digits for that grade on the correct row. For example, 9|2 represents a grade of 92.

Complete the chart as follows.

Exam Grades in Statistics

```
5 | 2 5 9
6 | 3 1 1 0 8 8
7 | 5 3 0 5 8 1 0 9 5 3
8 | 2 4 3 4 8 9 2
9 | 2 1 0 3
```

The stem-and-leaf display in Example 4 allows us to see that most grades centered around the 70s. If grades are assigned on a 10-point scale, it will be easy for the teacher to see that there were four A's (90s) and three F's (50s). The individual grades are also displayed for further use, if needed.

Practice Set 8-3

1. In the past year, 900 cases of accidental in-home poisonings were reported to authorities in a large western city. The location within the home where the poisoning occurred is given, along with the number of cases reported. Use these data to draw a circle graph.

Location in Home	Number of Cases
Kitchen	369
Bathroom	189
Bedroom	108
Other location	234
Total	900

2. A sample of 1260 people were asked the following question: "What meal are you most likely to eat in a fast-food restaurant?" Use these data to draw a circle graph.

Meal	Number of Responses
Breakfast	97
Lunch	616
Dinner	398
Snack	126
None of the above	23

3. A local community college is considering a new student teacher evaluation form. One concern is the amount of time necessary to complete the evaluation. A random sample of 25 students were asked to complete the evaluation. The time in minutes it took them to complete the evaluation were as follows:

19	23	35	16	26
12	17	15	12	18
18	19	8	12	18
22	30	21	16	44
20	15	17	19	23

   (a) Construct a stem-and-leaf display for these data.
   (b) What conclusions can you draw from the graph?

4. A survey of 20 patients was done at a local hospital. The number of pills each patient was given per day was recorded as follows.

10	6	8	15
2	7	14	10
12	8	11	9
1	7	9	5
13	15	3	6

   (a) Construct a stem-and-leaf display for these data.
   (b) What conclusions can you draw from the graph?

5. The average monthly rainfall amounts in Honolulu, Hawaii, are listed below.

Month	Jan.	Feb.	March	April	May	June
Rainfall (inches)	4.20	2.26	3.08	1.16	.86	.32

Month	July	Aug.	Sept.	Oct.	Nov.	Dec.
Rainfall (inches)	.60	.76	.67	1.41	2.89	3.64

   (a) Construct a bar graph of this information.
   (b) There is a rainy season in Honolulu and a dry season. Which six months make up the rainy season?
   (c) If you are traveling in the winter months, (November, December, January, February), which month would be the best time for your trip?

6. A survey was taken among 200 high school students. They were asked which of the following careers they admired most. The following chart summarizes the results.

Career	Number of students
Doctor	60
Scientist	50
Military Officer	50
Lawyer	30
Athlete	10

   (a) Construct a bar graph of this information.
   (b) Construct a circle graph of this information.

7. Following are the ages of students in a math class at a local community college.

$$18, 18, 19, 19, 20, 20, 22, 22,$$
$$22, 23, 23, 25, 27, 42, 43, 57$$

   (a) Construct a graph that is appropriate for these data.
   (b) Discuss the "spread" of this graph as it pertains to the average age of a student in the class.

8. Following are the grades on a final exam in a statistics class.

$$77, 89, 84, 83, 80, 80, 83, 82, 85, 92,$$
$$87, 88, 87, 86, 99, 93, 79, 100, 83, 54$$

   (a) Construct a graph that is appropriate for these data.
   (b) Does the graph resemble any of the curves that we have studied?
   (c) What conclusions can you draw from the graph?

9. The U.S. Bureau of Citizenship and Immigration Services (USCIS) keeps records of where persons entering the United States and seeking to stay come from originally. The following graph shows where immigrants came from during the 1990s.

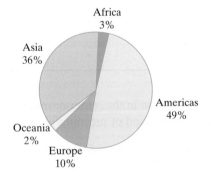

   (a) What percentage of immigrants entered the United States from Africa?
   (b) Did more than one-third of immigrants come from Asia?
   (c) If there were approximately 7.5 million immigrants to the United States during the 1990s, about how many came from the Americas?
   (d) Where did the least number of immigrants come from?

10. A survey was taken among a group of students to determine their favorite sport. The following table shows their responses.

Sport	Number of responses
Football	100
Baseball	30
Tennis	20
Basketball	200
Soccer	50

   (a) What percentage of students preferred football?
   (b) What percentage of students preferred soccer?
   (c) Construct a circle graph of this information.

Section 8-4 Descriptive Statistics

In this section, we will look at some of the common manipulations that can be done with quantitative data once they have been collected. The main topic will be averages for data. As we shall see, there can be several different averages for a given set of data. This means that we cannot just use the word **average** to stand for all of them. All averages are called **measures of central tendency**. There are three commonly used measures of central tendency: the **mode**, the **median**, and the **mean average**.

Definition **The Mode**

The mode of a set of data is the most often repeated observation(s) or item(s).

Example 1: Finding the Mode of Qualitative Data

In a recent survey of the senior class at Jones High School, 60 students had brown eyes, 30 had blue eyes, 5 had green eyes, and 10 had hazel eyes. What is the mode of this set of data (i.e., the modal eye color)?

The most often repeated observation is brown eye color; therefore, the mode = brown.

Example 2: Finding the Mode of Quantitative Data

Find the mode of the following sets of numbers:

(a) 2, 4, 6, 8, 8, 10, 12
(b) 2, 2, 3, 4, 4, 4, 5, 6, 6
(c) 2, 2, 3, 4, 4, 4, 5, 5, 6, 6, 6, 7, 7
(d) 1, 1, 2, 2, 3, 3, 4, 4, 5, 5, 6, 6

(a) In this set of numbers, 8 is the only repeated number, so the mode = 8.
(b) The numbers 2, 4, and 6 are all repeated, but 4 is repeated more than any of the others, so the mode = 4.
(c) There are three 4s and three 6s. These numbers are repeated more than any other numbers in the list and are repeated the same number of times. So there are two modes, 4 and 6. (This is called *bimodal*.)
(d) There are repeated numbers, but none is repeated more than any other. There is no mode.

This example illustrates that a set of data may have no mode, one mode, or more than one mode.

Definition: The Median

The **median** of a set of observations is the observation in the center or middle of the list of items after they have been placed in some kind of meaningful order. This average has its own symbol, $\tilde{x}$, called "*x*-tilde."

Example 3: Finding the Median of Qualitative Data

In a recent survey of the senior class at Jones High School, 60 students had brown eyes, 30 had blue eyes, 5 had green eyes, and 10 had hazel eyes. What is the median of this set of data?

For a set of data to have a median, it must be data that can be placed in a meaningful order. There is no meaningful order for eye color, so there is no median.

Example 4: Finding the Median of Quantitative Data

Find the median of the following sets of data:

(a) 1, 2, 3, 3, 5, 6, 7, 9, 9
(b) 3, 6, 3, 1, 4, 8, 9, 5, 6
(c) 2, 6, 4, 7, 8, 1, 2, 9

(a) Because these numbers are already in a meaningful order, simply find the middle of the list. There are nine numbers in the list, so the fifth one (counting from either end) would be in the middle of the list. The median = $\tilde{x} = 5$.

(b) First, sort the numbers into numerical order as follows:

$$1, 3, 3, 4, 5, 6, 6, 8, 9$$

Now find the middle number.

$$\tilde{x} = 5$$

(c) Again, sort the numbers into numerical order.

$$1, 2, 2, 4, 6, 7, 8, 9$$

Finding the number in the middle is not as easy this time. There are eight numbers in this list. Counting four from each end places the middle *between 4 and 6*. When this happens, the median must be calculated by adding the two numbers on either side of the middle and then dividing by two.

$$4 + 6 = 10$$
$$10 \div 2 = 5$$
$$\tilde{x} = 5$$

Whenever a set of numerical data has an even number of items, the median will be between two of the items. If there is an odd number of items in the list, then the median will be one of the items, the one in the middle.

Definition: The Mean Average

The **mean average** is found by totaling the observations in a set of data and then dividing the total by the number of items in the original list. This average has its own symbol, $\bar{x}$, called "*x*-bar."

$$\bar{x} = \frac{\Sigma x}{n}$$

Example 5: Finding the Mean of Qualitative Data

In a recent survey of the senior class at Jones High School, 60 students had brown eyes, 30 had blue eyes, 5 had green eyes, and 10 had hazel eyes. What is the mean average of this set of data?

For a set of data to have a mean average, it must be numerical data so that the amounts may be added. Therefore, there is no mean average.

Example 6: Finding the Mean of Quantitative Data

Find the mean average of the following sets of data:

(a) 3, 4, 5, 5, 7, 8, 9, 11, 0, 15   (b) 2.3, 6, 7.3, 4, 6, 7, 6.3

(a) First, add the numbers and then divide by 10, which is the number of items in this list of numbers.

$$3 + 4 + 5 + 5 + 7 + 8 + 9 + 11 + 0 + 15 = 67$$
$$67 \div 10 = 6.7$$

The mean average = 6.7.

(b) Again add and this time divide by 7.

$$2.3 + 6 + 7.3 + 4 + 6 + 7 + 6.3 = 38.9$$

$$38.9 \div 7 = 5.557142857\ldots$$

The mean average is about 5.6.

When you are asked to find some kind of average of a list of numbers or a group of items, you must first determine which of the several averages may apply. This depends on the kind of data with which you are working. As shown in the preceding examples, not all types of data have a median or a mean average.

If a set of data has several valid averages, as most numerical data will, the question then becomes, "Which is the *best average*?" If a set of data has only one valid average (see Examples 1, 3, and 5), then this question is easy to answer. However, if there are several averages available, some thought must be given to this question.

Example 7

Which Average Is the Best?

Find the mode, median, and mean of the following sets of data. Then determine the "best" average for each set.

(a) 2, 2, 3, 4, 5, 6, 7, 8, 8
(b) 2, 2, 3, 4, 5, 6, 7, 8, 80

(a) The modes are 2 and 8. Because the numbers are already in order, the middle of the list is easy to find. The median = 5. To find the mean average, add and divide:

$$2 + 2 + 3 + 4 + 5 + 6 + 7 + 8 + 8 = 45$$

$$45 \div 9 = 5$$

The mean average is 5.

The averages are not all the same number, so which is "best"? First, reconsider what averages are supposed to tell us about a set of data. They are *measures of central tendency*. They are supposed to tell us about the value of items near the middle of the set of items once they are arranged properly. Are 2 and 8 near the middle of the list? No. How about 5? Yes, it is near the middle of the list. Therefore, either the mean or the median could be chosen. If you *must* pick one, the average of choice is the *mean*, because all numbers in the list contribute to its calculation.

(b) This list is almost identical to list (a), with one major difference: one extremely large data item. This will have an effect on the averages. The mode = 2. The median = 5. The mean = $117 \div 9 = 13$.

Now all three of the averages are different from each other. Again, which number is closest to the middle of this list? In this case, 5 is, not 13. The value of one of the numbers, 80, is so extreme compared to the rest of the list that it skews the mean average. Because the median is not affected by the size of the numbers, its value is the same as it was in list (a). The *median* is the "best" average.

As a matter of practice in everyday life, we use the word "average" to refer to all three measures of central tendency. However, a builder who builds houses for the "average American family" is interested in the most common family size or the mode. The "average cost of college tuition" is best described using the median cost of college tuition, because tuition at a few very expensive colleges may distort or skew the mean average of college tuition. When computing a "grade point average," the mean average is used in order to reflect *all* grades made by that student during the course. Our correct use of the statistical terms *mean*, *median*, and *mode* will reflect the differences in meaning of these measures of central tendency.

The Weighted Mean

In some situations, data items vary in their degree of importance. For example, a final exam might be 25% of your final average in a particular course, whereas each test counts 20% and homework 15%. Because of these differences in values, computing the final average would not be done by just adding grades and dividing. The following formula is needed to calculate the **weighted mean** by assigning different weights (w) to different values (x).

$$\text{Weighted mean:} \quad \bar{x} = \frac{\Sigma(w \cdot x)}{\Sigma w}$$

Example 8 — The Final Course Average: An Example of a Weighted Mean Average

Suppose your grades in a course are as follows: tests—80, 85, 75; homework average—91; final exam—78. If homework is 15% of your final grade, tests are 20% each, and the final exam is 25%, we can calculate your final average as follows:

$$\bar{x} = \frac{\Sigma(w \cdot x)}{\Sigma w}$$

$$\bar{x} = \frac{(15 \times 91) + (20 \times 80) + (20 \times 85) + (20 \times 75) + (25 \times 78)}{15 + 20 + 20 + 20 + 25}$$

$$\bar{x} = \frac{8115}{100} = 81.15$$

College grade point averages are also computed with a weighted mean. Each letter grade is assigned the appropriate number of points (A = 4, B = 3, C = 2, etc.), and each course is assigned a weight equal to the number of credit hours for the course. For a good exercise in calculating a college grade point average, do Lab Exercise 5, titled GPA Mania, at the end of this chapter. The pattern used in Example 8 may be used to calculate the GPA.

Practice Set 8-4

1. Write a brief paragraph discussing possible sources of confusion about the word *average*.

2. Define and distinguish among the mean, the median, and the mode.

Find all possible averages (mean, median, and mode) of the following sets of data. Then choose the "best" average from among them.

3. 2, 3, 4, 6, 2, 3, 8, 2, 7, 9

4. 2.3, 4.5, 6.8, 2.2, 5.8, 9.0, 2.5

5. 3, 5, 6, 72, 4, 8, 9, 9, 3, 9, 2, 5, 7, 77, 4, 3, 8, 2, 9, 6

6. 200, 258, 265, 236, 298, 195, 300, 199, 245, 245, 227

7. 1.23, 2.85, 3.21, 1.54, 2.65, 0.35, 6.88, 4.25, 2.88, 2.69, 3.27, 11.99

8. 23.6, 25.1, 22.5, 20.4, 26.9, 24.7

9. According to the U.S. Census Bureau, the average (median) household income in the United States in 2012 was $48,451, whereas the average (mean) household income was $60,528. This result is about $12,000 higher than the median household income. Why do you think there is a difference?

10. Following are the ages of students in a math class at a local community college.

$$18, 18, 19, 19, 19, 20, 20, 22, 22,$$
$$22, 23, 23, 25, 27, 42, 43, 57$$

Which average would you use if asked to give the average of this data set? Briefly justify your answer with a brief discussion or explanation.

11. Would you use the mean, median, or mode to give a fair representation of the "average" in the following situations?
    (a) Builders planning houses are interested in the average size of an American family.
    (b) The grade point average for students at a community college
    (c) The average cost of a house in Cook County is $95,000.
    (d) The average American's share of the national debt is $8000.

12. For each of the following statements, determine which average, the mean, median, or mode, was likely to have been used to draw the conclusion that was reached.
    (a) Half of the workers at PrimoWidgets make more than $6.32 per hour and half make less than $6.32 per hour.
    (b) The average number of children in each family renting apartments in Yonkers Heights is 1.82 children per family.
    (c) In a recent survey of people purchasing convertibles, it was found that most prefer the color red.
    (d) The average age of professors in the University of Texas system is 43.7 years.
    (e) The most common fear today is the fear of having to speak at a public gathering.

13. The following is a list of daily high temperatures as measured in Uranium City, Saskatchewan, Canada, for several consecutive days last January. The temperatures are measured on the Celsius scale.

$$2, 5, 3, 0, 0, -3, -7, 1$$

Find the mean, median, and mode for these temperatures.

14. The following body temperatures (measured on Fahrenheit scale) were recorded for 10 male patients. Find the mean, median, and mode.

$$96.3, 98.2, 97.9, 98.7, 98.3,$$
$$98.2, 98.8, 98.3, 98.2, 97.8$$

15. Mary is taking three courses this semester. Her biology course is a four-hour course and she received an A in the course. She made Bs in her history course and her English course, both of which are three-hour courses. What is her GPA for this semester? (Use Example 8 as your guide. A = 4 points, B = 3 points, C = 2 points, etc.)

16. Sonya's Dress Shop sold 23 dresses yesterday as shown on the chart that follows.

Brand	Price	Number Sold
A	$62	3
B	$45	7
C	$38	5
D	$29	2
E	$22	6

  (a) What was the mean price per brand?
  (b) What was the mean price per dress sold?
  (c) What was the median price of the dresses sold?
  (d) What was the modal selling price of a dress?

17. In a recent trade magazine, the average price of a new car in the United States was said to be about $19,000.00. Discuss how you think the person writing the magazine article got this value, and whether you think it represents a mean, median, or mode, and why.

18. A doctor's office is advertising for a receptionist. If there are two doctors, four nurses, and a nurse's aide on staff, and the mean average salary of the staff in this office is $60,000.00 per year, can you expect to earn this amount if you apply for the receptionist job? Why is this average skewed?

19. Sonya has a mean average of 77.6 on the first five of six tests. To get an overall average of 83.0, what must Sonya score on the sixth test?

20. Zack has grades of 65, 84, and 76 on three tests in Biology. What grade does he need on the last test to have an average of exactly 80?

21. The mean is the "most sensitive" average because it is affected by any change in the data.
    (a) Determine the mean, median, and mode for 1, 2, 3, 4, 4, 7, 11.
    (b) If you now change the 7 to a 10 in the same list, which average(s) will be affected by this change? What is (are) the new average(s)?

22. Two students have the following grades at the end of a semester. Lori: 90, 80, 80, 20. Luann: 90, 80, 80, 80. Calculate the mean, median, and mode for each student. Which "average" is best for each girl?

23. Rubber Baby Buggy Bumper Manufacturing Company has only three employees. Their salaries are $25,000, $30,000, and $35,000 per year. If each employee were to receive a $2000 raise, how would this affect the mean of these salaries?

24. Referring to problem 23, if all the employees received a 4% raise, how would that affect the mean average of the salaries?

25. In Greenwich, a small town of 300 residents, the mean annual salary is $60,000. The town's wealthiest person earns $10 million annually. Do you believe that this mean salary is typical of the average annual salary of the residents of Greenwich? Why or why not?

26. Using the data in problem 25, calculate the average salary of the residents of Greenwich excluding the wealthiest resident.

27. The mean of five numbers is 8. If one of the numbers is removed from the list, the mean becomes 7. What is the number that was removed?

28. Create a set of eight test scores so that the mean score is higher than the median score.

29. Find the mean, median, and mode of the ages of cars driven to school by you and your classmates in this class.

30. Find the mean, median, and mode of the manufacturers of the cars driven to school by you and your classmates in this class.

Section 8-5 Variation

Averages do not tell all that one might like to know about a set of numbers. For example, consider the set of statistics derived from the sets of numbers shown in Table 8-1. Looking at this table, you could easily assume that the sets of numbers were identical and not really three different sets at all. However, if you look at the actual sets of numbers shown in Table 8-2, it is obvious that there are differences.

TABLE 8-1 Averages

Number Set	Mean Average	Median
A	5	5
B	5	5
C	5	5

TABLE 8-2 Numbers for the Averages in Table 8-1

Number Set	Numbers				
A	5	5	5	5	5
B	6	5	5	5	4
C	7	6	5	4	3

Averages (measures of central tendency) try to tell us about numbers near the middle of a set. If more information is needed, then other statistics must be calculated based on some other criteria.

Suppose we had a question about how similar the numbers in a group are to each other (or how different they are). An average would not tell us anything about that question. In statistics, the difference between and among numbers in a set of data is called **variation**. Measures of variation (or **dispersion**) tell us about similarities and dissimilarities between numbers within a set of observations. There are two commonly used measures of variation, the **range** and the **standard deviation** of the set of data.

Definition	**Range**
The range of a set of data is the difference between the highest and the lowest number in the data set.
$$R = \text{(highest number)} - \text{(lowest number)}$$

Example 1 — **Finding the Range of Data Sets**

Referring to Table 8-2, what is the range of each set of data?

For set A: $R = 5 - 5 = 0$
For set B: $R = 6 - 4 = 2$
For set C: $R = 7 - 3 = 4$

It can be seen from the calculations in this example that all three sets of numbers are *not* the same. All three have different ranges. When the value of the range is reported along with the measures of central tendency, it is easy to see that there are differences in the sets of numbers.

The range of a set of data is easy to calculate and gives us some sense of the variability of our data. However, one problem with range is that it only uses two data items—the highest and the lowest—to give an answer. An extreme value, either high or low, can create a large range value even if the majority of the numbers is relatively consistent. Consider the set of grades consisting of 80, 82, 20, and 86. Although the range is 66 points, three of the four grades are fairly close together. Because of this problem with range, another method of variability, called standard deviation, is more commonly used in describing the dispersion of a set of data.

Definition	**Standard Deviation**
A rough measure of the average amount by which the observations in a set of data deviate from the mean average value of the group. This deviation may be either above or below the value of the mean.

The standard deviation measures how much the data differ from the mean. It is symbolized by either the Greek letter σ (sigma) or the letter s. The σ is used when the standard deviation of a population of data is being calculated. The letter s is used for the standard deviation when the data being analyzed represent a sample taken from a population. For purposes of this book, we will assume that the data we are analyzing represent sample data.

When the standard deviation of a set of data is small, most of the data items are close to the mean. As the spread of the data around the mean becomes larger, the standard deviation number becomes larger. Look at these two sets of data.

$$8, 8, 9, 10, 10 \quad\quad 6, 7, 9, 11, 12$$

The mean of both sets of data is 9. However, the numbers in the first set of data are all closer to 9; therefore, the first set of numbers will have a smaller standard deviation value.

The desirability of having a large or small standard deviation number in a set of data varies according to the problem being assessed. In IQ scores, for example, a relatively large standard deviation would be expected. However, a small standard deviation is important in quality control procedures used by many industries. If a company manufactures bags of potato chips, it strives to have consistency in its product. For example, although each individual bag of potato chips might have a slightly different weight, all products should be extremely close in weight to the value stated on the bag. This requires a small standard deviation in the weights of the bags.

The calculation of the standard deviation can be complicated, especially with a larger data set. However, to completely understand the definition of standard deviation as an average deviation from the mean, we will illustrate the step-by-step process of finding the standard deviation.

STEPS FOR CALCULATING A STANDARD DEVIATION

To Find the Sample Standard Deviation of a Set of Data

1. Find the mean $\bar{x}$ of the data.
2. Make a chart having three columns:

   Data: x Data − Mean: $x - \bar{x}$ (Data − Mean)^2: $(x - \bar{x})^2$

3. List the data vertically under the column marked "Data."
4. Complete the "Data − Mean" column for each piece of data.
5. Square the values obtained in the "Data − Mean" column and record these values in the "(Data − Mean)^2" column.
6. Find the sum of the "(Data − Mean)^2" column.
7. Divide the sum in step 6 by $n - 1$ (the number of data items minus 1).
8. Find the square root of the quotient in step 7. This is the standard deviation number.

These steps are summarized by the standard deviation formula. (*Note:* Σ is the mathematical symbol for summation—"add them all up.")

$$s = \sqrt{\frac{\Sigma(x - \bar{x})^2}{n - 1}}$$

Example 2 **Calculating the Standard Deviation of a Data Set**

Find the standard deviation of the following set of data:

$$8, 6, 0, 2, 9$$

(a) Find the mean:

$$\bar{x} = \frac{8 + 6 + 0 + 2 + 9}{5} = 5$$

(b) Chart the data:

Data (x)	Data − Mean ($x - \bar{x}$)	(Data − Mean)² [$(x - \bar{x})^2$]
8	8 − 5 = 3	$(3)^2 = 9$
6	6 − 5 = 1	$(1)^2 = 1$
0	0 − 5 = −5	$(-5)^2 = 25$
2	2 − 5 = −3	$(-3)^2 = 9$
9	9 − 5 = 4	$(4)^2 = 16$
Total	0	$\sum(x - \bar{x})^2 = 60$

$$s = \sqrt{\frac{\sum(x - \bar{x})^2}{n - 1}} = \sqrt{\frac{60}{4}} = \sqrt{15} = 3.87$$

Calculating the standard deviation by hand involves a lot of manipulation. Now is a good time to use the capabilities of our calculators to do some complicated calculating.

CALCULATOR MINI-LESSON
Finding a Standard Deviation on a Calculator

A *standard scientific calculator* can do the job, but there are many different ways of entering the initial data. It would be best to consider your calculator's instruction booklet to learn how your version works.

A *graphing calculator* (TI-84 Plus) will do the job efficiently as follows: Press STAT. Select EDIT. You should now be looking at a screen that will allow you to enter lists of numbers. Enter each number, then press the ENTER key. After entering the last number, press STAT. Use the arrow key to move over to CALC. Choose 1-Var Stats by pressing ENTER. Press ENTER a second time. You should now see a list of calculated statistical values. $\bar{x}$ = the mean average, s_x = the sample standard deviation, σ_x = the population standard deviation, n = the number of items in the list, Med = the median, and min X and max X are the smallest and largest numbers that you entered.

Example 3 Using a Calculator to Compute a Standard Deviation

Use your calculator to obtain the standard deviation for the following numbers.

$$8, 6, 0, 2, 9$$

With a regular scientific calculator, use the $\Sigma+$ or the M+ button to enter the data. You may need to put your calculator into statistics mode. (See your instructions for specific information.)

The function labeled s or σ_{n-1} is the sample standard deviation number. It should be 3.8729833..., which rounds to the same answer obtained in Example 2.

When you have finished a problem, you must remember to clear the data from your calculator to start a new problem. This is usually done with SHIFT CE/C or using a button labeled CSR.

If you are using a TI-84 Plus calculator to do your calculations, use the button labeled STAT to access a menu that will let you enter data EDIT and clear old data CLRLST. The CALC function at the top of the screen is used to calculate statistical values of the data. The 1-Var Stats menu option is the one used to find the standard deviation of the list you have entered. You must tell the calculator to calculate for L_1, L_2, or whichever list you have used for the problem.

Problem Set 8-5

1. What does it mean if the standard deviation of a set of data is 0?

2. Without actually calculating the standard deviation, can you tell which of the following sets of data has the greater standard deviation? Explain.

    3, 6, 9, 12, 15   213, 216, 219, 222, 225

Using your calculator, find the mean, range, and standard deviation of each of the following data sets.

3. 0, 1, 2, 3, 4, 5, 6, 7, 8, 9
4. 2, 4, 6, 8, 10, 12, 14, 16, 18, 20
5. 0, 0, 2, 5, 6, 9, 11, 15, 18, 18
6. 2, 2, 2, 2, 2, 2, 2, 2, 2, 2
7. 6, 6, 10, 12, 3, 5
8. 120, 121, 122, 123, 124, 125, 126
9. 4, 0, 3, 6, 9, 12, 2, 3, 4, 7

10. Find the mean average, range, and standard deviation of each of the following sets of numbers. What are the similarities and differences between the data sets? Explain these.

    Data set A: 10, 10, 10, 10
    Data set B: 1, 6, 13, 20
    Data set C: 1, 1, 18, 20
    Data set D: 2, 2, 20, 20

11. Two different sections of a math class take the same quiz and the scores are recorded in the following table. Find the range and standard deviation for each section. What do the range values lead you to conclude about the variation in the two sections? Why is the range misleading in this case? What do the standard deviation values lead you to conclude about the variation in the two sections?

    Section 1: 10 90 90 90 90 90 90 90 90 90 90
    Section 2: 20 35 40 40 45 70 75 80 85 95 100

12. The same biology test is given to two different sections of the class. The scores are recorded in the following table. Find the range and standard deviation of each section. What do the range values tell you about the variation in the two sections? Why is the range misleading in this case? What do the standard deviations lead you to conclude about the variation in the two sections?

    Section 1: 15 95 95 85 80 91 72 78 95 73 64 82 97
    Section 2: 20 35 40 40 47 70 78 80 80 88 92 95 100

13. The Forever and Enduring brands of D-cell batteries are both labeled as lasting 150 hours. In reality, they both have a mean life of 155 hours, but the Forever batteries have a standard deviation of 5 hours, whereas the Enduring batteries have a standard deviation of 20 hours. Which brand is the better choice? Why?

14. The Utah Department of Transportation is trying to decide whether to install high-pressure sodium vapor lights or low-pressure sodium vapor lights at various points along its interstate highway system. High-pressure bulbs have a mean lifetime of 7000 hours with a standard deviation of 200 hours. Low-pressure sodium bulbs have a mean lifetime of 6400 hours with a standard deviation of 50 hours. Which bulbs should it choose? Why?

15. If you manufacture potato chips that are sold by weight, not by volume, would you want a small standard deviation or a large standard deviation in your quality control measurements? Give a reason for your answer.

16. Assume that the duration of a normal human pregnancy can be described by a mean of nine months (270 days) and a standard deviation of one-half month (15 days).
    (a) Most but not all babies will arrive within two standard deviations of the mean. Therefore, a mother would not expect a baby to arrive sooner than _____ days or later than _____ days.
    (b) In a paternity suit, the suspected father claims that because he was overseas during the entire 10 months before the baby's birth, he could not possibly be the father. Any comments?

17. The annual salaries of three employees at Maxton Office Resources are $20,000, $25,000, and $30,000.
    (a) If each of these employees receives a $1000 raise, how does that affect the mean of these salaries? How does it affect the standard deviation?
    (b) If each of these employees receives a 3% raise, how does that affect the mean of these salaries? How does it affect the standard deviation?

18. Carole has taken three tests in her civics class. She now has a mean average score of 80 and a median grade of 79. If the range of her scores is 13 points, what are her three test grades?

19. If you add the same amount to each piece of data in a data set, what happens to the value of the mean, range, and standard deviation of the set?

20. Without actually doing the calculations, decide which of the following two sets of data will have the greater standard deviation. Explain.

    5, 8, 9, 10, 12, 16   8, 9, 9, 10, 10, 11

21. Refer to problem 23 in Practice Set 8-2 and answer this question, "If each employee were to receive a $2000 raise, how would this affect the standard deviation of these salaries?"

22. Refer to problem 24 in Practice Set 8-2 and answer this question: "How does the 4% raise affect the standard deviation of the salaries?"

Section 8-6 Normal Curve

In this section, we will demonstrate some applications of the statistics that we have learned to calculate. When data sets are collected and then analyzed, histograms or other types of graphs of the data are often constructed. Many types of data will form a graph that is shaped like a **normal curve** (or **bell curve**). The normal curve has applications in many areas of business and science. A sample graph is shown in Figure 8-7.

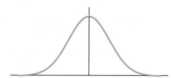

FIGURE 8-7
A normal curve.

Distributions of data that are collected can differ in many ways. As shown in Figure 8-8, data may differ in the location of the average value, in the overall spread of the data, and in shape. In this section, only data that are normally distributed will be considered. Distributions that are not normal require a more extensive study of statistics than is possible in this textbook.

What types of data might be "normally distributed"? Suppose that you were to ask a large number of men how tall they were. You would expect to find a few very

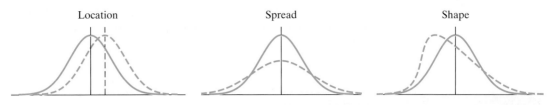

FIGURE 8-8
Normal curve variations.

tall men and a few very short men. However, you would expect to find that the vast majority of men were of approximately the same height. If you were to build a graph of this data set, you would plot each height onto the graph. The shape of the graph would slowly emerge as more and more data were added. Eventually, there would be enough data plotted to see the true "shape" of the data. (Refer to Figure 8-9.)

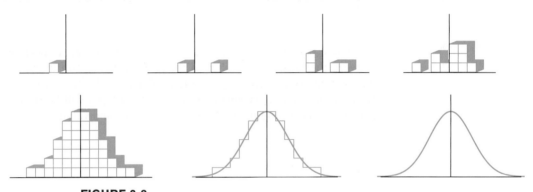

FIGURE 8-9
Building a normal curve.

If we are sure that the data that have been collected are approximately **normally distributed** (i.e., a graph of the data would be shaped like the normal or bell curve), then we can use some characteristics of the normal curve to analyze the data.

To set up a normal curve for a particular set of data, first calculate the measures of central tendency and dispersion (mean average, median, mode, range, and standard deviation) as described in the preceding sections of this chapter. Because each data set will probably have a different mean average and standard deviation, the normal curve for each will be slightly different.

Figure 8-10 shows a family of normal curves with the same standard deviation number but different mean averages. Figure 8-11 shows a series of normal curves with the same mean average but different standard deviations. (*Note:* The larger the standard deviation number, the broader the normal curve will be.)

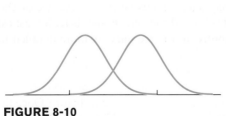

FIGURE 8-10
Normal curve family.

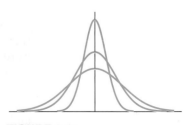

FIGURE 8-11
Another normal curve family.

Once a set of data has been collected and is known to be approximately normally distributed, then the **empirical rule** can be applied to the analysis of the data.

Theorem | **The Empirical Rule**

For a distribution that is symmetrical and normally distributed:

1. Approximately 68.2% of all data values will lie within one standard deviation on either side of the mean.
2. Approximately 95.4% of all data values will lie within two standard deviations on either side of the mean.
3. Approximately 99.7% of all data values will lie within three standard deviations on either side of the mean.

These percentages are based on the amount of **area under the normal curve** between the locations specified in the empirical rule. In theory, the curve extends infinitely in both directions, and the total area under the whole curve is 1 (100%). Because the curve is **symmetrical**, these percentages will be split evenly on either side of the mean. Figure 8-12 shows these areas and values.

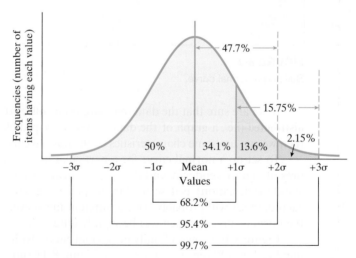

FIGURE 8-12
Standard normal curve.

As you may have already surmised, the use of the normal curve assumptions (the empirical rule) and the percentages is really just estimation. In fact, the proper word is **probability**, not percentage. A more detailed method of analysis is needed for most studies of data.

Table 8-3 is a **table of areas** based on the normal curve and **standard scores**, commonly called z-scores. To use the table, you must first calculate the mean average and standard deviation for your data as before. Once this has been done, calculating a z-score is simple. The z-score will be used to locate a position on the table of areas.

TABLE 8-3 Table of z-Scores: Proportion of Area under the Standard Normal Curve for Values of z

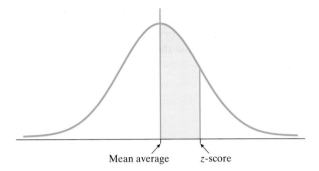

Values are for areas beginning at the mean and going out to the z-score of choice.

z	A	z	A	z	A	z	A	z	A	z	A	z	A	z	A	z	A
0.00	0.000	0.37	0.144	0.74	0.270	1.11	0.367	1.48	0.431	1.85	0.468	2.22	0.487	2.59	0.495	2.96	0.499
0.01	0.004	0.38	0.148	0.75	0.273	1.12	0.369	1.49	0.432	1.86	0.469	2.23	0.487	2.60	0.495	2.97	0.499
0.02	0.008	0.39	0.152	0.76	0.276	1.13	0.371	1.50	0.433	1.87	0.469	2.24	0.488	2.61	0.496	2.98	0.499
0.03	0.012	0.40	0.155	0.77	0.279	1.14	0.373	1.51	0.435	1.88	0.470	2.25	0.488	2.62	0.496	2.99	0.499
0.04	0.016	0.41	0.159	0.78	0.282	1.15	0.375	1.52	0.436	1.89	0.471	2.26	0.488	2.63	0.496	3.00	0.499
0.05	0.020	0.42	0.163	0.79	0.285	1.16	0.377	1.53	0.437	1.90	0.471	2.27	0.488	2.64	0.496	3.01	0.499
0.06	0.024	0.43	0.166	0.80	0.288	1.17	0.379	1.54	0.438	1.91	0.472	2.28	0.489	2.65	0.496	3.02	0.499
0.07	0.028	0.44	0.170	0.81	0.291	1.18	0.381	1.55	0.439	1.92	0.473	2.29	0.489	2.66	0.496	3.03	0.499
0.08	0.032	0.45	0.174	0.82	0.294	1.19	0.383	1.56	0.441	1.93	0.473	2.30	0.489	2.67	0.496	3.04	0.499
0.09	0.036	0.46	0.177	0.83	0.297	1.20	0.385	1.57	0.442	1.94	0.474	2.31	0.490	2.68	0.496	3.05	0.499
0.10	0.040	0.47	0.181	0.84	0.300	1.21	0.387	1.58	0.443	1.95	0.474	2.32	0.490	2.69	0.496	3.06	0.499
0.11	0.044	0.48	0.184	0.85	0.302	1.22	0.389	1.59	0.444	1.96	0.475	2.33	0.490	2.70	0.497	3.07	0.499
0.12	0.048	0.49	0.188	0.86	0.305	1.23	0.391	1.60	0.445	1.97	0.476	2.34	0.490	2.71	0.497	3.08	0.499
0.13	0.052	0.50	0.192	0.87	0.308	1.24	0.393	1.61	0.446	1.98	0.476	2.35	0.491	2.72	0.497	3.09	0.499
0.14	0.056	0.51	0.195	0.88	0.311	1.25	0.394	1.62	0.447	1.99	0.477	2.36	0.491	2.73	0.497	3.10	0.499
0.15	0.060	0.52	0.199	0.89	0.313	1.26	0.396	1.63	0.449	2.00	0.477	2.37	0.491	2.74	0.497	3.11	0.499
0.16	0.064	0.53	0.202	0.90	0.316	1.27	0.398	1.64	0.450	2.01	0.478	2.38	0.491	2.75	0.497	3.12	0.499
0.17	0.068	0.54	0.205	0.91	0.319	1.28	0.400	1.65	0.451	2.02	0.478	2.39	0.492	2.76	0.497	3.13	0.499
0.18	0.071	0.55	0.209	0.92	0.321	1.29	0.402	1.66	0.452	2.03	0.479	2.40	0.492	2.77	0.497	3.14	0.499
0.19	0.075	0.56	0.212	0.93	0.324	1.30	0.403	1.67	0.453	2.04	0.479	2.41	0.492	2.78	0.497	3.15	0.499
0.20	0.079	0.57	0.216	0.94	0.326	1.31	0.405	1.68	0.454	2.05	0.480	2.42	0.492	2.79	0.497	3.16	0.499
0.21	0.083	0.58	0.219	0.95	0.329	1.32	0.407	1.69	0.455	2.06	0.480	2.43	0.493	2.80	0.497	3.17	0.499
0.22	0.087	0.59	0.222	0.96	0.332	1.33	0.408	1.70	0.455	2.07	0.481	2.44	0.493	2.81	0.498	3.18	0.499
0.23	0.091	0.60	0.226	0.97	0.334	1.34	0.410	1.71	0.456	2.08	0.481	2.45	0.493	2.82	0.498	3.19	0.499
0.24	0.095	0.61	0.229	0.98	0.337	1.35	0.412	1.72	0.457	2.09	0.482	2.46	0.493	2.83	0.498	3.20	0.499
0.25	0.099	0.62	0.232	0.99	0.339	1.36	0.413	1.73	0.458	2.10	0.482	2.47	0.493	2.84	0.498	3.21	0.499
0.26	0.103	0.63	0.236	1.00	0.341	1.37	0.415	1.74	0.459	2.11	0.483	2.48	0.493	2.85	0.498	3.22	0.499
0.27	0.106	0.64	0.239	1.01	0.344	1.38	0.416	1.75	0.460	2.12	0.483	2.49	0.494	2.86	0.498	3.23	0.499
0.28	0.110	0.65	0.242	1.02	0.346	1.39	0.418	1.76	0.461	2.13	0.483	2.50	0.494	2.87	0.498	3.24	0.499
0.29	0.114	0.66	0.245	1.03	0.349	1.40	0.419	1.77	0.462	2.14	0.484	2.51	0.494	2.88	0.498	3.25	0.499
0.30	0.118	0.67	0.249	1.04	0.351	1.41	0.421	1.78	0.463	2.15	0.484	2.52	0.494	2.89	0.498	3.26	0.499
0.31	0.122	0.68	0.252	1.05	0.353	1.42	0.422	1.79	0.463	2.16	0.485	2.53	0.494	2.90	0.498	3.27	0.500
0.32	0.126	0.69	0.255	1.06	0.355	1.43	0.424	1.80	0.464	2.17	0.485	2.54	0.495	2.91	0.498	3.28	0.500
0.33	0.129	0.70	0.258	1.07	0.358	1.44	0.425	1.81	0.465	2.18	0.485	2.55	0.495	2.92	0.498	3.29	0.500
0.34	0.133	0.71	0.261	1.08	0.360	1.45	0.427	1.82	0.466	2.19	0.486	2.56	0.495	2.93	0.498	3.30	0.500
0.35	0.137	0.72	0.264	1.09	0.362	1.46	0.428	1.83	0.466	2.20	0.486	2.57	0.495	2.94	0.498	3.31	0.500
0.36	0.141	0.73	0.267	1.10	0.364	1.47	0.429	1.84	0.467	2.21	0.487	2.58	0.495	2.95	0.498	3.32	0.500

Definition: z-Score (Standard Score)

A *z*-score is the number of standard deviations (*s*) that a particular piece of data (*x*) is from the mean average ($\bar{x}$) for the set of data. If a *z*-score is positive, then the piece of data in question is above the mean average, and if the *z*-score is negative, then it is below the mean average. A *z*-score is calculated as follows:

$$z = \frac{x - \bar{x}}{s}$$

(See Figure 8-13.)

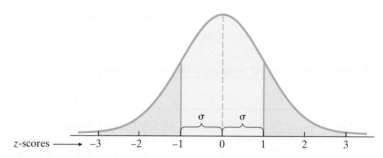

FIGURE 8-13
Normal curve and *z*-scores.

Example 1: Calculating z-Scores

Given a mean test score of 80 with a standard deviation of 5, calculate the following *z*-scores.

(a) What would your *z*-score be if you scored 85 on the test?
(b) What would your *z*-score be if you scored 70 on the test?
(c) What would your *z*-score be if you scored 93 on the test?
(d) What about for a test score of 90?
(e) If you had a *z*-score of -1.5 on this test, what numerical grade did you have?

(a) $$z = \frac{x - \bar{x}}{s} = \frac{85 - 80}{5} = \frac{5}{5} = 1$$

A *z*-score of 1 means that a score of 85 on this test is one standard deviation above the mean average for all students who took this test.

(b) $$z = \frac{x - \bar{x}}{s} = \frac{70 - 80}{5} = \frac{-10}{5} = -2$$

A *z*-score of -2 means that a test score of 70 is two standard deviations below average on this test.

(c) $z = \dfrac{x - \bar{x}}{s} = \dfrac{93 - 80}{5} = \dfrac{13}{5} = 2.6$, or 2.6 standard deviations above average

(d) $z = \dfrac{x - \bar{x}}{s} = \dfrac{90 - 80}{5} = \dfrac{10}{5} = 2$, or 2 standard deviations above average

(e) Here we will be solving the z-score formula for the value of x, the data value corresponding to your test grade.

$$z = \frac{x - \bar{x}}{s}$$

$$-1.5 = \frac{x - 80}{5}$$

$$-7.5 = x - 80$$

$$72.5 = x$$

Therefore, your score was 72.5 on the test.

A z-score is often used to compare an individual data point to the general population to which it belongs and to compare different groups with each other. Consider the following examples.

Example 2 Comparing Heights by Using z-Scores

Assume that the average height of a male in the United States is 70 in., with a standard deviation of 2.0 in., and that the average height of a female is 67 in., with a standard deviation of 2.0 in. If one of your male classmates is measured to be 75 in. tall and a female in your class is also measured to be 75 in. tall, who is the "taller" compared to the group to which they belong?

To make a valid comparison between males and females, it is important to realize that the two groups do not have the same mean average height. There is a true difference in the average heights of males and females. So the correct answer to who is "taller" would be the one that is higher above average for his or her particular group.

$$z_{male} = \frac{75 - 70}{2.0} = 2.5 \text{ standard deviations above average}$$

$$z_{female} = \frac{75 - 67}{2.0} = 4.0 \text{ standard deviations above average}$$

The female is much further above average height for a female than is the male student. Therefore, the female is the "taller" as compared to others in her group.

Once you know how to calculate a z-score, then you are ready to use the table of areas under the standard normal curve (Table 8-3). This table will be used to determine the **probability of events** based on data collected; it will also show you the origin of the percentages given in the empirical rule.

Table 8-3 gives the proportion of the area under the normal curve between the mean of the data and the chosen z-score. This is pictured in Figure 8-14.

The absolute value of the z-score is used so that areas to the left and right of the center of the normal curve are all given based on positive z-scores. The table is only for half of the curve. Remember that in a theoretical normal curve, 50% of the area is below the mean and 50% is above the mean.

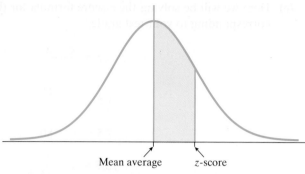

FIGURE 8-14
Area under the normal curve.

Example 3 — Areas Applied to Grade Point Averages

Admission to a state university depends in part on the applicant's high school grade point average (GPA) at graduation. Suppose that the mean average GPA of all applicants to this university is 3.20, with a standard deviation of 0.30, and the GPAs of applicants are approximately normally distributed.

(a) If all applicants who have a GPA of 3.50 or higher are automatically admitted, what proportion of applicants will be automatically admitted?
(b) Applicants with GPAs below 2.50 will be automatically denied admission. What proportion will be automatically denied?
(c) Applicants with GPAs of 3.75 or higher may apply for special scholarships. What proportion of applicants will be eligible for these scholarships?
(d) What proportion of applicants have GPAs between 3.00 and 3.50?
(e) What is the probability that you would randomly select an applicant to the university who has a GPA between 3.00 and 3.50?

(a) To solve these types of problems, you should sketch a normal curve with the given mean and standard deviation. Next, locate the value of special interest on your sketch and shade the area of interest.

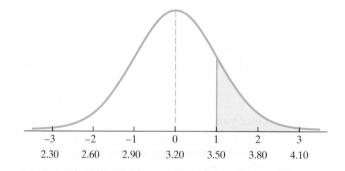

Calculate the z-score for a GPA of 3.50.

$$z_{3.50} = \frac{x - \bar{x}}{s} = \frac{3.50 - 3.20}{0.30} = 1.00$$

Now find $z = 1.00$ on the area table. For $z = 1.00$, the proportion of area under the curve between the mean and 1.00 is 0.341. Because we are interested in the proportion of students with GPAs *higher* than this, we will subtract this number from 0.50, resulting in 0.159. Thus, approximately 16% of the applicants will be automatically admitted.

(b) As in part (a), locate the area specified. This area will begin at a GPA of 2.50 and extend into the left-hand tail of the normal curve.

$$z_{2.50} = \frac{2.50 - 3.20}{0.30} = \frac{-0.70}{0.30} = -2.33$$

Now find $|z| = 2.33$ on the area table. The number from the table is 0.490. Again, we will need to subtract this proportion from 0.50 to find the proportion lower than this value. Calculating this tells us that fewer than 1% of applicants will be automatically rejected.

(c)
$$z_{3.75} = \frac{3.75 - 3.20}{0.30} = 1.83$$

The proportion of the area under the curve between the mean and $z = 1.83$ is 0.466 or (46.6%). Subtracting from 50%, we find that slightly more than 3% of applicants will be eligible for these special scholarships.

(d) The area between GPAs of 3.00 and 3.50 is not a tail-end area. It overlaps the center line (mean) of the normal curve. We will need to find two z-scores, one above the mean and one below the mean. The proportions corresponding to these two z-scores will then be added to answer the question.

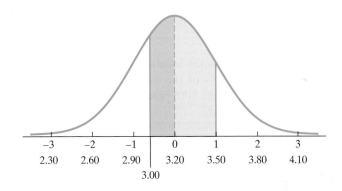

$$z_{3.00} = \frac{3.00 - 3.20}{0.30} = -0.67 \quad \text{(proportion from table: 0.249)}$$

$$z_{3.50} = \frac{3.50 - 3.20}{0.30} = 1.00 \quad \text{(proportion from table: 0.341)}$$

Adding the two proportions will give the total proportion of the area shown in our sketch: $0.249 + 0.341 = 0.590$ (or about 59%) of applicants.

(e) This is the same area as specified in part (d) of this problem. Thus, the probability that you would randomly pick an applicant with a GPA between 3.00 and 3.50 is 0.590 (or a 59% chance). ∎

Often, in testing such as achievement tests given in elementary school and related areas, it is useful to measure the position of an individual score relative to all other scores in the whole distribution. Ranking scores or values or data points is also useful when comparing differing groups (refer to Example 2 on comparing heights). One common way of doing this is to calculate the **percentile rank** for the score. A *percentile rank* indicates the *percentage of scores falling below* the given score. This can be done fairly easily using z-scores and finding the total area to the left of the given score.

Example 4 — Calculating a Percentile Rank

Refer to Example 3, part (c).

(a) What must the percentile rank of an applicant's GPA be to get one of the special scholarships?
(b) What is the percentile rank of the median GPA of 3.20?

(a) The GPA must be at least 3.75 to be eligible for these scholarships, and the z-score was calculated to be 0.466 in Example 3(c). This is the proportion of the area under the normal curve between the mean and the required minimum GPA of 3.75 ($z = 1.83$). To find the total area to the *left* of $z = 1.83$, we must add 0.466 and 0.500. So, $0.466 + 0.500 = 0.966$ or about 97% of all GPAs are less than 3.75.

Thus, the percentile rank for a GPA of 3.75 would be 97—in other words, the 97th percentile.

(b) $$z_{3.20} = \frac{3.20 - 3.20}{0.30} = 0.00$$

This is the location of the peak of the normal curve and, by definition, exactly 50% of the data are located to the left of this point. Thus, a GPA of 3.20 is at the 50th percentile. In fact, *the median score is always at the 50th percentile rank*.

Practice Set 8-6

Value Watt lightbulbs have an average lifetime of 750 hr of use before burning out, with a standard deviation of 80 hr, according to the quality control department. Assume these times are normally distributed.

1. What percent of bulbs will last between 750 and 830 hr?

2. What percent of bulbs will last longer than 910 hr?

3. What percent of bulbs will burn out before being used 670 hr?

4. If a company were to buy 1500 of these lightbulbs, how many may burn out before burning for 830 hr?

Suppose that men's shirts have an average neck size of $15\frac{1}{4}$ in., with a standard deviation of $\frac{1}{2}$ in. Assume that neck sizes are normally distributed.

5. What percent of men have a neck size of 15 in. or larger?

6. What percent of men's neck sizes are between $14\frac{1}{2}$ and $16\frac{1}{2}$ in.?

7. If a company is going to manufacture 75,000 men's shirts, how many should have a neck size between $14\frac{1}{2}$ and $16\frac{1}{2}$ in?

Scores on the mathematics portion of the Scholastic Aptitude Test (SAT) are normally distributed, with a mean average math score of 500 and a standard deviation of 100. Use this information to answer the following questions.

8. What proportion of high school students taking the math portion of the SAT test will score 675 or higher?

9. What proportion will score less than 350?

10. What proportion will score between 350 and 675?

The average burning times of American Standard lightbulbs (made in China) are normally distributed, with a mean of 1200 hr and a standard deviation of 120 hr, according to quality control studies. A company installs a large number of these bulbs at the same time.

11. Before what amount of time should we expect 1% of them to burn out? (*Hint:* This will split the total area under the normal curve into an area of 0.0100 on the left-hand tail and 0.9900 on the other.)

12. Before what amount of time should we expect 10% of them to fail?

13. Before what amount of time should we expect 95% of them to fail?

14. If the company installed 2000 lightbulbs, about how many would they have to replace after 1500 hr of use?

A researcher in the town of Killeen did a survey with children. He discovered that they watched TV an average of 20 hr per week (standard deviation = 2.0 hr, and the data are normally distributed).

15. If a child is chosen at random in the town of Killeen, what is the probability that he or she watches TV for more than 22 hr per week?

16. If a child is chosen at random in the town of Killeen, what is the probability that he or she watches TV for less than 14 hr per week?

17. If a child is chosen at random in the town of Killeen, what is the probability that he or she watches TV for between 15 and 25 hr per week?

18. If a child is chosen at random in the town of Killeen, what is the probability that he or she watches TV less than 25 hr per week?

Beech Mountain Ski Resort has measured the depth of the snowpack (in inches) each February for the past 20 years. The results are given in the following table.

37	52	25	48	26	41	22	15	52	40
50	16	58	59	51	26	39	12	37	20

19. Find the mean average and standard deviation of the snow depths.

20. Assuming that the snowpack depths are approximately normally distributed, what are the maximum and minimum snow depths that define the 68% area under the normal curve? (In other words, what range of snowpack depths will occur 68% of the time at Beech Mountain?) Below what amount will the snowpack depth be 95% of the time?

Assume that the mean average height of a male in the United States is about 72 in., with a standard deviation of 5 in., and that the height of males in the United States matches a normal distribution very well. Find the percent of males in the following categories.

21. Males taller than 65 in.

22. Males shorter than 63 in.

23. Males between the heights of 64 and 74 in.

24. Men with heights below 60 in. are not eligible for military service. What percent of men will be rejected because they are too short?

25. Suppose that you took a mathematics achievement test to try and skip some basic math courses when you enrolled in a college. If you scored at the 85th percentile, what percentage of persons who have taken the same test scored below your level?

26. On most tests, scores can range from 0 to 100. If you got a grade of 90 on such a test, would this mean that your score was at the 90th percentile? Why or why not?

27. Ray and Susan are in different sections of Algebra II at East High School. Ray scored 82 on the first test in his class. In his class, the mean average was 65, with a standard deviation of 12. Susan scored a 78 on the first test in her section of the class. In her section, the mean was 65, with a standard deviation of 10. Whose grade was better if they were graded on a curve referenced to the normal curve?

28. Mario scored 82 on his history final. The class mean was 78 with a standard deviation of 3 points. He scored 72 on his biology final and the mean grade for that class was 65 with a standard deviation of 5 points. Which grade was better as compared with the class mean?

29. Given a mean test score of 78, with a standard deviation of 5, calculate the z-score for a test score of 82.

30. Given a mean test score of 84 and a standard deviation of 7, calculate the z-score for a grade of 90 on the test.

31. Suppose that your z-score for the test in problem 29 was a negative number. What does that mean about your actual test score? Suppose your z-score was -1.6; what grade did you make on the test?

32. If your z-score on the test in problem 30 was 0.4, what was your actual test score?

Problems 33 to 38 refer to a normal distribution of IQ scores that have a mean average of 100 and a standard deviation of 15. You may round your answers to the nearest whole number or whole percentage.

33. What percent of IQs are below 110?

34. What percent of IQs are above 115?

35. What IQ score corresponds to the 5th percentile?

36. What percent of IQs fall between 130 and 145?

37. What pair of IQs separate the middle 95% from the remainder of the distribution?

38. If the z-score for your IQ is 1.3, what is your IQ?

39. A banker studying customer needs finds that the numbers of times people use automated teller machines (ATMs) in a year are normally distributed, with a mean of 40.0 and a standard deviation of 11.4 (based on data from McCook Marketing Research, Inc.). Find the percentage of customers who use ATMs between 30 and 50 times. Among 5000 customers, how many are expected to have between 30 and 50 uses in a year?

40. The average Internet connection time for customers of connectme.com is 75 seconds with a standard deviation of 10 seconds. What percent of customers are able to connect between 63 and 83 seconds?

41. The Beanstalk Club has a minimum height requirement of 5 ft 10 in. for women to join the club. If women in the general population have a mean height of 5 ft 5.5 in. and $s = 2.0$ in., approximately what percentage of women would be eligible for membership?

42. If the height requirement were increased by 1 inch to 5 ft 11 in., what percentage of women would then be eligible to join the Beanstalk Club? (See problem 41.)

Section 8-7 Scatter Diagrams and Linear Regression

In statistical analysis, there are many problems that require a comparison of two variables. We often want to know, "Are these two variables related?" or "Is there a strong correlation between these two factors?" We are not examining cause and effect but looking at mathematical relationships that will allow us to predict the behavior of one variable based on knowledge about another variable. For example, is a voter's opinion of a president's defense policy related to his or her political party affiliation? In this section, we will introduce one method of examining these relationships.

One way to see quickly if two variables are closely related is to use a graph called a **scatter diagram**. The data being analyzed are written in the form of ordered pairs, (x, y), where x is the independent variable and y is the dependent variable. For example, if we were interested in examining the relationship between a person's height and weight, x would be a person's height and the corresponding y would be

Example 1

Drawing a Scatter Diagram: Do Heavier Cars Get Poorer Gas Mileage?

Weight of Car (lb)	Gas Mileage (mpg)
1600	46
4100	15
2100	31
2300	20
3400	15
3300	20
2700	25
1800	31
2100	20
2050	36
3800	15
3500	20

Draw a scatter diagram that relates the weight of various cars with their gas mileage for city driving.

We will let the weight of the cars be the x (independent) variable and gas mileage (miles per gallon of gas, mpg) the y (dependent) variable. We make this choice of variables because we suspect that the gas mileage may *depend* on the weight of the car. A plot of these ordered pairs is shown in Figure 8-15.

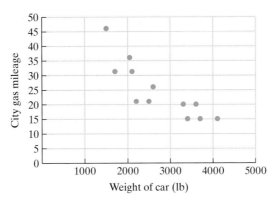

FIGURE 8-15
Weight versus mileage for selected cars.

We can now look at the plot of the data to determine whether there appears to be a relationship between the weight of a car and its gas mileage in the city. Generally, it looks as though gas mileage decreases as the weight of a car increases. Because the points trend downward from left to right, we might say that there appears to be a **negative correlation** between the weight of a car and the gas mileage that it gets. (Remember from Chapter 2 that a line that slopes downward from left to right has a negative slope.)

In generalizing about the relationship between two variables, you can use a scatter diagram to help you draw some preliminary conclusions about your data. For example, you might try to answer the following questions about the data after you have drawn a scatter diagram.

(a) Does the pattern of the scatter diagram roughly follow a straight line (or other shape such as a parabola)?

(b) If it is somewhat linear, do the points slope upward (positive slope) or downward (negative slope)?

(c) Is the pattern of dots on the scatter diagram fairly tightly grouped or widely scattered?

(d) Are there any points that look significantly out of place (a significant deviation)?

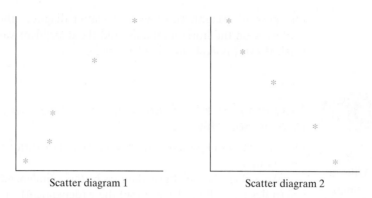

FIGURE 8-16
Scatter diagram comparison.

Look at the scatter diagrams in Figure 8-16. Notice that the points in both scatter diagrams line up very well. They may not lie in a perfectly straight line, but they are very close. Such a pattern indicates a strong relationship between the two variables. In scatter diagram 1, the line of points is "uphill," from left to right (a positive slope). Thus, there is a *positive relationship* (sometimes called a *positive correlation*) between the x and y variables. This can also be called a *direct variation*. See Chapter 4 to review this kind of relationship.

In scatter diagram 2, the slope is "downhill," from left to right (a negative slope). Thus, there is a *negative relationship* (sometimes called a *negative correlation*) between x and y.

If a scatter diagram shows that a set of data is fairly linear, then the next logical step is to set up a model for the data. The usual model would be to determine the equation of the line that best fits the data points that you have. The idea is to choose a line through the data points that by visual inspection seems to best fit the data.

This **best-fit line** can be calculated based on the data that you have. There is an area of statistics that deals with this process of determining the best-fit line. It is called **linear regression analysis**. The name is intimidating and so are the necessary formulas. We will not go into the formulas in this book, but a graphing calculator can be used to do the job for us, as we will see later. The line that best fits the data as determined by linear regression techniques is called the **least-squares line**. The regression equation is in the form $y = ax + b$. This equation should remind you of the slope-intercept form of a linear equation, which you studied previously. The equations determined by this method are models of the data that can be used as **prediction equations**. Given the regression equation for a situation involving two variables, one can predict the value of the dependent variable based on chosen values of the independent variable.

An important note needs to be made before we go too far with this topic. A high degree of correlation (either positive or negative) between two quantities does *not* imply (or prove) that one is the cause of the other (causation). As an example, the size of children's shoes and their measured reading ability will show a fairly high degree of correlation. That does not mean that having big feet causes a child to read well. Having larger feet and reading better are both a consequence of growing older, but neither *causes* the other.

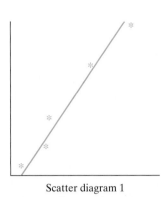
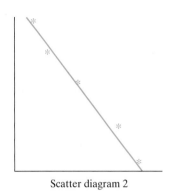

Scatter diagram 1   Scatter diagram 2

FIGURE 8-17
Scatter diagrams with least-squares lines.

In Figure 8-17 are the same two scatter diagrams that were shown previously, but these include the least-squares lines for the data. Notice that the lines do not go through all the points.

Example 2 An Application of a Prediction Equation

A study was done of the heights and weights of a sample of women ages 20–30. Use the resulting prediction equation (least-squares line) to predict the weight of a 26-year-old woman who is 65 in. (5 ft 5 in.) tall.

$$\text{Prediction equation:} \quad y = 4.71x - 186.5$$

where y is weight and x is height in inches.

$$\text{Solution:} \quad y = 4.71(65) - 186.5 = 119.65 \text{ or about } 120 \text{ lb}$$

As previously mentioned, deriving a prediction equation is a complicated process and beyond the scope of this textbook. However, we can use a graphing calculator to help us derive prediction equations. General directions are given in the following example, but you will need to refer to your calculator manual for specific instructions.

Example 3 Using a Calculator to Derive a Prediction Equation

The following table gives the cooking time for a turkey based on its weight. Use this table to find a prediction equation for cooking time based on weight.

If you are using a TI-84 Plus graphing calculator, enter the data into the calculator. STAT, EDIT, L_1 (weight), L_2 (cooking time). Then go to the STAT, CALC menu and find the LinReg (ax + b) $(ax + b)$ option. When you press ENTER for this option, the values of a and b will be displayed. Use these in the form $y = ax + b$ and write the prediction equation for this problem.

Weight (lb)	Cooking Time (hr)
5	3.5
7	4.0
10	4.5
14	5.5
18	6.75
22	8.5

$$a = 0.2840557276 \approx 0.28$$
$$b = 1.860294118 \approx 1.86$$

$$y = 0.28x + 1.86$$

where x = weight and y = cooking time.

Now predict the cooking time for a 20-lb turkey.

$$y = 0.28x + 1.86 = 0.28(20) + 1.86 = 7.46$$

or about 7.5 hr. (If you are following along with this example in your calculator, do not erase the numbers you have listed until you finish reading this section. We are going to refer back to this example shortly.)

Once the regression line is calculated, another piece of information is needed to determine just how good the model (equation) will be at making predictions about potential data. Look at the two scatter diagrams in Figure 8-18 and their associated best-fit lines. You will note that one of the sets of data is much closer to fitting the line than the other. In other words, scatter diagram 1 is nearer to actually being a line than scatter diagram 2. There is a number that can be calculated to indicate the degree to which a particular regression line fits the actual data. This number is the **correlation coefficient** (r) and is sometimes called the Pearson correlation coefficient, after Karl Pearson, who developed this method in the 1890s. The value of r is a number between 1 and -1 ($-1 \le r \le 1$). If $r = 1$ or $r = -1$, then the data are perfectly linear in nature. The sign of the number for r indicates whether the correlation is positive or negative, as has already been discussed. A value for r that is not equal to 1 or -1 indicates that the data are not exactly linear. The closer the value is to 0, the less linear the data. The following list gives the most commonly used degrees of correlation based on the value of r.

$0 < |r| < 0.33$ shows a *weak correlation* (or none at all when r is near 0)

$0.34 < |r| < 0.67$ shows a *moderate correlation*

$0.68 < |r| < 1.00$ shows a *strong correlation*

Refer back to the data you entered into your calculator in Example 3. In addition to calculating the coefficients for the least-squares line, the calculator will also calculate a value for the correlation coefficient, r. On the screen with the values for a and b (STAT , CALC , LinReg (ax + b)), you will find the value of r. (If the value for r is not visible, use the CATALOG function and select the option Diagnostic On .) For the data in Example 3, the value of r is 0.9879245263. This value is very close to 1 and indicates that there is a high positive correlation between the weight of a turkey and the time it takes to cook it.

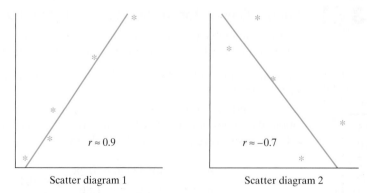

FIGURE 8-18

Scatter diagrams with least-squares lines and correlation coefficients.

Practice Set 8-7

1. In general, is the correlation between the age of a secondhand car and its price positive or negative? Why?

2. In general, is the correlation between current age and years of life expectancy positive or negative? Why?

3. In general, is the correlation between the weight of a car and its gas mileage positive or negative? Why?

4. In general, is the correlation between people's height and weight positive or negative? Why?

5. The correlation coefficient r of the relationship between years of education and personal income is 0.75.

   (a) What are the independent and dependent variables in this problem?
   (b) How would you describe the strength of this relationship?
   (c) What does the positive sign of the correlation coefficient indicate in this problem?

6. The correlation between the number of police officers patrolling a public park and the number of muggings is calculated as $r = -0.97$.

   (a) What are the independent and dependent variables in this problem?
   (b) How would you describe the strength of this relationship?
   (c) What does the negative sign of the correlation coefficient indicate in this problem?

7. The following least-squares prediction equation is based on college GPAs (y) and SAT scores (x) in English:

   $$y = 0.005x + 0.40$$

   If a student scores 600 on the SAT, predict the student's GPA.

8. A scientist developed a prediction equation to predict the number of seedpods that would develop on a plant based on the number of branches on the plant. The equation is

   $$y = 18x - 208$$

   Predict the number of seedpods that would be produced by a plant with

   (a) 16 branches
   (b) 20 branches

9. Using a sample of 50 college graduates, a regression equation for estimating the final college GPA (y) based on a student's SAT score (x) was derived. Use the equation $y = 0.002x + 0.667$ to estimate the final college GPA of a student with an SAT score of 1060.

10. Using a sample of 15 houses in one neighborhood, a salesperson studied the relationship between the size of a house (in square feet) and the selling price of the property (in thousands of dollars). She developed a prediction equation, $y = 0.0728x - 25.0$, to assist in the prediction of the selling price of a house based on its square footage. Use this equation to predict the selling price of a 1500-square-foot house in this neighborhood.

11. The following data were recently collected. We will use these data to determine if there is a relationship between age and the length of hospital stay for surgical patients.

Age	Days in Hospital
40	11
36	9
30	10
27	5
24	12

    (a) Draw a scatter diagram. (*Note:* Let x = age and y = days in hospital.)
    (b) Does there appear to be a correlation between age and number of days in the hospital? Find the r-value for this problem to verify your answer.
    (c) What is the regression equation for the data?
    (d) Use the equation to predict the length of stay for a patient who is 32 years old. Do you believe this prediction is reliable? Why or why not?

12. Is there a correlation between a man's height and his shoe size? Can a man's shoe size be predicted from his height? How reliable is this prediction? Use the data in the table collected from adult males to answer the following questions.

Height (in inches)	68	73	67	74	65	70	75
Shoe Size	10	11	10	13	8	9	13

Height (in inches)	70	67	72	64	68	74	69
Shoe Size	10	9	12	8	9	12	10

(a) Draw a scatter diagram using this data set. Be sure to label it properly.
(b) What is the correlation coefficient for these data? Is there a strong, weak, or moderate correlation?
(c) What is the regression equation for the data?
(d) Predict the shoe size of an adult male whose height is measured at 5 ft $7\frac{1}{2}$ in.

13. Medical research indicates that there is a positive correlation between the weight (x) of a 1-year-old baby and the weight (y) of that baby as a mature adult (30 years old). A random sample of medical files produced the following information for 11 females:

x (lb)	21	25	23	24	20	15
y (lb)	125	125	120	125	130	120

x (lb)	25	21	17	24	26
y (lb)	145	130	115	130	140

(a) Draw a scatter diagram of these data.
(b) Is there a strong, weak, or moderate correlation?
(c) What is the prediction equation?

(d) What is the correlation coefficient of these data?
(e) Predict the weight of a 1-year-old 20-lb baby at age 30.

14. Sociologists are interested in the correlation between years of education and yearly income. The following data were collected from eight men with full-time jobs who had been employed for 10 years.

Years of Education	12	13	10	14	11	14	16	16
Yearly Income (in thousands)	34	45	36	47	43	35	50	42

(a) Use these data to find a prediction equation for income (y) based on years of education (x).
(b) Predict the income after 10 years of a man with 15 years of education.
(c) What is the degree of correlation?

15. Ten students in a physics class have each taken two brief quizzes. Each quiz has a maximum score of 10. The scores of all 10 students for tests 1 and 2 are listed next. (*Note:* The scores are given in the same order by student in both lists so that the first score in both lists is for student 1, the second in both for student 2, etc.)

Quiz 1 scores: 6, 5, 8, 8, 6, 7, 7, 9, 10, 4
Quiz 2 scores: 8, 7, 7, 10, 8, 5, 6, 8, 10, 6

(a) Construct a scatter diagram of the scores, letting quiz 1 scores be the xs.
(b) Does there appear to be any correlation between the scores on the quizzes? If so, what kind of correlation is it?

16. Find a prediction equation for the data in problem 15. Let x = quiz 1 scores and y = quiz 2 scores. Use this equation to predict the quiz 2 score of a student who scored 8 on the first quiz.

17. A real estate salesperson in a small town is studying the relationship between the size of a home (in square feet) and the selling price of the property. A random sample of 15 homes is obtained.

Area (in sq ft)	Selling price (in thousands)
1400	88
1600	102
1450	109
1500	115
1400	96
1800	193
1560	107
1540	171
1490	86
1700	125
1330	58
1810	175
1700	160
1750	145
1360	102

(a) Draw a scatter diagram of these data. Does there appear to be a correlation?
(b) What is the correlation coefficient for these data? Is there a strong, weak, or moderate correlation?
(c) What is the regression equation for these data?
(d) If a home has 1575 square feet, use the regression equation to estimate the selling price. Do you believe this estimate is reliable? Why or why not?

18. Refer to the following table for homes sold in Halifax County.

List Price	Selling Price
150,000	142,000
142,995	139,500
305,000	289,000
160,000	152,800
225,000	207,000
90,000	88,500
157,000	149,500
153,900	151,000
199,000	185,000
242,000	231,500
170,000	167,000
193,950	190,000

(a) Construct a scatter diagram for selling price versus list price.
(b) What is the correlation coefficient for these data? Is there a strong, weak, or moderate correlation?
(c) What is the regression equation for these data?
(d) Based on these data, what is the predicted selling price of a house listed at $235,500? Do you believe that your prediction is reliable? Why or why not?

Chapter 8 Summary

Key Terms, Properties, and Formulas

area under the normal curve
average
bell curve
best-fit line
bias
circle graph
cluster sampling
convenience sampling
correlation coefficient (r)
data
descriptive statistics
dispersion
empirical rule
frequency distribution
frequency table
inferential statistics
least-squares line
linear regression analysis
mean average
measures of central tendency
median
midpoint of the class
modal class
mode
negative correlation
normal curve
normally distributed
percentile rank
pie graph
population
prediction equation
probability
probability of events
qualitative data
quantitative data
random sampling
range
sample
scatter diagram

standard deviation   strata   table of areas
standard scores   stratified sampling   variation
statistics   symmetrical   weighted mean
stem-and-leaf display   systematic sampling   z-scores

Formulas to Remember

Mean average: $\bar{x} = \dfrac{\text{total of all numbers in the data set}}{\text{number of items in the data set}} = \dfrac{\Sigma x}{n}$

Range: $R = (\text{highest number}) - (\text{lowest number})$

Standard deviation for a sample: $s = \sqrt{\dfrac{\Sigma(x - \bar{x})^2}{n - 1}}$

z-score: $z = \dfrac{x - \bar{x}}{s}$

Chapter 8   Review Problems

1. Discuss the differences between descriptive and inferential statistics.

2. Discuss the differences between quantitative and qualitative data.

For each of the following situations, state which type of sampling plan (simple random, stratified, cluster, systematic, convenience) was used. Explain whether or not you think the sampling plan would result in a biased sample.

3. To survey the opinions of its customers, an airline company made a list of all of its flights and randomly selected 25 flights. All the passengers on those flights were asked to fill out a survey.

4. A large department store wanted to know if customers would use an ATM machine if it were installed inside the store. An interviewer was posted at the door and told to collect a sample of 100 opinions by asking the next person who came in the door each time she had finished an interview.

5. To find out how its employees felt about higher student fees imposed by the legislature, a community college divided employees into three categories: staff, faculty, and student employees. A random sample was selected from each group, and they were telephoned and asked for their opinions.

Identify possible misuses of statistics in problems 6 and 7.

6. The results of a study testing a new medication on a sample of 10 male heart patients were released today, and the company stated that this medication is a breakthrough for all heart patients.

7. Clients who try our weight-loss milk shake lose an average of 6.21 lb in 2 weeks.

Find the mean average, median, mode, range, and standard deviation of each of the following sets of data.

8. 10, 11, 12, 13, 14, 15, 16, 17, 18

9. 2, 4, 6, 8, 8, 9, 22, 25, 32, 57

10. Looking back at the sets of numbers in problems 8 and 9, which group has the most variation? Explain how you know this.

11. Looking back at the sets of numbers in problems 8 and 9, what is the "best" average for each set? Explain why you chose the value that you did.

The troubleshooter at a local computer center has recorded the number of calls to technical support from January 1 to February 19.

Number of Calls Daily, January 1 to February 19

6	5	3	7	4	8	10	3	4	5
9	8	3	5	7	4	7	6	7	10
7	3	4	5	5	6	6	8	6	3
3	8	10	5	12	9	6	11	3	8
11	12	14	7	13	14	7	8	11	14

12. Construct a frequency table of the preceding data (use six classes).

13. Construct a histogram. Describe the distribution shape.

A survey is conducted, and it is discovered that the average adult male's waist size is 35 in., with a standard deviation of 2 in. Assume a normal distribution.

14. What percent of men have a waist size greater than 37 in.?

15. What percent of men have a waist size less than 35 in.?

16. What percent of men have a waist size between 31 and 37 in.?

17. You took two tests yesterday. On the math test, you scored 88. The class average for this test was 80, with a standard deviation of 5. You also took a history test and scored 88. The class average for the history test was 85, with a standard deviation of 3. Compared to the other students in these classes, on which test did you do better?

Tough Tread Tire Company manufactures tires for automobiles. The company quality control department has determined that its best tire has a tread life of 45,000 miles, with a standard deviation of 8200 miles, and the data are normally distributed.

18. If the company guarantees these tires for 30,000 miles, what proportion can it expect to replace under this guarantee?

19. What if the company guarantees the tires for 40,000 miles?

A student of criminology, interested in predicting the age at incarceration (y) using the age at first police contact (x), collected the following data.

Age at First Contact	Age at Incarceration
11	21
17	20
13	20
12	19
15	18
10	23
12	20

20. Construct a scatter diagram of these data. Does there appear to be a strong correlation?

21. Find the regression equation for this set of data.

22. Predict the age at incarceration for an individual who at age 14 had his first contact with police.

23. What is the linear coefficient of correlation for these data?

24. Interpret the meaning of the coefficient of correlation in this situation.

Chapter 8 Test

If you were to collect the following data, would it be quantitative or qualitative data?

1. The bacteria count in your house drinking water

2. The occupation of your parents' friends

3. The favorite TV show of members of your immediate family

4. Identify the possible misuse of statistics in this statement: "100% of students who took the Real Estate licensing exam this year passed it."

Consider the following sets of data:

Data set A: 2, 4, 6, 8, 10

Data set B: 1, 3, 5, 7, 19

Without calculating any numerical values, answer the following questions.

5. Which group will have the larger variation?

6. What measure of central tendency would be the "best" for each group?

Calculate the mean average, median, mode, range, and standard deviation of each of the following sets of data.

7. 32, 33, 33, 35, 35, 36, 36, 36, 40, 42, 44

8. H, M, L (High, Medium, Low)

Many airline passengers seem weighted down with their carry-on luggage. Just how much weight are they carrying? The carry-on luggage weights for a random sample of 40 passengers returning from a vacation to Hawaii were recorded in pounds.

Weight of Carry-on Luggage (lb)

30	27	12	42	35	47	38	36	27	35
22	17	29	3	21	10	38	32	41	33
26	45	18	43	18	32	31	32	19	21
33	31	28	29	51	12	32	18	21	26

9. Construct a stem-and-leaf display of these data.

10. Does this display approximate any of the shapes we have discussed?

11. A certain community college gives a numerical-skills placement test to all incoming students each year. The average score made by students on this test is 41, with a standard deviation of 4. If a score of 37 is needed to place out of the first required math course, what percent of incoming students must take this course? Assume a normal distribution of scores.

A recent statistics test had a class average score of 80, with a standard deviation of 5, and the grades are normally distributed.

12. If a student scored 65 on this test, what would that student's z-score be?

13. If your z-score was 1.75, what was your grade on the test?

14. If 70 is the minimum passing grade, what proportion of students passed?

15. If a grade of B is earned by scoring between 80 and 89, what proportion of students made a B?

16. If your height were two standard deviations above that of the average person of your sex, what would the z-score for your height be?

17. A company manufactures automobile batteries and wants to guarantee them for 60 months (5 years). Quality control testing determines that its average battery has a life of 64 months, with a standard deviation of 3 months. What proportion of batteries is the company likely to be replacing under this guarantee? Assume a normal distribution.

18. A regression equation for predicting weight in pounds (y) from height in inches (x) is

$$y = 4x - 130$$

Predict the weight of a person who is 69 in. tall.

19. Following is a list of the ages of 10 employees at a local company and their current salaries. Determine the linear coefficient of correlation of this set of data and then interpret its meaning in this situation.

Age	Salary ($)
25	22,500
55	46,000
27	41,500
30	31,500
35	36,000
58	71,500
19	17,500
33	51,000
40	37,000
42	38,500

The following table lists the number of Hardy Burger restaurants that were in operation worldwide in the year indicated.

2003	2004	2005	2006	2007
6385	6648	7121	7547	8030

2008	2009	2010	2011	2012
8696	9400	9835	10,526	11,150

20. Draw a scatter diagram of these data.

21. What is the prediction equation for the number of restaurants that Hardy Burger has in operation?

22. Based on these data and your equation in problem 21, predict the number of restaurants that will be in operation in 2010.

Suggested Laboratory Exercises

Lab Exercise 1 — A Statistical Analysis of M&Ms

As a class, buy a number of small packages (all of the same weight) of plain chocolate M&Ms. Open each package and weigh the contents of each separately. Also, count the number of M&Ms of each color in each package. Keep a complete record for each package separately.

The M&M/Mars company mixes colors in a certain percentage that it has found to be pleasing to its customers. The current (2004) mix is approximately 13% brown, 14% yellow, 13% red, 24% blue, 20% orange, and 16% green.

Calculate the average package weight and standard deviation. Using your class standard deviation, and assuming that the package weights are normally distributed, determine the proportion that meet the package claims of weight. Discuss your results.

Is the color mix that you measured approximately what was intended by the manufacturer? Discuss your results.

Assuming you washed your hands and used clean holders for weighing your candy, eat all of 'em.

Lab Exercise 2 — College Placement Test Statistics

Gather information from your college student services office or registrar's office about math placement testing at your college. Determine the average test score on a particular test—introductory algebra, for example. Determine the proportion of students who place out of the basic algebra course at your college. Construct a graph of the data and discuss what you see. Other questions may be answered about these scores if your instructor wishes.

Lab Exercise 3 — Radio Advertisements

Contact a local radio station and find out the amount of advertising time per hour of broadcast it claims to follow. Get the average and standard deviation if at all possible. Listen to that station for 1 hour and, with a stopwatch in hand, measure the amount of time devoted to advertising. Does the time you measured match the station claims?

Lab Exercise 4 — Predicting Growth for the CVS Drug Store Chain

CVS is a chain of drug stores scattered across the United States. According to its website (www.cvs.com) it began in 1963 with one store and has now grown to over 5700 by opening new stores and buying out other smaller stores and chains. The following table contains data gained from the CVS website. It lists the number of stores operated by CVS during several years. This information is rounded to make calculations easier but the actual information is in a history section on their site. Using a statistical analysis process known as regression analysis, a linear equation modeling the rate of growth of the chain can be derived. The equation based on the pairs of numbers in this chart, $(x, y) = (1988, 750)$ for example, is $y = 264.0x - 523{,}914$.

1. Complete the following table of values using the given equation where $y =$ the model number of stores. For example: $y = 264.0(1988) - 523{,}914 = 918$.

Year	Actual Stores	Model Stores	Actual Stores – Model Stores
1988	750	918	−168
1990	1250		
1997	3750		
1998	4100		
2004	5000		
2007	5700		
2009	7000		

2. There will be some negative numbers in the last column. What do they mean?
3. By what percent does the model over/underestimate the number of stores in 2004?
4. Below are two more equations that can be generated using this same set of data. Are either of these better than the original model at predicting the number of stores in the chain?
    (a) $y = 264.0x - 22{,}273$, where the pairs of numbers are like (88, 750) instead of 1988 or (104, 5000) instead of 2004.
    (b) $y = 264.0x + 696.8$, where the pairs are like (1, 750) and (3, 1250).
5. Which of the three models does the best overall job of predicting the number of stores in the chain? Use that equation to predict the number of stores in the chain in 2010 or 2011, and then check the website for CVS to see how well the equation worked. For what purposes might a company like CVS try to model both past and future aspects of their business?

Lab Exercise 5 GPA Mania

If you are like most students, you are manic about your grade point average or GPA (or sometimes a QPA for quality point average). Also, you are probably not sure exactly how a GPA is calculated. Well, why not do this little exercise and check to be sure that the college has properly calculated your GPA? Of course, we must first write an expression (model) for the calculation of a GPA.

At most colleges (but not all, so you should check to be sure your college uses this model), each course earns a specific number of *credit hours* (2, 3, 4, etc.). Your college then assigns a number of *grade points* (or quality points) to each letter grade. For example, a grade of A usually earns 4 grade points per credit hour, a B earns 3 grade points, a C earns 2, a D earns 1, and an F earns 0 points.

Suppose you and four friends all took a history course that was a 3-credit-hour course and you each made a different grade. The one who made the

A would receive (3 credit hours)(4 points per credit hour for an A) = 12 grade points

B would receive (3 credit hours)(3 points per credit hour for a B) = 9 grade points

C would receive (3 credit hours)(2 points per credit hour for a C) = 6 grade points

D would receive (3 credit hours)(1 point per credit hour for a D) = 3 grade points

F would receive (3 credit hours)(0 points per credit hour for an F) = 0 grade points

The number of grade points for every credit class that you take is calculated this way, and then the total number of grade points earned is divided by the total number of credit hours of all the classes involved. This is a cumulative operation, semester after semester.

1. Write an equation to calculate your overall GPA.
2. Using a copy of your current transcript, calculate your own GPA using your formula (model) as determined in part 1.
3. This is not a normal mean average because each hour of credit earned may not be worth the same number of points. What kind of average is this?

Lab Exercise 6

Cookie Monster

This activity is essentially a brief look at product testing and comparisons. The guidelines given here are brief and not nearly as detailed or rigorous (statistically or procedurally) as the "real thing" would be, if it were done by a manufacturer. The good part is that you get to eat chocolate chip cookies and call it a math lab.

Are all chocolate chip cookies equal? Do they all taste good? Are they all the same size and texture? Do they all cost the same? Do the more expensive ones really taste better (by whatever standard you wish to measure "better")? Let's do some investigating to find out. What follows are some basic guidelines for performing this investigation. You may wish to modify these questions for your particular class.

1. What cookies to buy? Buy as many different brands of chocolate chip cookies as the class budget will allow. Purchase only "hard" chocolate chip cookies or only "soft" chocolate chip cookies and not a mixture. A mixture will impact the results of questions involving the texture of the cookies. Be sure to record the brand name, actual price (no tax), and weight of each package. From this, you can determine the cost per ounce of cookies for each brand.
2. Next, open each package and have someone who will not actually participate in the testing (perhaps your instructor) label some plain brown lunch bags with letters A through enough to match up with the number of different brands purchased. Then have this person place cookies of one brand in each bag. This is the only person who should know which brand is in which bag. Make a list for future reference.
3. Now the taste-testers get to eat some cookies. (Got milk?) Remember, we're after opinions here, not skipping lunch. What opinions do we want? Well, let's rate the cookies as to their taste, texture, and appearance and give an overall rating of the brand. Perhaps a rating sheet patterned as follows would be helpful.

Rating Chocolate Chip Cookies

Name of tester: _____.
My favorite brand of chocolate chip cookie before this test was: _____.
("no preference" is a valid response here)
I am   Male   Female   (please circle one)
Rating Scale:
P = poor, F = fair, G = good, V = very good, and E = excellent
Using this scale, please rate each brand of cookie (A–?) by circling your opinion in each of the following categories:

Brand	Taste	Texture	Appearance	Overall
A	P F G V E	P F G V E	P F G V E	P F G V E
B	P F G V E	P F G V E	P F G V E	P F G V E
etc.	P F G V E	P F G V E	P F G V E	P F G V E

4. Now for the mathematical fun. Let's use a 5-point scale, with P = 1 and E = 5, and convert the entire class data set into number ratings.
5. Using the numbers for taste and the cost per ounce numbers, is there a relationship between the two? What kind?
6. Make any other comparisons that your instructor may choose. For example, compare the overall ratings with cost.
7. Have a general discussion of this testing procedure. What is good, and what is bad about it? If you had it to do over again, what would you change? And so forth.

Lab Exercise 7

Heart Rate versus Body Temperature

We can evaluate the linear correlation coefficient for any two sets of paired data using the correlation coefficient r, and then proceed to analyze the resulting values. Could a high correlation between two variables be caused by an outside factor that is influencing each of the two original variables?

The table gives the body temperature and heart rate for ten men and ten women involved in a recent study. Answer the following questions using this data set.

Men	Body Temperature	Heart Rate	Women	Body Temperature	Heart Rate
Mario	98.0	74	Jennifer	98.2	73
Frank	97.1	82	April	98.4	79
John	97.2	64	Connie	98.7	72
Anthony	98.6	77	Drema	99.0	79
Geraldo	99.0	81	Jessica	98.6	82
Ty	98.4	82	Clara	98.7	59
Shane	98.2	72	Sherri	99.3	68
Cameron	98.6	66	Anne	98.2	65
Marcus	97.7	77	Maria	98.8	84
Chad	97.8	74	Alexandra	98.6	77

1. Prior to analyzing the data, give your opinion about the strength of correlation between the two variables, body temperature and heart rate. Do you believe there should be a high correlation? Why or why not?
2. Compute the correlation coefficient between heart rate and body temperature using the entire data set. Based on this number, are the two quantities highly correlated?
3. Determine the line of best fit (regression equation) for the heart rate-body temperature data using temperature as the independent variable (x).
4. Next determine the correlation coefficient and the regression equation using only the data corresponding to female subjects.
5. Use both linear formulas to predict the heart rate of a female with a body temperature of 99°F.
6. Which prediction do you think is better? Why?
7. What other factors do you believe might influence these variables?
8. Try this experiment with your classmates by having them gather this information at home and bring it to class for analysis.

Lab Exercise 8

Deriving an Exponential Function from a Data Set

We have looked at linear regression models in this chapter. However, not all data will fit the linear model. Some data are best modeled with an exponential function of the form $y = ab^x$. (Recall your work with these functions in Chapter 4.) In this problem, we will look at the per capita income in the state of Texas for years 1970–2010 to create a function that models the growth in per capita income.

The table lists the per capita (per person) income in the state of Texas for the last four census reports: 1970, 1980, 1990, 2000 and 2010. The data source is the U.S. Bureau of Economic Analysis.

Per Capita Income (in dollars) for the State of Texas

Years Since 1970	Income
0	$ 3646
10	$ 9439
20	$16,747
30	$27,752
40	$39,493

(www.infoplease.com)

1. Enter the data points into your graphing calculator. Put the years in L_1 and the income in L_2.
2. To fit an exponential function to the data, press STAT and choose the CALC option. Now select option 0 ExpReg. Next, press VARS, choose the Y-VARS option, then choose option 1, Function, and then choose option 1, Y_1. When you press ENTER, you will see the values of a and b in the equation $y = ab^x$. By doing these keystrokes, you have also copied the regression equation into the Equation Editor® screen as Y_1. What is the regression equation for these data? Round your values to one decimal place.
3. Now use the calculator to make a scatter diagram of this data set. Go to the STATPLOT function, choose option 1, and turn on the first plot. Arrow down to choose the picture of a scatter diagram (the first graph shown), and be sure that the screen shows that the data will be taken from L_1 and L_2. Now press the ZOOM button and select option 9, ZOOMSTAT. The graph should be displayed on your screen. Sketch it below. Is this a linear function?

4. Use the equation to predict the per capita income in Texas in the year 2001 ($x = 31$). What amount did you get? The reported per capita income in 2001 was $28,581. How close was your prediction? Now predict the per capita income in the year 2010. How accurate do you believe this prediction to be?

1 Commonly Used Calculator Keys

KEY   FUNCTION*

SHIFT or 2nd — Allows access to the calculator functions printed above the keys

+/− — Used to change the sign of a number or enter a negative number into the calculator

$\sqrt{x}$ — Used to take the square root of a number

x^2 — Used to raise a number to the second power (square a number). Enter 10^2 as: 10 x^2

y^x or x^y — Used to raise a number to an exponential power. Enter 5^3 as: 5 y^x 3 =

EXP or EE — Used to enter a number written in scientific notation into the calculator. Enter 3×10^2 as: 3 EE 2

$a^b/_c$ — Used to enter a fraction into the calculator. Enter $\frac{1}{2}$ as: 1 $a^b/_c$ 2

x^{-1} or $\frac{1}{X}$ — Used to find the reciprocal of a number

π — Gives the value of pi to nine decimal places

$\bar{x}$ — Gives the arithmetic mean of a set of data when the calculator is in the statistics mode

σ or σ_n — Gives the population standard deviation for a set of data when the calculator is in the statistics mode

σ_{n-1} or s — Gives the sample standard deviation for a set of data when the calculator is in the statistics mode

Calculator Practice

Purpose: To familiarize students with the functions found on a scientific calculator

Apparatus: Scientific calculator, pencil

Introduction: Scientific calculators are used frequently in our study of mathematics to facilitate the arithmetic calculations required by our problems. Though different manufacturers include varying functions, many of the same functions are found on all scientific calculators. This exercise will allow you to explore your own

---

*The keystrokes and keys on individual calculators vary according to brands. This list is a general resource to help you quickly find an explanation of certain keys. For complete information, consult the manual that accompanied your particular calculator.

calculator and become familiar with its operation. For instructions specific to your brand of calculator, see your instruction manual.

Procedure:

SHIFT or 2nd

1. This key allows you to access the functions written above each key.

$-$ $+/-$

2. This key is used to change the sign of the number that appears on your screen. To enter a negative number, many calculators require that you enter the number first, followed by the $+/-$ button.

Activity: **(a)** Enter the number 5 and then push the $+/-$ key. A negative sign should appear in front of the 5. To add two negative numbers in the calculator, look at the following example. Calculate $-5 + (-3)$ by entering 5 $+/-$ $+$ 3 $+/-$ $=$. The answer on your screen should be -8.

(b) Try these problems in your calculator.
  1. $-6 + (-3) =$
  2. $-8 - (-5) =$
  3. $5 \times (-4) =$
  4. $(-10) \div (-2) =$

$a\text{b/c}$

3. This key is used when entering a fraction into the calculator.

Activity: **(a)** To enter the fraction $\frac{2}{3}$ use this sequence: 2 $a\text{b/c}$ 3. The screen display will look like 2 ⌋ 3. To reduce a fraction such as $\frac{5}{10}$ enter 5 $a\text{b/c}$ 10 $=$. To enter a mixed fraction such as $1\frac{1}{4}$ enter 1 $a\text{b/c}$ 1 $a\text{b/c}$ 4. The screen display reads 1_1 ⌋ 4.

(b) Use the fraction key to solve the following problems.
  1. $\frac{1}{4} + 2\frac{1}{2} =$
  2. $(-\frac{1}{2})(1\frac{3}{4}) =$
  3. $-\frac{3}{4} + 1\frac{1}{2} - (-\frac{1}{4}) =$
  4. $(3\frac{1}{4}) \div 5 =$

x^2 y^x

4. These keys are exponent keys.

Activity: **(a)** The x^2 key is used specifically for squaring numbers. To solve the problem 5^2, enter 5 x^2. The answer appears on the screen immediately. The y^x key allows you to raise any base to any given exponent. To calculate the problem $(-5)^3$, enter 5 $+/-$ y^x 3 $=$. The screen should display the an-swer -125.

(b) Use the appropriate exponent key to solve the following problems.
  1. $0.8^2 =$
  2. $2^5 =$
  3. $(\frac{3}{4})^2 =$
  4. $(-3)^4 =$

$\sqrt{x}$

5. This key is a square root key. This key allows you to reverse the process of squaring done by the x^2 key.

Activity: **(a)** If you are asked to find $\sqrt{16}$ you need to find the number that was originally squared to get 16. To do this on the calculator, enter 16 $\sqrt{x}$ and the answer 4 will appear on the screen.

(b) Calculate the following values. (Round to the nearest tenth, if necessary.)
  1. $\sqrt{1.25} =$
  2. $\sqrt{8} =$
  3. $\sqrt{12,100} =$
  4. $\sqrt{4000} =$

( )

6. Parentheses keys are used for calculating problems containing grouping symbols.

Activity: **(a)** If a problem contains grouping symbols, they can be entered into the calculator as you enter the numbers. If a problem has a number immediately preceding the parentheses indicating the operation of multiplication, you will probably need to include the $\times$ button in your entry. For example, to solve $3(5 + 2)$, enter 3 $\times$ $(5 + 2)$ $=$ and the display should show 21 as the correct answer.

(b) Calculate the answers to each of the following problems.

1. $2(15 - 6) =$
2. $(-4 + 3) - (6 - 4) =$
3. $5(8 - 3) + 6 =$
4. $-2(4 + 7) - 3^2 =$

EXP or EE **7.** Your calculator will have one (but not both) of these keys. This key is used to enter values into the calculator that are given in scientific notation form.

Activity: **(a)** A number written in the form 5×10^3 is said to be written in scientific notation. To enter this number into the calculator, enter 5 EE 3. The display shows 5^{03}. If you now press $=$, the number will be changed to the decimal equivalent 5000. The $\times$ and the 10 are not displayed on most screens when a number is entered in scientific notation form.

(b) Use the scientific notation key to solve the following problems.

1. $2.5 \times 10^{-2} =$
2. $(6 \times 10^{-4}) \times (2.1 \times 10^2) =$
3. $5.14 \times 10^4 =$

2 Formulas Used in This Text

GEOMETRY FORMULAS

Circumference of a circle: $C = 2\pi r = \pi d$

Area formulas:
- square $\quad A = s^2$
- rectangle $\quad A = lw$
- triangle $\quad A = 0.5bh$
- circle $\quad A = \pi r^2$

Hero's formula for the area of a triangle:

$$A = \sqrt{s(s-a)(s-b)(s-c)} \quad s = \frac{a+b+c}{2}$$

Pythagorean theorem: In a right triangle, $a^2 + b^2 = c^2$, where a and b are the lengths of the legs and c is the length of the hypotenuse.

BUSINESS FORMULAS

$$\text{Percent increase/decrease} = \frac{\text{new value} - \text{original value}}{\text{original value}} \times 100$$

Simple interest: $I = Prt$

Maturity value: $M = P + I$

Interest compounded quarterly: $M = P\left(1 + \dfrac{r}{4}\right)^{4t}$, where r = annual rate, t = number of years

Interest compounded monthly: $M = P\left(1 + \dfrac{r}{12}\right)^{12t}$, where r = annual rate, t = number of years

Interest compounded daily: $M = P\left(1 + \dfrac{r}{365}\right)^{365t}$, where r = annual rate, t = number of years

Straight-line depreciation: $\dfrac{\text{original value} - \text{residual value}}{\text{number of years}}$

Fixed-rate mortgage monthly payment formula: $P = A\left[\dfrac{\dfrac{r}{12}\left(1 + \dfrac{r}{12}\right)^{12t}}{\left(1 + \dfrac{r}{12}\right)^{12t} - 1}\right]$

where A = amount borrowed
t = number of years
r = annual rate

ALGEBRA FORMULAS

Direct variation: $y = kx$

Inverse variation: $y = \dfrac{k}{x}$

Joint variation: $y = kxz$

Slope of a line: $m = \dfrac{y_2 - y_1}{x_2 - x_1}$

Linear equation: $Ax + By = C$
Slope-intercept form: $y = mx + b$
Point-slope form: $y - y_1 = m(x - x_1)$
Quadratic formula: If $ax^2 + bx + c = 0$, $a \neq 0$, then
$$x = \dfrac{-b \pm \sqrt{b^2 - 4ac}}{2a}$$

Exponential growth: $y = Ae^{rn}$
- Power function: $f(x) = cx^k$
- Exponential function: $f(x) = b^x$

Cramer's rule: To find the solution of a system of equations $ax + by = c$, $dx + ey = f$:

$$x = \dfrac{\begin{vmatrix} c & b \\ f & e \end{vmatrix}}{\begin{vmatrix} a & b \\ d & e \end{vmatrix}} \qquad y = \dfrac{\begin{vmatrix} a & c \\ d & f \end{vmatrix}}{\begin{vmatrix} a & b \\ d & e \end{vmatrix}}$$

PROBABILITY FORMULAS

Empirical Probability:

$$P(E) = \dfrac{\text{number of times event } E \text{ has occurred}}{\text{total number of times the experiment has been performed}}$$

Theoretical Probability:

$$P(E) = \dfrac{\text{number of ways event } E \text{ can occur}}{\text{total number of possible outcomes}}$$

Odds in favor: event occurs : event does not occur
Odds against: event does not occur : event occurs

STATISTICS FORMULAS

Mean average: $\bar{x} = \dfrac{\Sigma x}{n}$

Range: $R = $ highest number $-$ lowest number

Sample standard deviation: $s = \sqrt{\dfrac{\Sigma(x - \bar{x})^2}{n - 1}}$

z-scores: $z = \dfrac{x - \bar{x}}{s}$

3 Levels of Data in Statistics

In addition to being classified as either quantitative or qualitative, data may also be classified based on how much real information they provide. There are four levels of data (or measurement): nominal, ordinal, interval, and ratio. As the level of the data collected increases from the lowest level (nominal) to the highest (ratio), so does the amount of information that is available from the data. We will now look at each level separately, starting with the lowest level.

Nominal-level data consist of *names only*. The data may be sorted into named groups or categories but no other comparisons are possible with this low level of data. No calculation or other mathematical manipulation may be done with nominal-level data. The "names" do not contain any inherent or implied order. Nominal-level data are *qualitative* in nature.

Ordinal-level data do contain more information than nominal-level data. Data of this level may be sorted into categories like nominal-level data but, in addition, these categories may be placed into a *meaningful order*. Generally speaking, ordinal-level data are qualitative in nature.

EXAMPLE 1   CLASSIFYING QUALITATIVE DATA

If you were to ask someone what color his or her eyes were, you would get data such as blue, brown, green, and hazel. There is no way to place eye color in a meaningful order that has to do with eye color. (Alphabetical order has to do with the alphabet, not eye color. In fact, translating these colors into another language could change their alphabetical order. For example, in Portuguese these colors are *azul*, *maroon*, *verde*, and *avela*. Placing these color names in alphabetical order would place *avela*, which is hazel, first in the list.) This means that eye color is nominal-level data. If you were to ask someone what their level of anxiety was in regard to taking a test on this chapter, you would get answers like high, medium, and low. These answers do have a meaningful order and thus are ordinal-level data.

Interval-level data can be placed in a meaningful order like ordinal-level data but also have the distinction of having equal intervals. Interval measurements allow us to make meaningful claims about measurable differences in amount between observations. Most measurement scales, such as for measuring temperature, are clearly interval-level data. Because interval measurements are numerical, this level of data is *quantitative* in nature. Also, "zeros" here are not "real zeros" in the sense of meaning "nothing" or "none of."

Ratio level is the highest level of data. Ratio-level data allow us to introduce the idea of one measurement exceeding another, not by just a certain amount as in interval-level data, but by a certain ratio such as "twice as much." Zeros here are "real," and this level of data is *quantitative* in nature.

EXAMPLE 2 CLASSIFYING QUANTITATIVE DATA

Consider measuring a temperature (we will use the Celsius scale) and measuring a height (we will use meters). First, look at the "zeros." There is a zero in both measurements but only one is a "real" zero. A temperature of 0°C is not a real zero in that it does not mean that there is no temperature at all or that there is no temperature below it. A measurement of 0 m is a real zero. It means no height and there is no measurement shorter than 0 m. A temperature of 40°C is not twice as hot as a temperature of 20°C but a height of 40 m is twice a height of 20 m. Thus, the Celsius temperature scale is interval level and heights in meters are ratio level.

The following table is a brief summary of the four levels of measurement that have been discussed.

Levels of Measurement

Level	Properties	Items Show	Examples	Type of Data
Nominal	Classifications	Differences in kinds	Sex Political party Colors	Qualitative
Ordinal	Classification order	Differences in degree	Letter grades Graded attitudes S, M, L H, M, L Scale of 1–10	Qualitative
Interval	Classification order Equal intervals	Measurable differences in amount	Celsius and Fahrenheit temperature scales GPAs	Quantitative
Ratio	Classification order Equal intervals True zero	Measurable differences in total amount	Income Weights Lengths Times	Quantitative

Now decide the proper level of data for each of the following:

(a) years of education that you have had
(b) your favorite TV show
(c) your daily caloric intake
(d) your place of birth
(e) your blood type

Answer Key

CHAPTER R
A REVIEW OF ALGEBRA FUNDAMENTALS

Practice Set R-1
1. $-2.6, -\sqrt{5}, -1, 0, \frac{7}{13}, 3\frac{1}{2}, 4$

3. (a) $<$ (b) $>$ (c) $>$

5. 2   7. -11 9. undefined   11. 2   13. $-\frac{11}{24}$
15. 4   17. $-\frac{3}{32}$ 19. 12   21. $-\frac{14}{3}$
23. -5 25. -3 27. -10 29. 0   31. 13
33. $\frac{5}{4} = 1.25$ 35. $-\frac{1}{4}$ 37. 0   39. 14.3125
41. $\frac{19}{8} = 2.375$ 43. 76   45. 0   47. 4   49. 0

Practice Set R-2
1. $x = -10$ 3. $x = -3.5$ 5. $x = 10$ 7. $x = 10$
9. $x = 4.5$ 11. $x = -2.5$ 13. $x = 11$
15. $x = -1$ 17. $x = 5$ 19. $x = -1.5$
21. $x = 1.25$ 23. $x = 95$ 25. $w = -\frac{1}{8} = -0.125$
27. $x = -\frac{45}{4} = -11.25$ 29. $x = -\frac{13}{3}$
31. $x = 0$ 33. $k = 120.54$ 35. $x = 24$
37. $x = -1.9$ 39. no solution   41. $a = -2.5$
43. $x = -6$ 45. $x = -4.5$ 47. $x = 12$
49. $x = -\frac{2}{5} = -0.4$

Practice Set R-3
1. $\frac{2}{25}$ 3. $\frac{180}{100} = \frac{9}{5}$ 5. $\frac{1}{3}$ 7. 0.05   9. 0.0005
11. 0.015   13. 1.25   15. 0.5%   17. $66\frac{2}{3}\%$
19. 150%   21. 11.25   23. 0.48   25. 1340
27. 50   29. 68%   31. $1\frac{1}{2}\%$ 33. 75%
35. $7.8 \approx 8$ free throws   37. $58.50   39. 46.8 lb
41. $3.83   43. 3%, 21 people
45. approximately 430 435   47. $8.06/hr
49. 12% discount   51. $6\frac{2}{3}\%$ increase

Practice Set R-4
1. 230,000,000   3. 3000   5. 0.605   7. 0.003
9. 7.5   11. 6.5×10^2 13. 3.2×10^9
15. 5×10^{-2} 17. 7.5×10^{-7} 19. 6.5×10^0
21. 1.47×10^8 km   23. 4.8×10^{-4}
25. 1.2495×10^{-9} 27. 4.375×10^{-10}
29. 4.6865×10^{21} 31. 3.057×10^{21}

Chapter R Review Problems
1. $-\frac{2}{3}, \frac{3}{2}$ 2. $9, -\frac{1}{9}$ 3. -2
4. $-7^2 + 9 = -49 + 9 = -40$ 5. 8.25
6. -6.865 (about)   7. 9   8. $-\frac{1}{3}$ 9. 42 500
10. 510   11. 0.0000175   12. 0.61
13. 1.508×10^6 14. 8.5×10^0
15. 2.78×10^{-1} 16. 1.08×10^{-4}
17. $x = \frac{16}{7}$ 18. $x = -3$ 19. $x = -\frac{4}{3}$
20. $n = 2$ 21. $x = 18$ 22. $x = 6$
23. $x = -6$ 24. $x = 0$ 25. 18.75
26. 12.5%   27. 1600   28. $320   29. 48%
30. $(1.5\%)(\$1800) = \$27, \$1800 + \$27 = \$1827$ new monthly salary   31. 1.6 kg   32. 1.5742 meters
33. 7.8%   34. 8%   35. $11.25   36. $999.98
37. $7.50   38. 125 lb   39. 6 ft $1\frac{1}{2}$ in.; 5 ft $6\frac{1}{2}$ in.
40. $25.48

Chapter R Test
1. The absolute value of a number is its distance from zero on the number line, and distances are always positive.
2. It is the "same" number with the opposite sign, or it is the number with the same absolute value, but on the opposite side of zero on the number line.
3. -16 4. 7   5. -31 6. 0.000208
7. 6.12×10^9 8. $y = -26$ 9. $x = 6$
10. $x = 1.8$ 11. $y = -25$ 12. $x = 1$
13. 12.75   14. approximately 22.3%   15. 650
16. $1.80   17. $6.87   18. 56.67% decrease
19. 113 V   20. max. = 0.454 cm, min. = 0.452 cm

A-1

CHAPTER 1
FUNDAMENTALS OF MATHEMATICAL MODELING

Practice Set 1-2
1. $t = 2.5$ hours   3. $500   5. 28.26 in.²
7. $h = 9$ inches   9. 30 in.²   11. 20°C
13. 14°F   15. $m = -\frac{1}{3}$ 17. $z = -\frac{1}{3}$ 19. $c = 5$
21. $7834.96   23. $x = -6$ or 1
25. 2,204,421 bacteria   27. $r = \frac{I}{pt}$ 29. $b = \frac{2A}{h}$
31. $L = \frac{P - 2W}{2}$ 33. $B = \frac{2A - bh}{h}$
35. $y = -\frac{2}{3}x + 2$ 37. $x = 2A - y$
39. $C = \frac{5}{9}(F - 32)$ 41. BMI = 22.4; normal

Practice Set 1-3
1. 3:8   3. 1:12   5. 2:3   7. 7:2   9. 1:40
11. 32 mpg   13. $0.35/min   15. $4.80/day
17. 1.6 lb/person   19. $x = 3.75$ 21. $x = 7$
23. $x = 12$ 25. $x = 6$ 27. $x = 3$ 29. $1.49/ft²
31. $0.56/oz   33. Brand Z   35. 160 calories
37. 6 dozen = 72 muffins   39. approx. 10.16 ft = 10 ft 2 in.   41. 54 volts   43. approx. 35.3 lb
45. 12.5 lb   47. $300   49. 45 toddlers
51. 22 ft long, 15.5 ft wide   53. 2359.72 mph
55. 2.5 days   57. 1.5 mL   59. 2.7 miles

Practice Set 1-4
1. $2x + 6$ 3. $7x - 2$ 5. $3(4 + x)$ 7. $\frac{1}{3}x - 5$
9. Ann = $x + 5000$, Bill = x
11. Width = x, Length = $2x + 5$ 13. $x = -5$
15. Emily, $24; Elena, $48   17. 20 m, 40 m
19. 86   21. about 470 kWh   23. $44
25. 20 points   27. 28, 29, 30   29. $-89, -91, -93$
31. lot = $23,333; house = $151,667
33. $12.32   35. approx. 2.94 hr = 2 hr 56 min
37. 16 male and 16 female   39. 36 bags
41. automotive div. = $378 million; financial div. = $105 million
43. Ohio State = 105,278; Penn State = 104,234
45. $149.95   47. $76   49. $1000
51. BMI = 23.4; normal

Chapter 1 Review Problems
1. $t = \frac{I}{Pr}$ 2. $y = -\frac{2}{3}x + 3$ 3. $d = \frac{C}{\pi}$
4. $c = P - a - b$ 5. 3:20   6. 4:3   7. 1:10
8. $12/hour   9. 5.5 bushels/tree   10. $3.66/lb
11. $x = \frac{12}{7}$ 12. $x = 6$ 13. $x = 9$ 14. $x = -1$
15. 6 in.²   16. 1440 liters   17. 69 dentists
18. $1.24 + (x - 4)(\$0.28) = \$3.76$; the call was 13 minutes long   19. 138 minutes   20. 2.5 hours
21. 15 in., 18 in.   22. 13 cm, 27 cm, 39 cm
23. $34.00   24. Let n be the number of nickels and so $2n - 2 =$ the number of dimes. There are 18 nickels and 34 dimes in the bank. $4.30
25. 3.5 hours   26. $N = 950$ units produced
27. First remove the parentheses by use of the distributive property, then add 2. $2 + 3(2x + 4) = 2 + 6x + 12 = 6x + 14$.
28. $5x + 3 = 6x$; $x = 3$ 29. 30°, 60°, 90°
30. 312 in. = 26 ft

Chapter 1 Test
1. $w = \frac{V}{lh}$ 2. $v = \frac{h + 16t^2}{t}$ 3. 1:6   4. 3:4
5. $103.30 per day   6. 1.25 lbs/person
7. 5 eggs per chicken   8. $x = \frac{35}{18}$ 9. $x = -\frac{60}{7}$
10. 18   11. 675   12. 64 ft
13. Let $x =$ the number of miles driven: $20.00 + $0.10/mile$(x) = $10.00 + $0.30/mile$(x)$; $x = 50$ mi
14. 750 passengers   15. Sarah is 9, and Michelle is 35
16. 5 ft, 16 ft   17. $2.52/gal
18. 35,898 people (rounding up to the nearest whole person) 8 years ago; down about 987 people per year
19. A grade of 99 is required, which is possible.
20. lot = $17,882, house = $134,118

CHAPTER 2
APPLICATIONS OF ALGEBRAIC MODELING

Practice Set 2-1
1. 1040 feet   3. 165 ft²; 18.3 yd²   5. 182.3 in²
7. approx. 158–160 plants   9. 35 bags
11. 35.6 in.²   13. no   15. 16-inch is the better buy.   17. 123.6 m²   19. (a) $0.75 per ft²
(b) $172.50 per front ft   (c) $32,670 per acre
21. (a) four times the area, twice the perimeter
(b) one-fourth the area, one-half the perimeter
(c) nine times the area, three times the perimeter
(d) four times the area, twice the perimeter
23. 78.5%

25. (a) $h \approx -1.3w + 10$ ft

(b)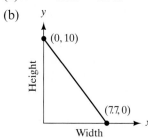

(c)

w	0	1	2	3
h	10	8.7	7.4	6.1

(d) theoretically, an infinite number

27. 50.3 ft²   29. $148.28

Practice Set 2-2
1. 90°, right triangle   3. 43°, obtuse triangle
5. 61°, right triangle   7. $A = 3.75$ cm²
9. $A = 30$ in.²   11. $A = 8750$ mm²
13. $A = 111.8$ in.²   15. $A = 2664$ cm²
17. $A = 15.02$ cm²   19. 45°, isosceles triangle
21. 116°, obtuse triangle   23. $h = 11$ in.
25. $A = 88.18$ ft²   27. $A = 21.3$ ft²
29. The area of one face is approximately 181,935 ft².

Practice Set 2-3
1. The triangle must be a right triangle, and you must know the length of 2 sides.
3. $c = 15$ 5. $a = 10.4$ 7. $b = 24$ 9. 24 ft
11. 9.4 m   13. 17.1 = 17-in. screen   15. 68.7 miles
17. 20.7 in.   19. 17.1 ft   21. $P = 60$ cm, $A = 120$ cm²   23. 8.9 units   25. No   27. 7.1 in.

Practice Set 2-4
1. to make drawings in 2-dimensions that are true representations of what we see in 3-dimensions with our eyes
3. The term *symmetry* implies that balance or a regular pattern exists.
5. horizontal symmetry: B, C, D, E, H, I, O, X
vertical symmetry: A, H, I, M, O, T, U, V, W, X, Y;
F, G, J, K, L, N, P, Q, R, S, and Z do not have reflection symmetry
7. H, I, O, S, X, Z   9. 120°   11. 72°
13.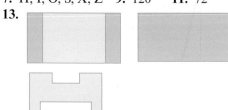

15. $\left(\dfrac{16 \text{ ft}}{1 \text{ in.}}\right)(21 \text{ in.}) = 336$ ft

17. $\left(\dfrac{3}{4}\right)(175 \text{ cm}) = 131.25$ cm

19. $\left(\dfrac{800 \text{ cm}}{1 \text{ cm}}\right)(42.2 \text{ cm}) = 33760$ cm = 337.6 m long

$\left(\dfrac{800 \text{ cm}}{1 \text{ cm}}\right)(8.2 \text{ cm}) = 6560$ cm = 65.6 m wide

21. $\left(\dfrac{5000}{1}\right)(8) = 40,000$ cm = 400 m

23. $\dfrac{573 \text{ ft}}{2.65 \text{ in.}} \approx 216.2$ ft/in.

25. $\left(\dfrac{0.5 \text{ in.}}{50 \text{ mi}}\right)(600 \text{ mi}) = 6$ in.

27. $(23 \text{ ft})(1.618) = 37.214$ ft = 37 ft 2.5 in.
29. $(45.5 \text{ cm})(1.618) = 73.6$ cm; $(45.4 \text{ cm})/(1.618) = 28.1$ cm

Practice Set 2-5
1. Pitch refers to how our ears interpret different frequencies of sound. The higher the frequency of a sound wave, the higher the pitch we would hear.
3. a recording in which the actual wave form is copied as closely as physically possible
5. Frequencies are measured numerically and recorded as numbers in short strings.
7. $(261.626 \text{ Hz})(2) = 523.252$ Hz is one octave higher. $(261.626 \text{ Hz})/2 = 130.813$ Hz is one octave lower.
9. $(392 \text{ Hz})(1.05946) = 415.3$ Hz   11. 2
13. 3   15. 1 quarter note   17. 1 eighth note
19. None are required.
21. similar to Figure 2.25(c)
23. similar to Figure 2.25(c)
25. (group project)

Chapter 2 Review Problems
1. $P = 70$ m; $A = 186$ m²
2. $12.02   3. 93°; obtuse triangle
4. $A = 1017.9$ cm; $C = 113.1$ cm
5. $A = 90$ in.²; $P = 63$ in.
6. 276.5 ft²   7. 21.3 in.²
8. (a) 1492.4 ft² (b) $3,693.50   9. 17 ft
10. 84.9 ft   11. yes, by the Pythagorean theorem
12. 21.6 in.
13. (a) vertical; (b) horizontal; (c) vertical; (d) neither
14. $\dfrac{50 \text{ cm}}{75 \text{ cm}} = \dfrac{2}{3}$

15. (32)(4.125 in.) = 132 in. = 11 ft
16. $\dfrac{32.4 \text{ cm}}{1.62} = 20$ cm
17. (35 mi)(1 in/20 mi) = 1.75 in
18. 1 eighth note and 1 half note
19. 27.5 Hz **20.** 6

Chapter 2 Test
1. $P = 60$ cm; 120 cm² **2.** 15.5 in.²
3. 56.5 ft **4.** 90°; right triangle
5. $A = 3$ ft **6.** 48 cm²
7. 4.4 ft² **8.** (a) $P = 80$ ft; $A = 370.125$ ft²
(b) $1,026.07 (or $1047.90 if cannot buy fractional part of a square yard)
9. 25.6 ft **10.** 2.9 cm
11. (a) 65.97 in. = 5.498 ft (b) approx. 961 times
12. O **13.** O, X
14. $(6250 \text{ m})\left(\dfrac{1 \text{ cm}}{2500 \text{ cm}}\right) = 2.5$ cm
15. $(75 \text{ ft})\left(\dfrac{1 \text{ ft}}{87 \text{ ft}}\right) = 0.862$ ft = 10.3 in.
16. No **17.** the legal pad-size paper
18. 3520 Hz **19.** 1 eighth note
20. The number of quarter notes per measure is different: 3 in 3/4 time and 4 in 4/4 time.

CHAPTER 3
GRAPHING

Practice Set 3-1
1.–12.

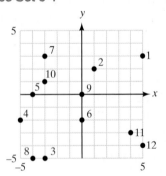

13. (0, 0) origin
15. (−3, 4) II
17. (−3, −3) III
19. (0, −3) y-axis
21. C **23.** F, C **25.** B, F **27.** no **29.** yes
31. (0, 8) (3, 5) (10, −2) (8, 0)
33. (0, −4) (3, −3) (6, −2) (12, 0)
35. (0, 0) (3, −0.75) (8, −2) (0, 0)
37. (0, 0) (3, 3) (−2, −2) (0, 0)
39. (0, −3) (3, −4.5) (−2, −2) (−6, 0)

Practice Set 3-2
1. Let the y value = 0 in the equation and solve for x.
3. Calculate the coordinates of another point on the line by assigning either x or y a nonzero value, and solving for the matching coordinate.
5. (6, 0) (0, −6)(4, −2) **7.** (3, 0) (0, 6) (4, −2)
9. (3, −2) (0, −2) (any real #, −2)
11. (5, 0) (0, 2) (10, −2)
13. x-intercept (−4, 0) y-intercept (0, −2)
15. x-intercept (−5, 0) y-intercept (0, −2.5)
17. x-intercept (3, 0) y-intercept (0, 2)
19. x-intercept (5, 0) y-intercept (0, −3)
21. x-intercept $\left(\dfrac{3}{4}, 0\right)$ y-intercept (0, −3)
23. x-intercept (0, 0) y-intercept (0, 0)
25. x-intercept (0, 0) y-intercept (0, 0)
27. x-intercept (−4, 0) y-intercept (0, −2)
29. x-intercept (2, 0) y-intercept (0, −1)
31. x-intercept (20, 0) y-intercept (0, −4)
33. x-intercept (3, 0) y-intercept (0, 6)
35. x-intercept (−3, 0) y-intercept none
37. x-intercept none y-intercept $\left(0, \dfrac{2}{3}\right)$
39. x-intercept (4, 0) y-intercept $\left(0, \dfrac{4}{3}\right)$

Practice Set 3-3
1. slope $= -\dfrac{5}{2}$
3. slope $= \dfrac{2}{11}$
5. slope $= \dfrac{5}{2}$
7. slope $= 8$ **9.** $x = 5$ **11.** $y = 22$
13. $y = -4x + 1$
  $m = -4$,
  y-intercept $= (0, 1)$

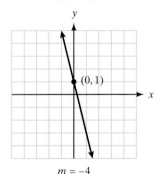

$m = -4$

Answer Key   A-5

15. $3x + 2y = 6$
$m = -\dfrac{3}{2}$,
y-intercept $= (0, 3)$

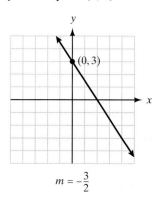

17. $2x - 3y = 4$
$m = \dfrac{2}{3}$,
y-intercept $= (0, -1.33)$

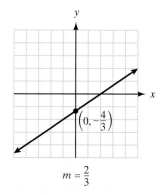

19. $4y - x = 10$
$m = \dfrac{1}{4}$,
y-intercept $= (0, 2.5)$

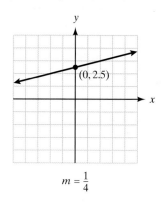

21. $\dfrac{3}{4}x - \dfrac{1}{2}y = \dfrac{5}{8}$
$m = \dfrac{3}{2}$
y-intercept $= (0, -1.25)$

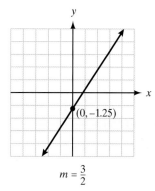

23. $y = -x$
$m = -1$
y-intercept $= (0, 0)$

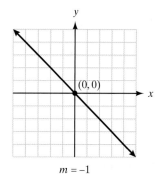

25. $4y = 12 + x$
$m = \dfrac{1}{4}$
y-intercept $= (0, 3)$

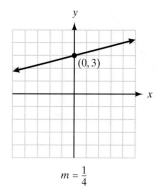

27. $m = \dfrac{3}{8}$ **29.** $m = \dfrac{1}{6}$ **31.** $m = 1$ **33.** $m = 0$
35. undefined   **37.** perpendicular   **39.** neither

Practice Set 3-4
1. $y = -2x + 8$, $2x + y = 8$
3. $y = \frac{1}{2}x - 5$, $x - 2y = 10$
5. $y = -2x - 25$, $2x + y = -25$
7. $y = \frac{5}{2}x + \frac{13}{2}$, $5x - 2y = -13$
9. $y = -\frac{1}{3}x + \frac{5}{3}$, $x + 3y = 5$
11. $y = 2x - 5$, $2x - y = 5$
13. $y = -2$, $y = -2$
15. $y = -x + 3$, $x + y = 3$
17. $y = \frac{1}{2}x - \frac{13}{6}$, $3x - 6y = 13$
19. $y = -2.5x - 7$, $5x + 2y = -14$
21. $x = 3$, $x = 3$ 23. $x = 1$, $x = 1$
25. $y = -2x - 5$, $2x + y = -5$
27. $y = -3x + 19$, $3x + y = 19$
29. $y = -x + 7$, $x + y = 7$
31. $y = -\frac{5}{2}x - \frac{21}{2}$, $5x + 2y = -21$
33. $x = 3$, $x = 3$ 35. $y = 3$, $y = 3$
37. $y = \frac{1}{3}x - 1$, $x - 3y = 3$
39. $y = -x$, $x + y = 0$

Practice Set 3-5
1. $y = -\frac{1}{2}x$ 3. $y = -2x + 2$ 5. $y = \frac{1}{2}x - 3$
7. $y = \frac{1}{3}x + 2$ 9. $y = x$ 11. $y = 3$

13. (a)

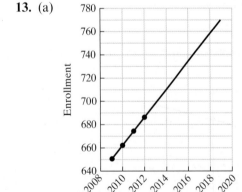

No, it will not exceed 800 by 2015.
(b) rate of change = 12 students per year
(c) $y = 12x + 650$ where x = number of years after 2009
(d) $y = 12(8) + 650 = 746$ students

15. (a)

Yes, it will be more than $35,000 before 2019.

(b) The rate of change is not constant between consecutive years.
(c) Since there is no steady rate of change, you cannot write a linear prediction equation.
(d) Using the graph to predict, the salary will be approximately $36,800. (Table represents a 2.5% increase in salary each year, so calculated amount is $36,841.45.)

17.

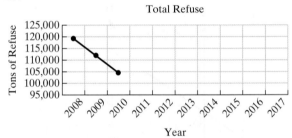

A reduction of 7253 tons per year is the rate of change. A constant rate for the next 7 years would be improbable since there is only so much refuse that can be recycled, thus reducing the total amount.

19. These rates do not have a linear trend since the changes in the price are not consistent throughout the years.

Chapter 3 Review
1.–4.

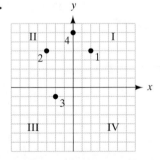

5. yes   6. $(-2, -12)$ 7. $(6, -40)$ 8. $(0, 4)$

9. (any real number, 7) **10.** $\left(0, -\dfrac{11}{3}\right)$

11.
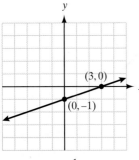
$m = \dfrac{1}{3}$

12.
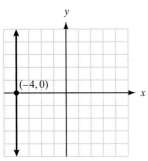
no y-intercept
$m =$ undefined

13.
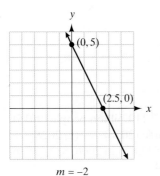
$m = -2$

14.
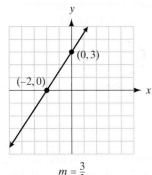
$m = \dfrac{3}{2}$

15.
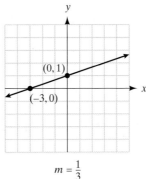
$m = \dfrac{1}{3}$

16.
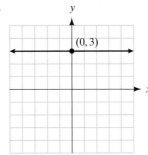
no x-intercept
$m = 0$

17. $m = 3$ **18.** $m = -2$
19. $m =$ undefined
20. $m = \dfrac{7}{2}$ **21.** $y = 2x - 3$
22. $y = 10x + 26$
23. $y = -2x + 10$
24. $y = -\dfrac{1}{2}x + 4$
25. $y = \dfrac{3}{4}x + 4$
26. $y = \dfrac{1}{2}x - 1$
27. $y = -\dfrac{1}{2}x + 4$
28. rate of change = 19,000

$y = 19{,}000x + 139{,}000$

29. (a) rate of change = 20 miles per gallon
(b) $y = 20x$ (c) 360 miles
30. (a) rate of change = $4 billion
(b) $y = 4x + 8$
(c) $72 billion

Chapter 3 Test
1.–3. See graph in back of textbook.

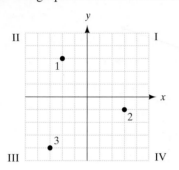

4. no

5.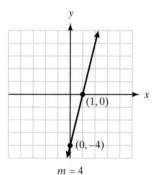

6.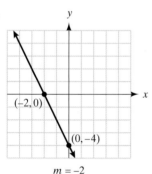

7.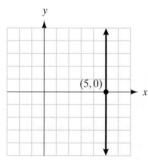
no y-intercept
$m =$ undefined

8.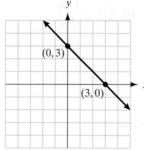
$m = -1$

9.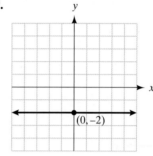
no x-intercept
$m = 0$

10. $m = \dfrac{1}{5}$ **11.** $m = \dfrac{1}{3}$ **12.** slope is undefined

13. $m = 1$ **14.** $y = -3x + 3$ or $3x + y = 3$

15. $y = \dfrac{1}{2}x + 4$ or $x - 2y = -8$

16. $y = 6x - 4$ or $6x - y = 4$

17. $y = -\dfrac{1}{2}x + 4$ or $x + 2y = 8$ **18.** $x = 2$

19. $y = -\dfrac{3}{4}x$ **20.** $y = \dfrac{3}{4}x$

21. (a) 0.072   (b) 7.113 billion people (2015); 8.193 billion people (2030)   (c) 2028

22. (a) 1698   (b) $45,828 in 2006; $37,338 in 2011   (c) 2014

23. $y = .614286x - 1220.6714$ approx. $17.7 billion

24. (a) $y = -4357.14x + 35,000$ (b) -4357.14 The office equipment depreciates in value by $4357.14 per year.   (c) $13,214.30   (d) approx. 8 years

CHAPTER 4
FUNCTIONS

Practice Set 4-1
1. A function is a relation or rule in which for each input value there is exactly one output value. The independent variable is the input and is the variable that we control. The dependent variable is

the output, and its value is a result of the original choice of the value of the independent variable.
3. (a) 68° (b) The high temperature is a function of the date. (c) temperature (d) dates
5. (a) The independent variable is the year, and the dependent variable is the world population.
(b) Domain includes the years 1980–2010, and the range is 4.2 billion–6.4 billion.
7. Grades are a function of student ID.
9. As the weight of a bag increases, the price of the bag increases, so price is dependent on weight.
11. As the slope of a hill increases, the speed of the scooter going uphill decreases, so the speed is dependent on the steepness of the hill.
13. As a person's age increases, his target heart rate when exercising decreases, so the target heart rate when exercising is dependent on a person's age.
15. 9:45 p.m. **17.** January 2 and December 30
19. 10:15 p.m. is the latest; 2:50 p.m. is the earliest
21.

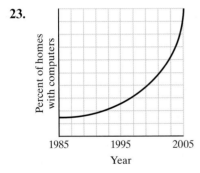

23.

25.

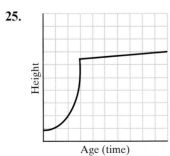

27. $d = 315$ ft; graph is not linear. The faster the speed, the longer the stopping distance.
29.

rate	5.5	5.75	6.0	6.25	6.5	6.75	7
payment	709.74	729.47	749.44	769.65	790.09	810.75	831.63

Practice Set 4-2
1. $g(0) = -5$ **3.** $g(10) = 25$
5. $f(x) = 2x + 1; f(-2) = -3$
7. $f(x) = 3x^2 - 1; f(-2) = 11$
9. $f(x) = \frac{5}{2}x - 4; f(-2) = -9$
11. $f(2) = 576$; 2 sec after the ball is thrown, it is 576 ft above the ground.
13. 173.41 cm **15.** 50.63 cm
17. (a) $2^{30} = 1,073,741,824$ (b) $10,737,418.24
19. (a) $960 (b) $M(n) = \$7.50n$
(c) domain = 0–200 people; range = $0–$1500
21. $f(20) = 170$ and $f(60) = 136$; as a person gets older, his target heart rate decreases.
23. $f(3) = \$29,080$; a person with 3 years of teaching experience will make an annual salary of $29,080; the independent variable is the number of years of experience and the dependent variable is the salary.
25. $f(4) = 0721$, which means that the sun rises at 7:21 a.m. on Jan 4th; $g(10) = 1654$, which means that the sun sets on Jan 10th at 1654 hours or 4:54 p.m.; the independent variable is the date; the dependent variables are the sunrise and sunset times.
27. (a) 75 mi (b) approx. 130 mi. from the graph (127.5 using the formula)
29. $f(100) = 30$ which means that it costs $30 to rent this car and drive 100 miles; the domain is numbers ≥ 0, and the range is numbers $\geq \$25$.

Practice Set 4-3
1. (a)

x	f(x)
1	$1.75
3	$5.25
5	$8.75
7	$12.25
9	$15.75

(b) Initial value is 0; rate of change is $1.75 per mile.
(c) $f(x) = \$1.75x$ where x = miles
(d) $f(8) = 14$, which means that an 8-mile taxi ride will cost $14.00.
3. (a) 5.5 million (b) 2% per year (c) 5.6 million
5. (a) $C(x) = \$525 + \$17.50x$ where x = number of guests (b) $4025 (c) 150 people

7. (a) −$340/year
(b) $V(x) = 1800 − 340x$ where $x =$ years
(c) $V(4) = 440$; after 4 years, the value is $440.
9. $f(x) = -\frac{1}{5}x$ 11. $f(x) = 5$
13. (a) rate of change $= +20$/year
(b) $f(x) = 450 + 20x$ where $x =$ years
(c) $f(15) = 750$; that in the year 2011, the predicted population is 750 people.
15. (a) $80,000/year
(b) $f(x) = \$600,000 + \$80,000x$ where $x =$ number of years after 2003
(c) in 11.25 years or during the year 2015
17. (a) $579/year
(b) $f(x) = \$3536 + \$579x$ where $x =$ number of years after 2008
(c)
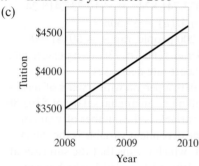
(d) $7010
19. (a) $f(x) = 100 + 5x$ where $x =$ inches > 60 in.
(b) $f(67) = 135$ lbs

Practice Set 4-4
1. direct   3. inverse   5. direct   7. $y = kz$
9. $a = kbc$ 11. $d = kef^3$ 13. $m = \dfrac{k\sqrt{n}}{p^3}$
15. $y = kx$ $k = \dfrac{1}{3}$ $y = 12$
17. $y = \dfrac{k}{x}$ $k = 54$ $y = 1.5$
19. $y = k\sqrt{x}$ $k = 6$ $y = 36$
21. $y = kxz^2$ $k = 2$ $y = 1536$
23. $p = \dfrac{kq}{r^2}$ $k = 32$ $p = 12$
25. 153.86 cm²   27. 12 ft   29. 36 in.   31. 4 in.
33. 142 lb.   35. 480 W   37. 0.72 in.
39. (a) direct variation   (b) $\left(\dfrac{4}{3}\right)\pi$ (c) 972π in.³

Practice Set 4-5
1. minimum $= -4$ roots $= -1$ and -5

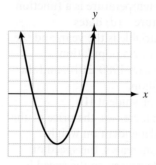

3. minimum $= -2.3$ roots $= -1$ and -4

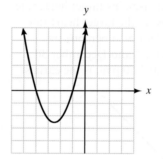

5. maximum $= 5.3$ roots $= 1.7$ and -1

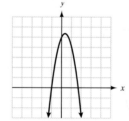

7. $f(x) = 3x^2$

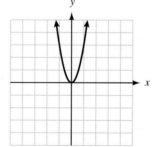

9. $f(x) = \frac{1}{3}x^3$

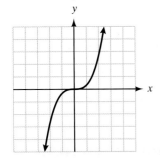

11. 9 sec, 5 sec   **13.** 6 sec, 8 sec
15. 3 sec   **17.** 0.2 sec, 4.6 ft
19. (a) 426 ft,   (b) 431.64 ft,   (c) 6.6 sec
21. $2070

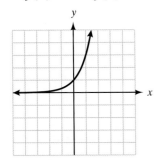

23. 5.00, −3.00   **25.** 2.18, 0.15   **27.** 3.00, −1.50
29. 2 sec, 5 sec   **31.** 33 sec   **33.** 72 units
35. 120 machines   **37.** (a) Neither is linear.
(b) Neither is linear.

Practice Set 4-6
1. $f(x) = 5^x$ $f(2) = 25$

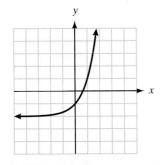

3. $f(x) = 3^x - 2$ $f(2) = 7$

5. $f(x) = \left(\frac{1}{2}\right)^x$ $f(2) = \frac{1}{4}$

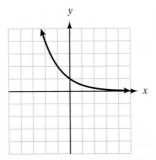

7. $f(x) = e^x$ $f(2) = 7.39$

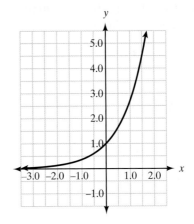

9. $f(x) = e^{-0.5x}$ $f(2) = 0.37$

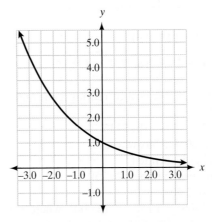

11. $f(10) = 15,767$ people   **13.** 28,733 people
15. 11,819 deer   **17.** 11,668,044 in 10 years;
8,643,900 in 50 years   **19.** 140,839,731 people
21. 46.7 g, 1.28 g   **23.** 218 g   **25.** $31,308.07
27. $2.38; 11.58 years

Chapter 4 Review
1. You will pay $0.72 tax on a $12 purchase.
2. Answers will vary.
3. Both the domain and the range values must be greater than or equal to 0.
4. As the age of a person increases, his height increases up to a certain age. Then it remains steady with some shrinkage in later years.
5. As time passes, the number of bacteria in a culture increases.
6. The faster the speed of a car, the less time it takes to drive from home to school.
7. independent variable: number of minutes dependent variable: cost
8. domain: ≥ 0 minutes; range: $\geq \$5.00$
9. $f(20) = 40$ ft; $f(40) = 120$ ft; $f(60) = 240$ ft; $f(80) = 400$ ft; the graph is nonlinear.

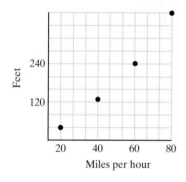

10. (a) -10 (b) -6 (c) 2 (d) -5
11. $f(x) = -3x + 2$ 12. $f(x) = -\frac{2}{3}x - 2$
13. $f(x) = x - 8$ 14. $f(x) = -\frac{3}{2}x + 3$
15. $f(x) = -4x + 1$ 16. 7, 1, -1, 1, 7
17. (a) -3 (b) 3 (c) 0
18. $C(120) = \$36.20$, the cost of renting a car and driving 120 miles
19. (a) $C(x) = 75 + 30x$ (b) $435
20. (a) $f(x) = 1000 + 5.50x$ where $x =$ number of books printed
(b) Domain is natural numbers ≥ 250.
(c) $f(375) = \$3062.50$
21. (a) $f(x) = 1400 + 0.15(x - 14{,}000)$, where x is taxable income   (b) $f(32355) = \$4153.25$
22. (a) $f(s) = \$1000 + 0.25s$ (b) $1625
23. $f(x) = 2x - 3$ 24. $f(x) = 0.5x + 4$
25. independent variable: time; dependent variable: length of fingernails; direct variation
26. independent variable: temperature; dependent variable; time; inverse variation

27. $s = kt$ 28. $z = \dfrac{k}{p}$
29. $m = krs$ 30. $x = \dfrac{k\sqrt[3]{z}}{y^2}$
31. $u = kv^3$ $k = 8$ $u = 64$
32. $x = \dfrac{k}{y^2}$ $k = 18$ $x = 0.72$
33. $s = ktg$ $k = 1.33$ $s = 56$
34. $N = \dfrac{kL^2}{M^3}$ $k = 2.25$ $N = 4.5$
35. $d = kt^2$ $k = 16.1$ $d = 144.9$ ft
36. $f = \dfrac{k}{d^2}$ $k = 7200$ $f = 18$ units
37. (a) minimum $= -8$; roots $= -1, 3$

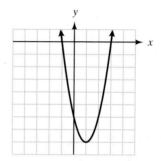

(b) maximum $= 9$; roots $= -4, 2$

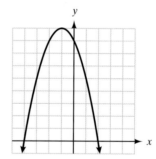

38. (a) $f(-3) = 40.5$

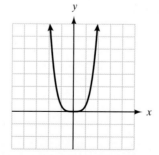

(b) $g(-3) = 54$

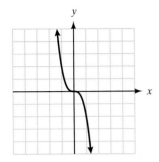

39. $10.4, -0.385$ **40.** 2 sec   **41.** 8

42. (a) $f(-2) = \dfrac{1}{9}$

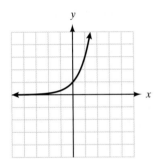

(b) $g(-2) = 4$

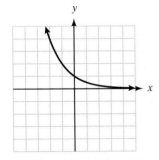

43. 122.9 million   **44.** 51,242 bacteria
45. 45,359 bacteria   **46.** 769,112 people

Chapter 4 Test
1. $25,000 initial value; $-$$1500 in value per year
2. The cost of a house is a function of its square footage. The larger the amount of square footage, the more a house usually costs. independent variable: the square footage, dependent variable: cost
3. (a) independent variable is time; dependent variable is the number of widgets produced
(b) initial value is 0; the function is not perfectly linear.
4. $f(x) = 2x - 4$ $f(-4) = -12$
5. $f(x) = 5$ $f(-4) = 5$

6. $f(x) = -3x^2 - 2$ $f(-4) = -50$
7. $V(t) = \$22{,}500 - \$2200t$
8. $C(12) = \$10.50$; a cab ride of 12 mi. is $10.50
9. (a) $f(x) = \$21 + \$28x$ where x is number of credit hours taken
(b) domain: set of whole numbers 12–15; range: {$357, $385, $413, $441}
(c) $f(14) = \$413$; cost of 14 credit hours
10. $f(x) = -3x + 2$
11. independent variable: speed; dependent variable: time to complete the race; inverse variation
12. independent variable: time; dependent variable: diameter of tree; direct variation
13. $t = \dfrac{kB}{P^2}$ $k = 5$ $t = 35$
14. $1139.69   **15.** 158.4 grams
16. maximum $= 9$; roots $= 3, -3$

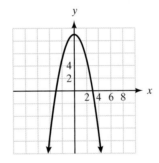

17. $f(4) = -16$

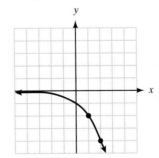

18. $11,484.09   **19.** 428.7 million people
20. $4.22, -0.55$ **21.** Answers vary.
22. 156 ft   **23.** 6.3 sec   **24.** 136,772 persons

CHAPTER 5 KEY
MATHEMATICAL MODELS IN CONSUMER MATH

Practice Set 5-1
1. (a) $1.92   (b) $11.68   (c) $19.21
3. state = $1600.94; county = $512.30
5. $25,120 in merchandise; $2009.60 in taxes

7. $199.95 **9.** $330 per year
11. $1207.14 **13.** 12% **15.** loss of $53,500
17. profit of $6500 **19.** loss of $8000
21. profit of $602,000 **23.** profit of $340
25. (a) $C(x) = 0.01x + 3500$ (b) $3705
(c) loss of $1245
27. $54.25 **29.** 80% markup

Practice Set 5-2
1. $I = \$45$; $M = \$545$ **3.** $I = \$216$; $M = \$1416$
5. 12% **7.** $11.25 **9.** $212 **11.** $162.50
13. (a) $420 (b) $2020 (c) 14%
15. interest = $85; rate = 10.2% **17.** $2148.85
19. $33,671.38
21. (a) $M = \$955.51$; $I = \$55.51$
(b) $M = \$955.61$; $I = \$55.61$
23. no; the account will contain $6104.98.
25. $2.17 more if compounded monthly
27. $29,045.86 **29.** $170,996.39

Practice Set 5-3
1. $5.26 **3.** $10.82
5. June: $5.25 finance charge; $342.10 unpaid balance
July: $342.10 beginning of month; $5.13 finance charge; $297.63 unpaid balance
August: $297.63 beginning of month; $4.46 finance charge; $416.24 unpaid balance
7. October: $0.90 finance charge, $371.51 unpaid balance
November: $371.51 beginning of month; $5.57 finance charge, $581.74 unpaid balance
December: $581.74 beginning of month; $8.73 finance charge; $926.89 unpaid balance
9. June: $19.43 finance charge; $1259.43 unpaid balance; July: $1259.43 beginning of month; $18.89 finance charge; $2,407.17 unpaid balance; August: $2407.17 beginning of month; $36.11 finance charge; $2349.35 unpaid balance
11. average daily balance: $177.28; finance charge: $2.66
13. average daily balance: $356.74; finance charge: $5.35
15. average daily balance: $694.49; finance charge: $10.42
17. 12.5% APR **19.** 11.0% APR
21. 9.0% APR
23. finance charge: $113.68; 8.5% APR
25. total interest $207.55
27. $236.68 **29.** $121.31

Practice Set 5-4
1. Answers will vary.
3. Lower monthly payments, repairs are under warranty (Answers will vary.)
5. Customize the car, sell the car at any time (Answers will vary.)
7. $48,394.50 **9.** $31,227.60 **11.** $60,887.28
13. interest = $5493.18; monthly payment = $722.96
15. interest = $2362.05; monthly payment = $503.03
17. interest = $5721.32; monthly payment = $776.51
19. Total paid: $11,430.00; finance charge: $2430
21. Total paid: $19,608.56; finance charge: $3158.56
23. 9% **25.** 6%
27. finance charge = $579.20; APR = 9.0%
29. $240.30 **31.** $21,803.79
33. amount to finance = $22,500,
payments = $523.13, total cost = $27,610
35. $936

Practice Set 5-5
1. Interest rate, type of loan, fees, points (Answers may vary.)
3. Subtract your monthly bills from your gross income and multiply by 36%.
5. Use the maximum amount of loan formula and the amortization table.
7. down payment = $15,325.00;
amount financed = $137,925.00
9. down payment = $44,975.00;
amount financed = $134,925.00
11. down payment = $94,650.00;
amount financed = $220,850.00
13. $306.05 **15.** $594.00 **17.** $1,098.00
19. $926.16 **21.** $559.89 **23.** $804.42
25. $425.85 **27.** $856.80 **29.** $1423.05
31. $167,037.86 **33.** $369,261.48 **35.** $125,246.55
37. monthly payment = $1,094.20
39. (a) down payment = $17,775.00;
monthly payment = $495.51 (b) $196,158.60
41. Answers will vary.

Practice Set 5-6
1. Insurance is an economical way of helping an individual deal with a severe financial loss by pooling the risk over a large number of people.
3. The amount of insurance specified by the policy is called the face value of the policy.
5. The payment that the insurance company makes to reimburse the policyholder is called an indemnity.
7. 20/40/15
9. Collision insurance pays for repairs to the vehicle of the insured when the policyholder is responsible for an accident.
11. $300.00 **13.** $204.60 **15.** $631.80
17. $211.40 **19.** $175.00 **21.** $264.60
23. $581.40
25. (a) $8,000 (b) $0

Practice Set 5-7
1. (a) $114,915.00 (b) $2298.30
3. (a) $1195.00 (b) $22.93
5. (a) $41,580 (b) $873.18
7. (a) $900.60 (b) $27.52
9. (a) $100 (b) 29,120,000 shares (c) $11.11
(d) $11.29 (e) -1.6% = down 1.6%
11. (a) $24.00 (b) 7,575,000 shares (c) $25.97
(d) $25.64 (e) up 1.3%
13. 250 shares **15.** 350 shares
17. Bonds are issued by companies or governments who are trying to raise money for projects. Bonds are generally safer investments than stocks. If a company goes bankrupt, the bondholders are paid before the stockholders receive any money. (answers may vary)
19. Invest more of your money in stocks and mutual funds since, over long periods of time, these investments usually grow at a better rate than bonds or savings accounts.

Practice Set 5-8
1. $1488.65 **3.** $3,520 **5.** $2297
7. $3125 **9.** $9.75 **11.** $5400 **13.** 8.5%
15. (a) $3199.65 (b) $3284.65
17. He will owe $49 less using the 4.5% rate.
19. $27,000 gross salary; $1045 tax
21. (a) Pay = 800 + 0.095(sales) (b) 0.095
(c) makes $9.50 for every $100 of sales
(d) yes (e) commission because it is dependent on sales (f) $18,157.89
23. total expenses = $1450; $1855 − $1450 = $405; 7%(1855) = $129.85 so he can save 7%
25. $296.70
27. (a) $1400.40 (b) fed/state tax $637.50; utilities $180.63; food $573.75; savings $233.75; entertainment $170; gas/car maint. $212.50; health ins. $85; misc. $396.47
29. Answers will vary.

Chapter 5 Review Problems
1. state tax: $564.66; county tax: $313.70
2. $3950 **3.** $2.23 **4.** 20%
5. $2350 **6.** $10,000 **7.** $1,425
8. Both plans would yield the same amount. **9.** 6.5%
10. $16,650.00 maturity value; $693.75 payment
11. $180 **12.** $8976.47 **13.** $477.36
14. $22,255.41 **15.** $39.38
16. $1355.09 avg daily bal.; $20.33 interest
17. $57.98 **18.** $28,347.90
19. $1254.83 interest; $485.67 payment
20. $7084.80 total paid; $584.80 interest; 3% interest rate
21. $20,232.91 **22.** $322.20 **23.** $491.67
24. $1615.57 **25.** $714.24 **26.** $843.48
27. $235.20 **28.** $331.80 **29.** $20,000
30. $79,863.49 **31.** (a) $12,276 (b) $306.90
32. 2500 shares **33.** 6.5% **34.** $712
35. tax = $1275 + 0.07(x − $21,250)
36. $1362.50 **37.** $357.90

Chapter 5 Test
1. $857.50 **2.** $27.90 **3.** $1550
4. $16,435 **5.** $353 **6.** $9,945
7. $4,891.61 **8.** 5%
9. $10,432.26 **10.** $35.00
11. average daily balance $189.50; interest $2.84
12. $56.38
13. monthly payment: $847.38; interest owed: $3125.79
14. $32,200 **15.** $550.82 **16.** $1597.50
17. $306 **18.** $0 **19.** $16,152.48
20. (a) $257.50 (b) $11.36 (c) $268.86

CHAPTER 6
MODELING WITH SYSTEMS OF EQUATIONS

Practice Set 6-1
1. $(3, -2)$ **3.** $(-3, -1)$ **5.** $\left(\dfrac{1}{2}, 1\right)$
7. $(1, 2)$ **9.** $(2, 0)$ **11.** $(4, 1)$ **13.** $(0, 1)$
15. infinite solutions—same line
17. no solution—parallel lines
19. $(-2, -4)$ **21.** $(-3, -1)$
23. infinite solutions—same line
25. no solution **27.** $(2, 0)$ **29.** $(2, -1)$
31. infinite solutions—same line

Practice Set 6-2
1. $(12, 6)$ **3.** $(2.5, -0.5)$ **5.** $(5, 11)$ **7.** $(2, -1)$
9. $(5, 1)$ **11.** $(5.5, 2.5)$
13. infinite solutions—same line **15.** 1
17. 0 **19.** 22 **21.** $(5, 1)$ **23.** $(7, 3)$
25. $(0, -4)$ **27.** $(-2, 6)$
29. infinite number of solutions
31. $(-2, -6)$ **33.** $(3.5, -1)$ **35.** $(4, -1)$
37. $(-6, -8)$ **39.** no solution—parallel lines
41. $\left(\dfrac{4}{5}, -\dfrac{1}{5}\right)$ **43.** no solution **45.** $(3, 4)$
47. $(2, 3)$ **49.** $(4, 3)$ **51.** $(13, 5)$
53. infinite solutions—same line
55. $(-5, 9)$ **57.** no **59.** 11 and 9 **61.** 12 and 18
63. 16 and 48 **65.** 21 and 28

Practice Set 6-3
1. 12 units @ \$350: 8 units @ \$425
3. 53 double, 27 single
5. \$8000 @ 4.5\%; \$4000 @ 5\%
7. tights = \$12.50, leotards = \$45.00
9. pizza = \$15; soda = \$4
11. Coca-Cola \$41.40 and Pepsi \$43.50
13. 14 mL of 10\%: 16 mL of 25\%
15. 12 g of 60\% and 8 g of 40\%
17. 20 mL of 10\%, 30 mL of 60\%
19. 18 pounds of peanuts; 8 pounds of cashews
21. 5 lb of each
23. \$66.67; where price equals demand
25. \$15   27. equilibrium price is \$5
29. 1500 backpacks
31. 160 miles; John's Rent A Car
33. \$18,750; straight 12\% commission
35. length = 5 ft, width = 2 ft
37. 12 inches by 18 inches
39. 4 yards by 10 yards

Practice Set 6-4
1. $\{(2, 4), (-5, 25)\}$ 3. $\{(-1, -1), (0.5, 0.5)\}$
5. $\{(6, 12), (1, 2)\}$ 7. $\{(3, 5), (-1, -3)\}$
9. $\left\{(1, 0), \left(\frac{1}{3}, -\frac{2}{3}\right)\right\}$
11. $\{(3, 2) (3, -2) (-3, 2) (-3, -2)\}$
13. The numbers are 3 and 4.
15. 3 and -7 or 7 and -3 17. 12 and 5
19. 15,000 pairs of sunglasses at \$37.50
21. 13 ft by 11 ft

Chapter 6 Review Problems
1. yes   2. yes   3. no   4. yes   5. $(7, 1)$
6. no solution   7. $(3, 2)$ 8. $(1, 3)$ 9. $(0, 2)$
10. $(2, -1)$ 11. $(3, -1.6)$ 12. $(1, -3)$ 13. $(3, 12)$
14. infinite number of solutions
15. $\left(\frac{1}{3}, -\frac{1}{2}\right)$ 16. no solution   17. $(6, 4)$
18. $(-4, 1)$ 19. $(5, 6)$ 20. $(-1, 3)$
21. no solution   22. $(1, -6)$ 23. $(1, 5)$
24. $(0, 8)$ 25. $\{(-3, -9), (1, -1)\}$
26. $\{(2.5, 13.5), (-1, 3)\}$
27. 34 boxes of donuts; 28 stadium cushions
28. 80 mL of 40\% acid; 40 mL of 70\% acid
29. 200 clocks   30. \$79   31. 5 nickels, 55 dimes
32. potatoes \$0.34/lb;   bananas \$0.69/lb
33. 1200 packages
34. length is 24 ft, and width is 18 ft
35. Answers will vary.

Chapter 6 Test
1. yes   2. 7   3. $(4, 1)$ 4. $(2, -1)$ 5. no solution
6. no solution   7. $(2, -1)$ 8. $(3, 1)$ 9. $(0.5, 1.5)$
10. $(2, -1)$ 11. $(3, 1)$ 12. $\{(-0.5, 0.5), (2, 8)\}$
13. $(0, 2), (-3.5, 3.75)$ 14. 500 calculators
15. 60 double rooms
16. 50 mL of 12\% solution and 150 mL of 8\% solution
17. \$4000 at 10.5\%: \$8000 at 12\%
18. length = 27 in.   19. \$6.00   20. 4250 leotards

CHAPTER 7
PROBABILITY MODELS

Practice Set 7-1
1. True   3. False   5. False   7. False   9. False
11. True
13. $A = \{x | x \text{ is a natural number less than } 6\}$
15. $\{2, 4, 6, 8, 10, 12, 14, \ldots, 40, 42, 44, 46, 48\}$
17. finite   19. infinite   21. $\{1, 2, 3, 4, 5, 6, 8\}$
23. $\varnothing$ or $\{\ \}$
25. $\{1, 4\}$

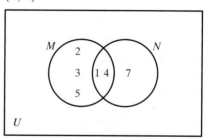

27. $\{1, 2, 3, 4, 5, 6, 8\}$

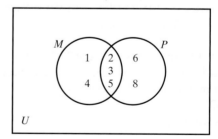

29. $\{2, 3, 5\}$

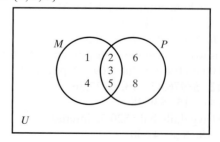

31. $\{1, 2, 3, 4, 5, 7\}$ 33. $\{1, 2, 3, 5, 7, 9\}$
35. $\{3, 5\}$ 37. $\varnothing$ or $\{\ \}$ 39. $\{5\}$

Practice Set 7-2

1. An experiment is something that you do in order to gain information (data); outcomes are all the possible results of a given experiment; and events are the individual possible outcomes that occur when the experiment is done.

3. The law of large numbers states that the more times an experiment is performed, the more accurately we can predict the probability.

5. independent events   7. dependent events

9. $\dfrac{155}{60000} = 0.3\%$

11. cardinal: $\dfrac{1}{6} = 16.7\%$, robin: $\dfrac{2}{3} = 66.7\%$

13. $\dfrac{5}{70} = 7.1\%$ 15. $\dfrac{41}{672} = 6.1\%$

17. $\dfrac{8}{72} = 11.1\%$ 19. 12.5%   21. 5%; 1250 people

23. $\dfrac{37}{55} = 67.3\%$ 25. $\dfrac{9}{55} = 16.4\%$ 27. 27%

29. 1 donor

31. (a) $\dfrac{95}{145} = \dfrac{19}{29} = 65.5\%$ (b) $\dfrac{50}{65} = \dfrac{10}{13} = 76.9\%$

(c) $\dfrac{35}{80} = \dfrac{7}{16} = 43.8\%$ (d) $\dfrac{15}{65} = \dfrac{3}{13} = 23.1\%$

33. (a) $\dfrac{120}{230} = \dfrac{12}{23} = 52.2\%$ (b) $\dfrac{50}{95} = \dfrac{10}{19} = 52.6\%$

(c) $\dfrac{60}{135} = \dfrac{4}{9} = 44.4\%$ (d) $\dfrac{75}{95} = \dfrac{15}{19} = 78.9\%$

Practice Set 7-3

1. Outcomes that have the same chance of occurring in an experiment.

3. No. A theoretical probability of one means that the event is certain to occur.

5. $P(s) = \dfrac{4}{11}$ 7. $P(a) = 0$

9. $\dfrac{10}{20} = \dfrac{1}{2}$ 11. $\dfrac{1}{20}$ 13. $\dfrac{4}{52} = \dfrac{1}{13}$

15. $\dfrac{13}{52} = \dfrac{1}{4}$ 17. 0   19. $\dfrac{12}{52} = \dfrac{3}{13}$ 21. $\dfrac{1}{13}$ 23. $\dfrac{1}{26}$

25. $\dfrac{3}{12} = \dfrac{1}{4}$ 27. $\dfrac{7}{12}$ 29. $\dfrac{1}{6}$ 31. $\dfrac{1}{2}$ 33. $\dfrac{1}{3}$

35. $\dfrac{1}{3}$ 37. $\dfrac{2}{3}$ 39. 13%   41. 0   43. $\dfrac{1}{10}$ 45. $\dfrac{4}{5}$

47. $\dfrac{1}{6}$ 49. $\dfrac{1}{4}$ 51. $\dfrac{35}{90} = \dfrac{7}{18}$ 53. $\dfrac{55}{90} = \dfrac{11}{18}$

55. $\dfrac{24}{100} = \dfrac{6}{25}$ 57. $\dfrac{84}{100} = \dfrac{21}{25}$

Practice Set 7-4

1. 1:1   3. 2:3   5. 1:12   7. $\dfrac{1}{2}$ 9. $\dfrac{3}{14}$

11. $\dfrac{1}{13}$ 13. 17:3   15. 3:17   17. (a) $\dfrac{1}{6}$ (b) 1:5

19. 1:1   21. 1:2   23. (a) $\dfrac{1}{4}$ (b) 1:3

25. 1:12   27. 10:3   29. 5:3   31. 999,999:1

33. $\dfrac{1}{6}$ 35. $\dfrac{2}{11}$ 37. $\dfrac{9}{14}$ 39. $\dfrac{7}{12}$ 41. $\dfrac{1}{7}$ 43. $\dfrac{2}{7}$

Practice Set 7-5

1. 4000   3. 125   5. 20   7. 8   9. $\dfrac{3}{8}$ 11. 12

13. $\dfrac{1}{12}$ 15. (a) 36   (b) $\dfrac{1}{6}$ (c) 0   (d) $\dfrac{1}{36}$

17. S = {HH1, HH2, HH3, HH4, HH5, HH6, HT1, HT2, HT3, HT4, HT5, HT6, TH1, TH2, TH3, TH4, TH5, TH6, TT1, TT2, TT3, TT4, TT5, TT6}

19. $\dfrac{1}{12}$

21.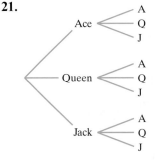

S = {AA, AQ, AJ, QA, QQ, QJ, JA, JQ, JJ}

23. $\dfrac{1}{9}$

25.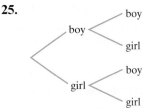

S = {bb, bg, gb, gg}

27. $\dfrac{3}{4}$ 29. 1,757,600   31. $\dfrac{1}{1{,}757{,}600}$ 33. 6

Practice Set 7-6

1. $\dfrac{8}{52} = \dfrac{2}{13}$ Yes; it is impossible to pick one card and get a king and a queen at the same time.
3. $\dfrac{26}{52} = \dfrac{1}{2}$ No; it is possible to pick one card and get a black spade because all spades are black!
5. $\dfrac{32}{52} = \dfrac{8}{13}$ No; it is possible to pick one card and get a black face card (jack of spades and clubs, queen of spades and clubs, king of spades and clubs).
7. $\dfrac{7}{8}$ 9. $\dfrac{5}{6}$ 11. 1   13. $\dfrac{1}{3}$ 15. $\dfrac{110}{205} = \dfrac{22}{41}$
17. $\dfrac{85}{205} = \dfrac{17}{41}$ 19. $\dfrac{135}{205} = \dfrac{27}{41}$ 21. $\dfrac{143}{228} = 62.7\%$
23. $\dfrac{183}{380} = 48.2\%$ 25. $\dfrac{37}{76} = 48.7\%$
27. $\dfrac{33}{50} = 66\%$ 29. $\dfrac{37}{60} = 61.7\%$

Practice Set 7-7

1. independent   3. dependent   5. independent
7. $\dfrac{1}{169}$ 9. $\dfrac{5}{18}$ 11. $\dfrac{1}{12}$ 13. $\dfrac{5}{46} = 10.9\%$
15. $\dfrac{85}{253} = 33.6\%$ 17. $\dfrac{108}{3125} = 3.5\%$
19. $\dfrac{32}{3125} = 1.0\%$
21. (a) $\dfrac{21}{99} = 21.2\%$ (b) $\dfrac{14}{165} = 8.5\%$
23. $\dfrac{1}{8}$ 25. $\dfrac{1}{8}$ 27. $\dfrac{44}{585}$ 29. $\dfrac{1}{133,225}$

Practice Set 7-8

1. 21   3. 42   5. 362,880
7. Only positive whole numbers have a factorial value; thus, no value here.
9. In a combination, order is unimportant, so that 1, 2 and 2, 1 are the same combination of the numbers 1 and 2. In a permutation, order is important, so that 1, 2 and 2, 1 are different permutations (orderings) of the numbers 1 and 2.
11. If Jane must be on the committee, then the other three members will be chosen from the nine remaining students, and order doesn't matter here. $_9C_3 = 84$
13. $9! = 362,880$ or $_9P_9 = 362,880$
15. A person may choose no toppings, or 1 or 2 or . . . no toppings + 1 topping + 2 toppings + 3 toppings + 4 toppings + 5 toppings
$1 + {_5C_1} + {_5C_2} + {_5C_3} + {_5C_4} + {_5C_5}$
$1 + 5 + 10 + 10 + 5 + 1 = 32$
17. 24   19. (a) 362,880   (b) 3,024
21. (a) 40,320   (b) 336   (c) 1680
23. 72   25. 1,000,000,000   27. 210   29. 84
31. $\dfrac{5}{42}$ 33. $\dfrac{1}{5}$ 35. $\dfrac{1}{14}$ 37. $\dfrac{3}{14}$
39. (a) 720   (b) 360
41. $\dfrac{1}{14}$
43. (a) $.940606 = 94.1\%$ (b) Answers will vary.
45. $\dfrac{1}{5}$ 47. $\dfrac{2}{39} = 0.05128$ 49. $\dfrac{1}{12}$
51. 1000   53. various answers

Chapter 7 Review Problems

1.

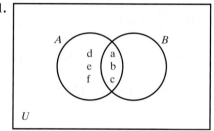

2. {a, b, c, e, g}   3. {f}   4. { } or $\varnothing$
5. {a, c, e, f, g, h}
6. No. All elements in A are NOT in B.
7. No. All elements in C are NOT in A.
8. Yes. There are six elements in set A.
9. No. There are two elements in set D – so it is finite.
10. A probability is a mathematical calculation of the likelihood of a given event occurring in preference to all other possible events that could occur in a given situation (or experiment). In other words, it is the proportion (or fraction) of times that a particular outcome will occur.
11. In problems where several outcomes are possible, use the addition rule if the word "or" is used to connect the events and use the multiplication rule if the word "and" is used.
12. A probability is the proportion of outcomes considered favorable compared to the total number of possible results. The odds of an event compares outcomes considered favorable to the number of outcomes considered unfavorable.
13. $\dfrac{5}{36}$ 14. 0   15. $\dfrac{5}{12}$ 16. $\dfrac{7}{20}$ 17. $\dfrac{13}{20}$
18. $\dfrac{1}{114}$ 19. $\dfrac{63}{200}$ 20. $\dfrac{6}{25}$ 21. $\dfrac{63}{113}$ 22. $\dfrac{24}{49}$
23. There are three face cards (K, Q, J) per suit (4 suits); thus, there are 12 face cards. This means that there are 40 nonface cards in the deck. Odds in favor of drawing a face card = 12:40 = 3:10.

24. $10^9 = 1$ billion   **25.** Yes, because by 2050 the population will be well over one billion.
26. $\frac{73}{75}$ or about a 97% chance
27. $\frac{92}{365}$ or about a 25% chance
28. $\frac{22}{43} = 51.2\%$ **29.** $\frac{26}{43} = 60.5\%$
30. $\frac{5}{23} = 21.7\%$ **31.** 210   **32.** 294,000   **33.** 455
34. 2 :373   **35.** 24   **36.** undefined   **37.** undefined
38. 35   **39.** 2520   **40.** 1771

Chapter 7 Test

1.
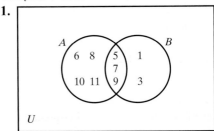

2. {5, 7, 9}   **3.** { } or ∅   **4.** {10, 11, 12, 13, 15}
5. S = {H1, H2, H3, H4, H5, H6, T1, T2, T3, T4, T5, T6}
6. $\frac{1}{12}$ **7.** $\frac{1}{4}$ **8.** $\frac{17}{30}$ **9.** $\frac{29}{60}$ **10.** $\frac{17}{45}$
11. 8,985,600   **12.** 1:1   **13.** $\frac{1}{2}$ **14.** 0.41   **15.** 6
16. 1   **17.** $\frac{1}{6}$
18. Only nine plates out of all possible plates will have all four numbers the same.
$9 \times 9 \times 9 \times 9 = 6561$ different possible plates
$\frac{9}{6561} = \frac{1}{729}$
19. $\frac{1}{2}$ **20.** $\frac{1}{13}$ **21.** $\frac{1}{2704}$ **22.** $\frac{11}{26}$
23. P(Ringo wins) = $\frac{1}{4}$ P(Bonnie wins) = $\frac{1}{16}$
P(Clyde wins) = $\frac{5}{13}$ P(Zero wins) = $\frac{2}{5}$
24. There are only two choices for children, boy or girl, each time, so for four births, $2 \times 2 \times 2 \times 2 = 16$ possible combination of boys and girls. But because order is not specified, many of the groupings are the same. If you list all possible combinations, 6 of the 16 will contain two boys and two girls. P(two girls and two boys) = $\frac{6}{16} = \frac{3}{8}$
25. 720   **26.** 0   **27.** $\frac{1}{1024}$ **28.** $\frac{2}{5}$ **29.** $\frac{3}{4}$ **30.** $\frac{3}{20}$

CHAPTER 8
MODELING WITH STATISTICS

Practice Set 8-1
1. inferential   **3.** descriptive   **5.** descriptive
7. descriptive   **9.** descriptive   **11.** quantitative
13. quantitative   **15.** qualitative   **17.** qualitative
19. quantitative   **21.** cluster sample
23. simple random sample   **25.** convenience sample
27. cluster sample   **29.** random sample
31. The numbering scale on graph exaggerates the differences.
33. Ask how many dentists were polled.
35. The number seems too precise for a general statement.
37. A misuse of a percent with a number greater than 100%.
39. The survey is not representative of the general population.

Practice Set 8-2
1. (a)

Classes	Frequency
50–54	1
45–49	2
40–44	3
35–39	4
30–34	6
25–29	20
20–24	12
15–19	2

(b)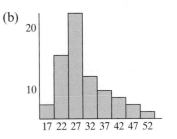

(c) skewed right, relatively few higher speeds

A-20 Answer Key

3. (a)

Classes	Frequency
500–549	1
450–499	0
400–449	3
350–399	3
300–349	4
250–299	2
200–249	8
150–199	7
100–149	10
50–99	10
0–49	6

(b)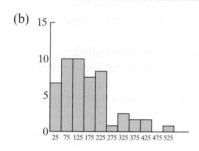

5. (a)

1	9899
2	078729299
3	1023951
4	1536
5	311584663
6	533

(b)

Classes	Frequency
60–66	3
53–59	7
46–82	3
39–45	4
32–38	3
25–31	9
18–24	7

(c)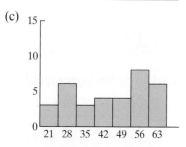

7. (a)

Classes	Frequency
52–57	1
46–51	1
40–45	4
34–39	5
28–33	14
22–27	21
16–21	4

(b)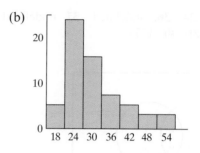

9. (a)

Number of people	Frequency
6–11	2
12–18	3
19–25	7
26–32	16
33–39	3
40–46	6
47–53	2
54–60	1

(b)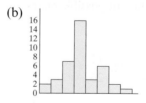

(c) normal

Practice Set 8-3

1.

(Pie chart: Kitchen, Bathroom, Bedroom, Other)

3. (a)
```
0 | 8
1 | 2 2 2 2 5 5 6 6 7 7 8 8 8 9 9
2 | 0 1 2 3 3 6
3 | 0 5
4 | 4
```
(b) Most students took between 12 and 19 minutes to complete the survey.

5. (a) horizontal axis—months
vertical axis—rainfall amounts
(b) Oct.—March
(c) February—less rainfall

7. (a)
```
1 | 8 8 9 9
2 | 0 0 2 2 2 3 3 5 7
3 |
4 | 2 3
5 | 7
```
(b) The majority of students are in their twenties.

9. (a) 3% (b) yes (36% > 33.3%)
(c) 3,675,000 (d) Oceania

Practice Set 8-4
1. Answers will vary.
3. mean = 4.6, median = 3.5, mode = 2, best avg. is the mean
5. mean = 12.6, median = 6, mode = 9, best is the median
7. mean = 3.65, median = 2.865, mode = none, best is the median
9. A relatively few individuals with huge incomes (like Michael Jordan and Bill Gates) "drag" the mean average off center. The median income is the best for most comparisons for persons and families of "normal" incomes.
11. (a) mode (b) mean (c) median (d) mean
13. mean = 0.125°C, median = 0.5°C, mode = 0°C
15. GPA = 3.4
17. Answers will vary.
19. 110
21. (a) mean = 4.57 median = 4 mode = 4
(b) The mean becomes 5 and the others remain unchanged.
23. The mean would increase by $2000.
25. Answers will vary. 27. 12
29. Answers will vary based on survey results.

Practice Set 8-5
1. All data items are identical.
3. mean = 4.5, R = 9, s = 3.0
5. mean = 8.4, R = 18, s = 6.9
7. mean = 7, R = 9, s = 3.3
9. mean = 5, R = 12, s = 3.6
11. for sec. 1: R = 80, s = 24; for sec. 2: R = 80, s = 27; Equal R's might lead you to conclude that the sets were identical, but the standard deviation numbers tell you that the sets are really different.
13. Forever batteries, because they have the smaller standard deviation
15. Small standard deviation indicates a consistent packing process.
17. (a) The mean increases by the amount added to all numbers but the standard deviation does not change. (b) Since 3% of each amount is different, different amounts are being added to each number; thus, both the mean and standard deviation will change.
19. The mean will increase by the amount added. The range and standard deviation will not change.
21. The standard deviation would not change.

Practice Set 8-6
1. 34.1% 3. 15.9% 5. 84.1% 7. 71,550
9. 0.067 (or about 7%) 11. about 920 hours
13. 1397 hours 15. 0.159 (or about a 16% chance)
17. 0.988 (or about a 99% chance)
19. mean = 36.3 in. s = 15.1 in.
21. 91.9% 23. 60% 25. 85%
27. Ray, because his z-score (1.4) is larger than Susan's (1.3)
29. z = 0.8 30. z = 0.86
31. It would mean that your test score was below the class average of 78; 70
33. 75% 35. 75 37. 71 and 129
39. about 3110 customers (62.2%) 41. about 1.2%

Practice Set 8-7
1. negative; as age increases, price decreases
3. negative; as weight increases, mileage decreases
5. (a) independent variable, education; dependent variable, personal income (b) strong positive correlation (c) As the years of education increase, personal income increases.
7. 3.4 9. 2.787
11. (a)

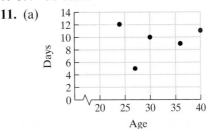

(b) no, $r = 0.16$
(c) $y = 0.065x + 7.35$
(d) 9.43 or 9 days; not reliable because of weak correlation

13. (a)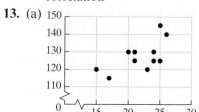
(b) a moderate correlation
(c) $y = 1.6x + 92.5$
(d) $r = 0.64$
(e) 124.5 lb

15. (a)
(b) moderate correlation; somewhat linear

17. (a)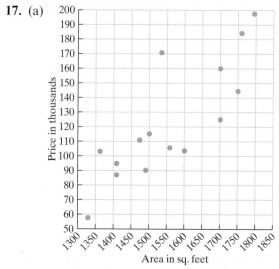
(b) $r = 0.82$; strong correlation
(c) $y = 0.197x - 185$
(d) 125.275 or $125, 275; fairly reliable since the correlation is strong

Chapter 8 Review Problems

1. Descriptive statistics makes no predictions or guesses. It summarizes a population of data. Inferential statistics takes descriptive data and uses them to help make "educated guesses" about the data or about the group from which the data were gathered.

2. Quantitative data are data that can be "quantified", i.e., they have amounts or numbers associated with them. Qualitative data are generally nonnumerical in nature.

3. cluster sample 4. convenience sample
5. stratified sample
6. Sample results cannot be applied to the general population since it was not representative.
7. Sample size is not given and the number is too precise for weight loss.
8. mean = 14, median = 14, mode = none, $R = 8$, $s = 2.7$
9. mean = 17.3, median = 8.5, mode = 8, $R = 55$, $s = 17.2$
10. The data in #9 have more variability. The larger value of the standard deviation tells us this.
11. For the data in #8, the mean average is best. The numbers are all relatively in the same size category. For data in #9, the median is the best average because the relatively large value of 57 drags the mean average off center for the group.

12.
Classes	Frequency
13–14	4
11–12	5
9–10	5
7–8	13
5–6	12
3–4	11

13.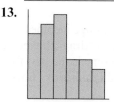
Positive skew

14. 15.9%   15. 50%   16. 81.8%
17. $z_{88 \text{ on math}} = 1.6$ and $z_{88 \text{ on history}} = 1.0$. The z-scores indicate that you would be farther above average for the math group than for the history group.
18. 0.034 (or about 3%)   19. 0.271 (or about 27%)
20.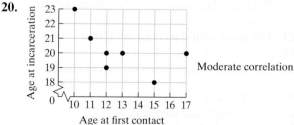
Moderate correlation

21. $y = 25.3 - 0.40x$ **22.** 19.7 years
23. $r = -0.61$
24. The r-value indicates a moderate negative correlation of these two variables.

Chapter 8 Test
1. quantitative   **2.** qualitative   **3.** qualitative
4. The number of students who took the test is not given.
5. data set B
6. for data set A, the mean average and for data set B, the median
7. mean = 36.5, median = 36, mode = 36, $R = 12$, $s = 3.9$
8. median = M
9.

0	3
1	2 7 0 8 8 9 2 8
2	7 7 2 9 1 6 1 8 9 1 6
3	0 5 8 6 5 8 2 3 2 1 2 3 1 2
4	2 7 1 5 3
5	1

10. normal distribution   **11.** approximately 16%
12. $z = -3.00$ **13.** grade = 88.75
14. 0.977 (or about 98%)
15. 0.464 (or about 46%)   **16.** $z = 2.00$
17. 0.092 (or about 9%)   **18.** 146 lb.
19. $r = 0.80$ (interpretations will vary)
20.

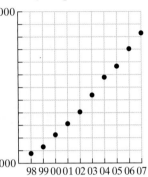

21. $y = 544.5x - 1,081,627.2$ **22.** 22,617

Index

A

absolute values, 3, 23
acute angles, 60, 89
acute triangles, 60, 89
adding notes, 83–88
addition
    indicator words, 41
    real numbers, 4
    rules, 322, 323, 338
additive inverses, 3, 23
additive opposites, 3
adjustable-rate mortgages (ARMs), 227, 250
advantages of leases, 219
algebraic solutions, 266–274
amortization tables, 224, 226, 250
analysis
    break-even, 279, 286
    linear regression, 386
    statistics, 395
And problems, 327–329
angles, 60, 61
annual percentage rate (APR), 212, 250
    automobiles, 219
    finding, 213
    payments, 214
applied variation, 164
applying
    Cramer's rule, 272
    cross-multiplication, 36–37
    functions, 141, 146–153
    graphs, 121–128
    intercepts, 102
    linear equations, 132
    relationships, 310
    modeling, 53–94
        music, 82–89
        patterns, 60–67
        plane geometry, 54–59
        right triangles, 67–72
    Pythagorean theorem, 68–72
    slopes, 131–132
architecture
    modeling in, 73–82
    scale, 77
areas, 89
    calculating, 56, 64, 92
    formulas, 56, 89
    irregular figures, 63
    measurements, 57–58
    Pythagorean theorem, 69
    tables, 376
    triangles, 61
arrays, 270
art, 73–82. *See also* drawing
astronomy, 82
atoms, 21–22, 23
automobiles, 217–223, 230
Avagadro's number, 21–23
average daily balance, 211–212, 250
averages, 363
    calculating, 36, 43
    variations, 369–374
axes, 96

B

balance, 78
Banker's Rule, 205, 250
banking, 202–210
bar graphs, 358–359
base commission, 242
beats, 83. *See also* music
bell curves. *See* normal curves
best-fit lines, 386
bias, 347
bimodal distribution, 356
BMI (body mass index), 31
body mass index. *See* BMI
bonds, 236–241, 250
break-even analysis, 279
budgets, 245, 246
business models, 196–201

C

calculators
    graphs, 264
        linear functions, 189
        nonlinear systems, 283
    mini-lessons
        combinations, 332
        exponential functions, 178
        exponents, 207
        factorials, 330
        histograms, 355
        max/min roots, 168
        order of operations, 7–8
        permutations, 332
        real numbers, 5–7
        scientific notation, 21
        standard deviations, 372
        prediction equations, 387
    TI-84 Plus, 132
calculus, 82
cards, drawing, 310
causal models, 28

charts, 123. *See also* graphs
checking solutions, 261
circles, graphs, 359–361
circumferences, 55, 89
classifications, 60
closed-end credit, 210, 250
closing costs, 223, 250
clusters of samples, 348
collision insurance, 233, 234
combinations, 330–337, 338
combining
    sets, 294
    variations, 162, 183
commissions, 242, 250
comparisons
    heights, 379
    leases to purchases, 221
    quantities, 45
    volumes, 92–93
composites, areas, 69
compound interest, 29, 203
    continuous, 208, 209
    equations, 205
    formulas, 31, 205
    quarterly, 208
comprehensive insurance, 233, 234
conditional probability, 301
consecutive integers, 43
consistency, 262, 286
constants
    proportions, 183
    of variations, 160
consumer credit, 210–216
consumer math, 195–258
Consumer Protection Act, 212
continuous compound interest, 250
convenience sampling, 348
converting notation, 20
coordinates, 96–100, 128
correlation coefficients, 388
costs
    linear systems, 274
    production, 198
counting
    numbers, 2, 23
    principle, 316, 330–338
Cramer's rule, 260, 270, 286
credit
    cards, 210
    consumer, 210–216
cricket function, 139–140
cross-multiplication properties, 36–37, 47
curves, normal, 374–384

D

data, 346
decay, 180–181, 193–194
decimals, 2
    notation, 24
    percentages, 14
    scientific notation, 20

decline, 180
decreases in values, 197–198
defining modeling, 28–29
demand, 278, 285, 286
denominators, 8, 35
dependence
    equations, 270, 286
    events, 301, 327, 338
    systems, 262, 263
    variables, 138, 141, 183
depreciation, 197, 250, 255
    linear functions, 156
    straight-line, 197–198
depth, sense of, 73
derivations, 87
descriptions, 138
    models, 28
    statistics, 346
determinants, 270, 286
    second-order, 271
    solutions, 272
deviations, 370–372
diagrams
    drawing, 66
    probabilities, 316–320
    scatter, 384–391
    tree, 338
    Venn, 294, 295, 321, 338
diameters, 58
dice, rolling, 306
digital
    digitizing, 86
    recordings, 86
    signal processing, 86
directly proportional, 160, 183
direct variations, 160–166
disjoint sets, 296, 338
dispersions, 370
distance
    formulas, 30
    functions, 146
    power functions, 175
    ratios, 34
distribution
    bimodal, 356
    frequency, 352, 353
    normal, 356, 375
distributive properties, 11, 23
dividends, 237, 250
divine proportion, 78
division
    indicator words, 41
    real numbers, 4, 5
domains, 148, 183
    definition of, 142
    identifying, 143
drawing
    cards, 310
    diagrams, 66
    graphs, 102
    histograms, 354
    objects, 323

perspective, 73
scale, 76–82, 89
scatter diagrams, 385
duration, 82

E

educated guesses, 28
electrical currents, 171–174
elements, 286, 294, 338
Elements, The, 54
empirical probability, 299, 338
empirical rules, 376
empty sets, 296, 338
equality, 294, 338
   indicator words, 41
   properties, 36
equally likely outcomes, 306, 338
equations
   areas of triangles, 61
   combinations, 331
   compound interest, 203
   continuous compounding, 208
   cross-multiplication properties, 36–37
   demand, 278
   dependent, 270
   equivalent, 9, 23
   exponential functions, 178
   fixed-rate mortgages, 227
   functions, 138, 146
   graphs, 260–266
   horizontal lines, 104
   increase/decrease, 17
   installment loans, 217
   linear, 98, 111, 128, 260
      applying, 132
      graphs, 101–105
      solutions, 9
   linear systems, 274–283
   lines, 120
   mathematical relationships, 309
   maximum amount of loan formulas, 225
   percentages, 14
   permutations, 331
   predictions, 386, 387
   quadratic functions, 168
   real numbers, 2
   rewriting, 32
   sales taxes, 196
   second-order determinants, 271
   simple interest formulas, 202
   slope-intercept forms, 110
   slopes of lines, 107
   supply, 278
   systems, 259–291
   variables, 162–163
   vertical lines, 104
   writing, 116–121, 122
equilateral triangles, 60
equilibrium points, 278, 286
equivalent equations, 9, 23
Euclid, 54
evaluating

   exponential functions, 178
   expressions, 6, 8
   functions, 147–150
events, 338
   addition rule, 322
   definition of, 298
   dependent, 327
   independent, 321, 327
   mutually exclusive, 321
   odds, 312
   order of, 331
   probabilities, 328, 379
experiments, 29, 298, 338
exponential functions, 178–182, 183, 192, 398–399
exponents, 23, 207
expressions
   evaluation, 6, 8
   fractions, 34
   operations, 7
   proportions, 36
   statements, 44
extremes, 36, 47

F

factorials, 330, 331, 338
factors, scale, 76
Fermat, Pierre, 293
finance charges, 213
finite sets, 294, 338
first harmonic, 84
first overtone, 84
fixed-rate mortgages, 226
flats, 85
forms
   general, 98, 128
   point-slope, 116, 128
   slope-intercept, 110, 125, 128, 260
formulas, 30–34
   areas, 56, 89
   BMI (body mass index), 31
   compound interest, 31, 205
   distance, 30
   empirical probability, 299
   functions, 138, 147, 149
   geometry, 54
   Hero's, 62
   installment loans, 217
   as models, 29
   perimeters, 30
   quadratic, 172–173
   real numbers, 2
   rewriting, 32
   simple interest, 202
   theoretical probability, 307
fractions
   linear equations, 10
   measurements, 34
   music, 83, 84
   percents, 14
frequencies, 84, 86, 87, 89
   distribution, 352, 353
   tables, 352–358

functions, 137–194, 183
    cricket, 139–140
    definition of, 139
    demand, 285
    distance, 146
    domains, 142, 143, 148
    evaluating, 147–150
    exponential, 178–183, 192
    formulas, 147
    graphs, 140–141
    linear, 154–157, 183, 189
    names, 190
    notation, 146–153, 183
    overview of, 138–146
    power, 166–177, 183
    profits, 170–171
    quadratic, 166–177, 183
    ranges, 142, 143, 148
    relationships, 141
    sales taxes, 146
    supply, 285
    transformations, 195
    variations, 160–166
    velocity, 146
    word problems, 149
fundamentals
    frequencies, 84, 86
    of modeling, 27–51

G

general forms, 98, 110, 128
geometry, 82
    linear equations, 280
    perspective, 73
    plane, 54–59
Golden Mean, 78
Golden Ratios, 89, 93–94
golden rectangles, 79, 93
graphing calculators, 330
graphs, 95–135
    applications of, 121–128
    calculators, 189, 264, 283
    circles, 359–361
    functions, 138–141, 150
    horizontal lines, 104
    linear equations, 101–105
    lines, 103, 109–122
    quadratic functions, 167
    reading, 358–363
    real numbers, 2
    rectangles, 96–100
    slopes, 105–115
    systems, 260–266
    vertical lines, 104
gross income, 198
growth, 180, 183

H

half-circle, 57
halves, symmetry, 75
Hambridge's Whirling Squares, 80
harmonics, 86

heights, 170, 190, 379
Hero's formula, 62
histograms, 352
home purchases, 223–229
horizon lines, 73
horizontal axis, 141
horizontal lines, 103–104, 109, 118, 128
hypotenuse, 67, 89

I

identifying
    coordinates of points, 97
    domains, 143
    ranges, 143
income
    gross, 198
    models, 241–250
inconsistent systems, 262
increase equations, 17
independence
    events, 301, 321, 327, 338
    outcomes, 321, 338
    variables, 138, 142, 183
indicator words, 41
inequalities, 290, 291
inferential statistics, 346
infinite
    sets, 294, 338
    solutions, 270, 286
installment buying, 210, 250
    automobiles, 217–223
    homes, 223–229
insurance, 230–235
integers, 2, 23, 43
intercepts
    applying, 102
    finding, 102
    sales predictions, 125
    slope-intercept forms, 110, 128, 260
interest, 202, 250
    automobile purchases, 218
    balances, 211–212
    compound, 29, 203
        continuous, 208, 209
        equations, 205
        formulas, 205
        quarterly, 208
    ordinary, 205
    short-term notes, 203
    simple, 33, 202
    unpaid balances, 211
interpreting graphs, 140–141
intersections, 105, 295, 338
    axes, 96
    finding, 296
inverses
    additive, 3, 23
    multiplicative, 5, 23
    proportions, 183
    variations, 160–166, 161
investments, 236–241
irrational numbers, 2, 23, 55

irregular figures, 63
isosceles triangles, 60, 89

J

joint variations, 162, 183

K

key statements, 41
key words, 40

L

laws
    of large numbers, 300, 338
    Ohm's law, 164
    of probability, 308, 338
leases, 219, 220, 250
least-squares lines, 386
legs, 67
liability insurance, 230, 232. *See also* insurance
linear
    combination method, 266, 268, 286
    equations, 98, 111, 128
        applying, 132
        equality, 10
        fractions, 10
        graphs, 101–105
        properties, 11
        solutions, 9–13
    functions, 183
        calculating, 155, 189
        definition of, 154
        depreciation, 156
        as models, 153–160
        tables, 156, 157
    regression, 384–391
    systems, 274–283
lines
    equations, 120
    graphing, 101
    graphs, 116–121, 122
    horizon, 73
    horizontal, 103–104, 109, 118, 128
    intercepts, 102
    numbers, 2
    origins, 103
    parallel, 112, 128
    perpendicular, 112–114
    slopes
        calculating, 106
        graphs, 109–112
    vertical, 103, 128
        graphs, 104
        slope-intercepts, 118
        slopes, 109
loans, 202
    annual percentage rate (APR), 213
    automobiles, 217–223
    maturity value of, 202
    monthly payments, 203
    short-term notes, 203
    simple interest, 203

locations
    of operations, 40
    points, 96
logarithmic spirals, 80
losses, 17
lowest common denominator (LCD), 112

M

marbles in a jar, 308
markup, 199, 250
masses, 34
MasterCard, 210
mathematical modeling, 28, 47. *See also* modeling
    automobiles, 217–223
    banking, 202–210
    business world, 196–201
    consumer credit, 210–216
    consumer math, 195–258
    home purchases, 223–229
    insurance, 230–235
    investments, 236–241
    personal income, 241–250
mathematical relationships, 309
maturity values, 202, 250
maximum height, 170
maximum house payments, 224
max/min roots, finding, 168
mean average, 363, 365
means, 36, 47, 78
measurements, 34
    angles, 60–61
    areas, 57–58
    ratios, 35
measures
    of central tendency, 363
    music, 83
medians, 363, 364
methods, samples, 348–349
midpoint of classes, 354
missing percents, finding, 17
misuse of statistics, 349–350
mixture problems, 276, 277
modal classes, 354
models, 143, 183
    applying, 53–94
    in art, architecture, and nature, 73–82
    defined, 28–29
    equations, 259–291
    exponential functions, 178–182
    formulas, 29, 30–34
    fundamentals of, 27–51
    linear functions, 153–160
    music, 82–89
    patterns in triangles, 60–67
    plane geometry, 54–59
    power functions, 166–177
    probability, 293–343
        combinations, 330–337
        odds, 312–316
        Or problems, 321–327
        outcomes, 306–312
        permutations, 330–337

models (*continued*)
    And problems, 327–329
    sets, 294–297
    tables, 324
    theory, 297–306
    tree diagrams, 316–320
    proportions, 34–40
    quadratic functions, 166–177
    ratios, 34–40
    right triangles, 67–72
    statistics, 345–399
    word problem strategies, 40–47
modes, definition of, 363
Modified Accelerated Cost Recovery System (MACRS), 255
mole of atoms, 21–22, 23
monthly loan payments, 203, 214
mortgages, 223, 250, 258
    adjustable-rate mortgages (ARMs), 227
    fixed-rate payments, 226
multiple points, 108
multiplication, 36–37, 47
    indicator words, 41
    real numbers, 4, 5
    rules, 327, 338
multiplicative inverses, 5, 23
music, 82–89, 93
mutual funds, 236–241, 250
mutually exclusive events, 321, 322, 338

N

names
    functions, 147, 190
    musical notes, 83
National Association of Securities Dealer Automated Quotations (NASDAQ), 236
natural exponential functions, 183
natural numbers, 2, 23
nature, modeling in, 73–82
negative
    correlation, 385
    skew, 356
    slopes, 107
net profits, calculating, 199
Newton, Sire Isaac, 82
New York Stock Exchange (NYSE), 236
nonlinear
    equations, 283–286
    functions, 166, 195
    system of equations, 286
non-mutually exclusive events, 323
normal curves, 374–384
normal distribution, 356, 375
notation
    decimals, 20, 24
    functions, 146–153, 183
    scientific, 19–22, 23, 24
    set-builder, 294, 338
    sets, 294
notes, 82, 83–88
null sets, 296, 338
numbers
    Avagadro's, 21–22, 23
    converting, 20
    counting, 23
    irrational, 23, 55
    law of large numbers, 300
    lines, 2
    natural, 23
    percents, 16
    rational, 23
    real, 2–5, 23, 96
    slopes, 105
    square arrays of, 270
    whole, 23
numerical codes, 346

O

objects, 73, 323
obtuse
    angles, 60, 89
    triangles, 60
octaves, 85
odds, 338
    definition of, 313
    probabilities, 312–316
Ohm's law, 164
open-end credit, 210, 250
operations
    locations of, 40
    order of, 6, 7–8, 23
    with real numbers, 2–5
opposites, additive, 3
options, insurance, 230–235
order
    of events, 331
    of operations, 6, 7–8, 23
    of subtraction, 108
ordered pairs, 96, 98, 260
ordinary interest, 205, 250
original values, 44
original wording, 41
origins, 96, 103, 128
Or problems, 321–327
outcomes, 338
    definition of, 298
    independent, 321
    probabilities, 306–312
    tree diagrams, 316–320
output, 141

P

pairs, ordered, 96, 98
parallel lines, 112, 113, 128
parentheses, 23
Pascal, Blasé, 293
patterns
    exponential functions, 192
    models, 60–72
payments
    calculating, 214
    credit cards, 210
    fixed-rate mortgages, 226
    maximum house, 224

Pearson correlation coefficients, 388
percentages, 13–19, 15, 23
   calculating, 16
   finding, 16
   increase/decrease, 17
   losses, 17
   missing, 17
   original values, 44
percentile rank, 382
perimeters, 54, 89
   formulas, 30
   rectangles, 54–55
permutations, 330–337, 338
perpendicular lines, 112–114
personal income, 241–250
perspective, 73, 74, 89
phi, 78
pitch, 69, 72, 85
planes
   coordinates, 96
   geometry, 54–59
planning budgets, 246
plotting points, 96, 101
points
   break-even, 279
   coordinates, 97
   equilibrium, 278, 286
   plotting, 96, 101
   slopes, 108, 116
   of view, 74
point-slope forms, 116, 128
policies. *See* insurance
polygons, 67
population, definition of, 347
portfolios, 239, 258
positive
   skew, 356
   slopes, 107
power functions, 166–177
   definition of, 174
   distance, 175
predictions, 28, 395
   equations, 386, 387
   law of large numbers, 300
   sales, 124
premiums
   comprehensive/collision insurance, 223
   liability insurance, 232
price-demand models, 29
prices, selling, 199
principal, 202, 250
probability models, 293–343
   calculating, 328
   combinations, 330–337
   counting, 330–337
   events, 379
   laws, 308, 338
   odds, 312–316
   Or problems, 321–327
   permutations, 330–337
   And problems, 327–329
   sets, 294–297
   tables, 324
   theory
      outcomes, 306–312
      overview of, 297–306
   tree diagrams, 316–320
processes, random, 298
production costs, 198
profits, 198
   functions, 170–171
   net, 199
   stocks, 237
projectiles, shooting, 171–172
properties
   cross-multiplication, 36–37, 47
   distributive, 11, 23
   equality, 36
proportions, 47, 89, 160
   constants, 183
   direct, 183
   divine, 78
   inverses, 183
   modeling, 34–40
   percents, 15
   scale, 76–82
   word problem solutions, 37–38, 39–40
purchasing stocks, 236–237
Pythagoras, 78
Pythagorean theorem, 67, 68–72, 89

Q

quadrants, 96, 128
quadratic functions, 166–177
   definition of, 166
   formulas, 172–173
   height, 170
   roots, 169
quadrivium, 82
qualitative data, 346–348
   bar graphs, 358–359
   means, 365
   medians, 364
   modes, 364
quantitative data, 346–348
   means, 365
   medians, 364
   modes, 364
quantities
   comparisons, 45
   linear systems, 274
quotients, 34. *See also* ratios

R

random
   processes, 298, 338
   sampling, 347
ranges, 148, 183
   definition of, 142, 370
   identifying, 143
rank, percentile, 382

rates, 47, 202, 250
  annual percentage rate (APR), 212
  of change, 105, 153
  insurance, 230–235
  units, 35
rational numbers, 2, 6, 23
ratios, 34–40, 47
  finding, 34
  Golden Ratios, 78, 89–94
  simplifying, 35
reading
  graphs, 358–363
  stock tables, 238, 239
real numbers, 2, 23, 96
  distributive properties, 11
  operations with, 2–5
reciprocals, 5, 23
recordings, digital, 86
rectangles, 29, 56
  coordinate systems, 96–100, 128
  golden, 79, 93
  perimeters, 54–55
reflections, symmetry, 75–76
regression, linear, 384–391
relations, 183
relationships, 138, 294
  functions, 141
  mathematical, 309
  slopes, 112
relative frequency, 299, 338
replacement, 318, 327, 328
residual values, 197, 250
rests, 83
revenues, sales taxes, 196
rewriting
  equations, 32
  formulas, 32
right angles, 60, 67, 68–72, 89
right triangles, 60, 67–72, 89
roots, 183
  max/min, 168
  quadratic functions, 169
rotational symmetry, 76
rules
  addition, 322, 338
    property of equality, 9
    of real numbers, 4
    of signed numbers, 23
  Banker's Rule, 205, 250
  counting principle, 316, 330
  Cramer's rule, 260, 270, 271, 286
  cross-multiplication property, 36–37
  division
    of real numbers, 4
    of signed numbers, 23
  empirical, 376
  functions, 138
  intercepts, 102
  laws of probability, 308
  multiplication, 327, 338
    property of equality, 9
    of real numbers, 4
    of signed numbers, 23

order of operations, 6
percents, 13, 14
probability laws, 308
ratios, 34
slopes
  of lines, 109
  of parallel lines, 113, 119
  of perpendicular lines, 114, 118
stem-and-leaf displays, 361
subtraction
  of real numbers, 4
  of signed numbers, 23
word problems, 40

S

sales
  predictions, 124
  taxes, 196
    calculating, 148, 197
    functions, 146
    revenues, 196
samples
  clusters, 348
  definition of, 347
  methods, 348–349
  random, 347
  space, 306, 338
saving accounts, 206, 254
scale, 76–82, 89
scalene triangles, 60, 89
scatter diagrams, 384–391
scientific notation, 19–24
scores, standard, 376, 378
second
  harmonic, 84
  overtone, 84
second-order determinants, 271
sections
  golden, 78
  spirals, 80
selections
  averages, 366
  direct/inverse proportions, 162
  sampling methods, 348–349
selling prices, finding, 199
semi-perimeters, 62. *See also* perimeters
semitones, 85
series of events, 328
set-builder notation, 294, 338
sets
  probabilities, 294–297
  unions, 322
  universal, 338
shapes, 56
shareholders, 236, 250
shares, 236, 250
sharps, 85
shooting projectiles, 171–172
short-term notes, 203
similar triangles, 77
simple interest, 33, 202, 203, 250
simplifying ratios, 35
sine waves, 87, 89

skews, statistics, 356
slope-intercept forms, 110, 128, 260
    equations, 116
    sales predictions, 125
slopes, 128
    applying, 131–132
    graphs, 105–115, 106–109
    linear functions, 153
    lines, 109–112
    parallel lines, 113
    perpendicular lines, 114
    points, 116
    relationships, 112
    vertical lines, 109
solutions, 98, 128, 286
    algebraic, 266–274
    checking, 261
    dependent systems, 263
    determinants, 272
    formulas, 32
    graphing calculators, 264
    graphs, 260
    infinite, 286
    linear equations, 9–13
    nonlinear equations, 284
    proportions, 15
    systems
        of equations, 260
        with no, 262
    variations, 163
    word problems, 37–43
specific points, finding, 101
spirals, logarithmic, 80
squares, 56
    arrays of numbers, 270
    Hambridge's Whirling Squares, 80
    root, 2, 23
standard
    deviations, 371, 372
    scientific calculators, 330
    scores, 376, 378
standard deviations, 370
state income taxes, 244
statements
    expressions, 44
    key, 41
statistics
    analysis, 395
    descriptive, 363–369
    misuse of, 349–350
    models, 345–399, 346–352
    normal curves, 374–384
    variations, 369–374
stem-and-leaf displays, 361
stocks, 236–241, 250, 258
    profits, 237
    purchasing, 236–237
    tables, 238, 239
straight commission, 242, 250
straight-line method of depreciation, 197–198, 250
strata, 347
strategies, word problems, 40–47
stratified sampling, 347
structures, 28
subsets, 296, 338
substitution method, 266, 267, 286
subtraction
    indicator words, 41
    order of, 108
    real numbers, 4
supply, 278, 285, 286
surveys, 342
symbols
    absolute values, 3
    common musical notes, 83
symmetry, 73, 75, 81, 89
    reflections, 75–76
    rotational, 76
systematic sampling, 347
systems
    of equations, 259–291, 286
    inequalities, 290, 291
    rectangular coordinates, 96–100, 128

T

tables
    amortization, 224, 226, 250
    areas, 376
    frequencies, 352–358
    functions, 149–150
    linear functions, 156, 157
    probability models, 302
    stocks, 238, 239
    of values, 138
taxes, sales. *See* sales taxes
tempos, 83, 88
terms, 36
theoretical probability, 307, 338
theory
    empirical rules, 376
    law of large numbers, 300
    probability models
        outcomes, 306–312
        overview of, 297–306
TI-84 Plus calculators, 132
time, 82
tones, 87
transformations of nonlinear functions, 195
tree diagrams, 316–320, 338
triangles, 56
    areas
        calculating, 61
        equations, 61
    models, 60–67
    scale, 77
    similar, 77
Truth-in-Lending Act, 212
two points, 108
types
    of angles, 60
    of averages, 366
    of lines, 103
    of sampling, 347
    of triangles, 60
    of variables, 138
    of variations, 160–166

U

unions, 294, 295, 322, 338
units
  measurements, 34
  rates, 35, 47
universal sets, 294, 338
unpaid balance method, 211, 250

V

values
  absolute, 3, 23
  functions, 147
  maturity, 202
  mixture problems, 277
  residual, 197, 250
  second-order determinants, 271
  tables of, 138
vanishing points, 73
variables
  dependent, 138, 141, 183
  equations, 162–163
  independent, 138, 142
variations
  applied problems, 164
  combined, 162, 183
  functions, 160–166
  height, 190
  inverse, 183
  joint, 162, 183
  normal curves, 375
  solutions, 163
  statistics, 369–374
velocity functions, 146
Venn diagrams, 294, 295, 296, 321, 338
verification solutions of linear equations, 98
vertical
  axis, 141
  lines, 103, 128
  graphs, 104
  slope-intercepts, 118
  slopes, 109
vibrations, 84
views
  perspective, 74
  point of, 74
Visa, 210
volume comparisons, 92–93
von Leibniz, Gottfeid Wilhelm, 82

W

waves, sine, 87, 90
weighted mean averages, 367
whole numbers, 2, 23
word problems, 48, 49–51
  formulas, 66
  functions as, 149
  indicator words, 41
  Pythagorean theorem, 71
  solutions, 37–43
  strategies, 40–47
writing, 116–122

X

x-axis, 96, 105, 128
x-coordinates, 96, 128
x-intercepts, 101, 102, 128

Y

y-axis, 96, 105, 111, 128
y-coordinates, 96, 128
y-intercepts, 101, 102, 112, 128

Z

z-scores, 376, 378

Formulas Used in This Text

Geometry Formulas

Circumference of a circle: $C = 2\pi r = \pi d$

Area formulas: square $\quad A = s^2$
rectangle $\quad A = lw$
triangle $\quad A = 0.5bh$
circle $\quad A = \pi r^2$

Hero's formula for the area of a triangle:

$$A = \sqrt{s(s-a)(s-b)(s-c)} \quad s = \frac{a+b+c}{2}$$

Pythagorean theorem: In a right triangle, $a^2 + b^2 = c^2$, where a and b are the lengths of the legs and c is the length of the hypotenuse.

The Golden Ratio: $\phi = \dfrac{a}{b} = \dfrac{a+b}{a}$

Business Formulas

Percent increase/decrease $= \dfrac{\text{new value} - \text{original value}}{\text{original value}} \times 100$

Simple interest: $I = Prt$

Maturity value: $M = P + I$

Interest compounded quarterly: $M = P\left(1 + \dfrac{r}{4}\right)^{4t}$, where $r =$ annual rate, $t =$ number of years

Interest compounded monthly: $M = P\left(1 + \dfrac{r}{12}\right)^{12t}$, where $r =$ annual rate, $t =$ number of years

Interest compounded daily: $M = P\left(1 + \dfrac{r}{365}\right)^{365t}$, where $r =$ annual rate, $t =$ number of years

Straight-line depreciation: $\dfrac{\text{original value} - \text{residual value}}{\text{number of years}}$

Fixed-rate mortgage monthly payment formula: $P = A \left[\dfrac{\dfrac{r}{12}\left(1 + \dfrac{r}{12}\right)^{12t}}{\left(1 + \dfrac{r}{12}\right)^{12t} - 1} \right]$
where $A =$ amount borrowed
$t =$ number of years
$r =$ annual rate